RF CIRCUIT DESIGN

Chris Bowick is presently employed as the Product Engineering Manager For Headend Products with Scientific Atlanta Video Communications Division located in Norcross, Georgia. His responsibilities include design and product development of satellite earth station receivers and headend equipment for use in the cable tv industry. Previously, he was associated with Rockwell International, Collins Avionics Division, where he was a design engineer on aircraft navigation equipment. His design experience also includes vhf receiver, hf synthesizer, and broadband amplifier design, and millimeter-wave radiometer design.

Mr. Bowick holds a BEE degree from Georgia Tech and, in his spare time, is working toward his MSEE at Georgia Tech, with emphasis on rf circuit design. He is the author of several articles in various hobby magazines. His hobbies include flying, ham radio (WB4UHY), and raquetball.

RF CIRCUIT DESIGN

by

Chris Bowick

Howard W. Sams & Co.
A Division of Macmillan, Inc.
4300 West 62nd Street, Indianapolis, IN 46268 USA

FIRST EDITION
FOURTH PRINTING—1987

International Standard Book Number: 0-672-21868-2
Library of Congress Catalog Card Number: 81-85517

Edited by: *Frank N. Speights*
Illustrated by: *T. R. Emrick*

Printed in the United States of America.

PREFACE

RF Circuit Design is written for those who desire a practical approach to the design of rf amplifiers, impedance matching networks, and filters. It is totally *user oriented.* If you are an individual who has little rf circuit design experience, you can use this book as a catalog of circuits, using component values designed for your application. On the other hand, if you are interested in the theory behind the rf circuitry being designed, you can use the more detailed information that is provided for in-depth study.

An expert in the rf circuit design field will find this book to be an *excellent reference manual,* containing most of the commonly used circuit-design formulas that are needed. However, an electrical engineering student will find this book to be a valuable bridge between classroom studies and the real world. And, finally, if you are an experimenter or ham, who is interested in designing your own equipment, *RF Circuit Design* will provide numerous examples to guide you every step of the way.

Chapter 1 begins with some basics about components and how they behave at rf frequencies; how capacitors become inductors, inductors become capacitors, and wires become inductors, capacitors, and resistors. Toroids are introduced and toroidal inductor design is covered in detail.

Chapter 2 presents a review of resonant circuits and their properties including a discussion of Q, passband ripple, bandwidth, and coupling. You learn how to design single and multiresonator circuits, at the loaded Q you desire. An understanding of resonant circuits naturally leads to filters and their design. So, Chapter 3 presents complete design procedures for multiple-pole Butterworth, Chebyshev, and Bessel filters including low-pass, high-pass, bandpass, and bandstop designs. Within minutes after reading Chapter 3, you will be able to design multiple-pole filters to meet your specifications. Filter design was never easier.

Next, Chapter 4 covers impedance matching of both real and complex impendances. This is done both numerically and with the aid of the Smith Chart. Mathematics are kept to a bare minimum. Both high-Q and low-Q matching networks are covered in depth.

Transistor behavior at rf frequencies is discussed in Chapter 5. Input impedance, output impedance, feedback capacitance, and their variation over frequency are outlined. Transistor data sheets are explained in detail, and Y and S parameters are introduced.

Chapter 6 details complete cookbook design procedures for rf small-signal amplifiers, using both Y and S parameters. Transistor biasing, stability, impedance matching, and neutralization techniques are covered in detail, complete with practical examples. Constant-gain circles and stability circles, as plotted on a Smith Chart, are introduced while rf amplifier design procedures for minimum noise figure are also explained.

The subject of Chapter 7 is rf power amplifiers. This chapter describes the differences between small- and large-signal amplifiers, and provides step-by-step procedures for designing the latter. Design sections that discuss coaxial-feedline impedance matching and broadband transformers are included.

Appendix A is a math tutorial on complex number manipulation with emphasis on their relationship to complex impedances. This appendix is recommended reading for those who are not familiar with complex number arithmetic. Then, Appendix B presents a systems approach to low-noise design by examining the Noise Figure parameter and its relationship to circuit design and total systems design. Finally, in Appendix C, a bibliography of technical papers and books related to rf circuit design is given so that you, the reader, can further increase your understanding of rf design procedures.

CHRIS BOWICK

ACKNOWLEDGMENTS

The author wishes to gratefully acknowledge the contributions made by various individuals to the completion of this project. First, and foremost, a special thanks goes to my wife, Maureen, who not only typed the entire manuscript *at least* twice, but also performed duties both as an editor and as the author's principal source of encouragement throughout the project. Needless to say, without her help, this book would have never been completed.

Additional thanks go to the following individuals and companies for their contributions in the form of information and data sheets: Bill Amidon and Jim Cox of Amidon Associates, Dave Stewart of Piezo Technology, Irving Kadesh of Piconics, Brian Price of Indiana General, Richard Parker of Fair-Rite Products, Jack Goodman of Sprague-Goodman Electronics, Phillip Smith of Analog Instruments, Lothar Stern of Motorola, and Larry Ward of Microwave Associates.

To my wife, Maureen, and daughter, Zoe . . .

CONTENTS

CHAPTER 1

CHAPTER 2

CHAPTER 3

CHAPTER 4

CHAPTER 5

CHAPTER 6

CHAPTER 7

APPENDIX A

APPENDIX B

APPENDIX C

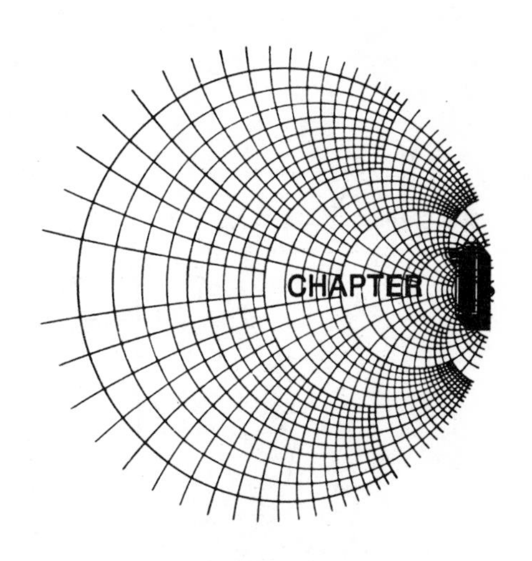

COMPONENTS

Components, those bits and pieces which make up a radio frequency (rf) circuit, seem at times to be taken for granted. A capacitor is, after all, a capacitor —isn't it? A 1-megohm resistor presents an impedance of at least 1 megohm—doesn't it? The reactance of an inductor always increases with frequency, right? Well, as we shall see later in this discussion, things aren't always as they seem. Capacitors at certain frequencies may not be capacitors at all, but may look inductive, while inductors may look like capacitors, and resistors may tend to be a little of both.

In this chapter, we will discuss the properties of resistors, capacitors, and inductors at radio frequencies as they relate to circuit design. But, first, let's take a look at the most simple component of any system and examine its problems at radio frequencies.

WIRE

Wire in an rf circuit can take many forms. Wire-wound resistors, inductors, and axial- and radial-leaded capacitors all use a wire of some size and length either in their leads, or in the actual body of the component, or both. Wire is also used in many interconnect applications in the lower rf spectrum. The behavior of a wire in the rf spectrum depends to a large extent on the wire's diameter and length. Table 1-1 lists, in the American Wire Gauge (AWG) system, each gauge of wire, its corresponding diameter, and other characteristics of interest to the rf circuit designer. In the AWG system, the diameter of a wire will roughly double every six wire gauges. Thus, if the last six gauges and their corresponding diameters are memorized from the chart, all other wire diameters can be determined without the aid of a chart (Example 1-1).

EXAMPLE 1-1

Given that the diameter of AWG 50 wire is 1.0 mil (0.001 inch), what is the diameter of AWG 14 wire?

Solution

AWG 50 = 1 mil
AWG 44 = 2 × 1 mil = 2 mils
AWG 38 = 2 × 2 mils = 4 mils
AWG 32 = 2 × 4 mils = 8 mils
AWG 26 = 2 × 8 mils = 16 mils
AWG 20 = 2 × 16 mils = 32 mils
AWG 14 = 2 × 32 mils = 64 mils (0.064 inch)

Skin Effect

A conductor, at low frequencies, utilizes its entire cross-sectional area as a transport medium for charge carriers. As the frequency is increased, an increased magnetic field at the center of the conductor presents an impedance to the charge carriers, thus decreasing the current density at the center of the conductor and increasing the current density around its perimeter. This increased current density near the edge of the conductor is known as *skin effect*. It occurs in all conductors including resistor leads, capacitor leads, and inductor leads.

The depth into the conductor at which the charge-carrier current density falls to 1/e, or 37% of its value along the surface, is known as the *skin depth* and is a function of the frequency and the permeability and conductivity of the medium. Thus, different conductors, such as silver, aluminum, and copper, all have different skin depths.

The net result of skin effect is an effective decrease in the cross-sectional area of the conductor and, therefore, a net increase in the ac resistance of the wire as shown in Fig. 1-1. For copper, the skin depth is approximately 0.85 cm at 60 Hz and 0.007 cm at 1 MHz. Or, to state it another way: 63% of the rf current flowing in a copper wire will flow within a distance of 0.007 cm of the outer edge of the wire.

Straight-Wire Inductors

In the medium surrounding any current-carrying conductor, there exists a magnetic field. If the current in the conductor is an alternating current, this magnetic field is alternately expanding and contracting and, thus, producing a voltage on the wire which opposes any change in current flow. This opposition to change is called *self-inductance* and we call anything that possesses this quality an *inductor*. Straight-wire inductance might seem trivial, but as will be seen later in the chapter, the higher we go in frequency, the more important it becomes.

The inductance of a straight wire depends on both its length and its diameter, and is found by:

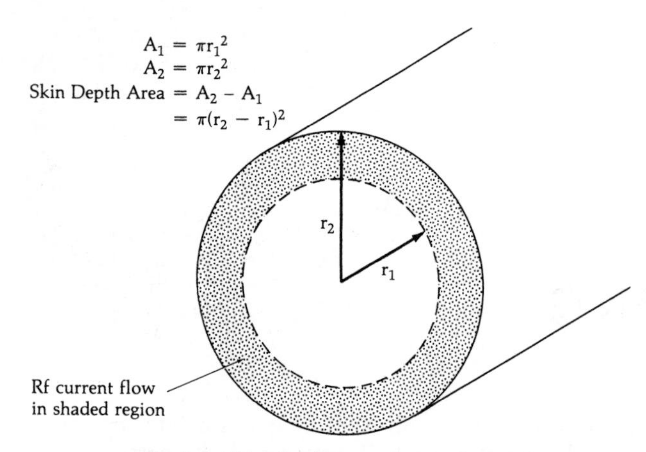

Fig. 1-1. Skin depth area of a conductor.

$$L = 0.002l\left[2.3 \log\left(\frac{4l}{d} - 0.75\right)\right] \mu H \qquad \text{(Eq. 1-1)}$$

where,

L = the inductance in μH,
l = the length of the wire in cm,
d = the diameter of the wire in cm.

This is shown in calculations of Example 1-2.

EXAMPLE 1-2

Find the inductance of 5 centimeters of No. 22 copper wire.

Solution

From Table 1-1, the diameter of No. 22 copper wire is 25.3 mils. Since 1 mil equals 2.54×10^{-3} cm, this equals 0.0643 cm. Substituting into Equation 1-1 gives

$$L = (0.002)(5)\left[2.3 \log\left(\frac{4(5)}{0.0643} - 0.75\right)\right]$$
$$= 57 \text{ nanohenries}$$

The concept of inductance is important because any and all conductors at radio frequencies (including hookup wire, capacitor leads, etc.) tend to exhibit the property of inductance. Inductors will be discussed in greater detail later in this chapter.

RESISTORS

Resistance is the property of a material that determines the rate at which electrical energy is converted into heat energy for a given electric current. By definition:

$$1 \text{ volt across } 1 \text{ ohm} = 1 \text{ coulomb per second}$$
$$= 1 \text{ ampere}$$

The thermal dissipation in this circumstance is 1 watt.

$$P = EI$$
$$= 1 \text{ volt} \times 1 \text{ ampere}$$
$$= 1 \text{ watt}$$

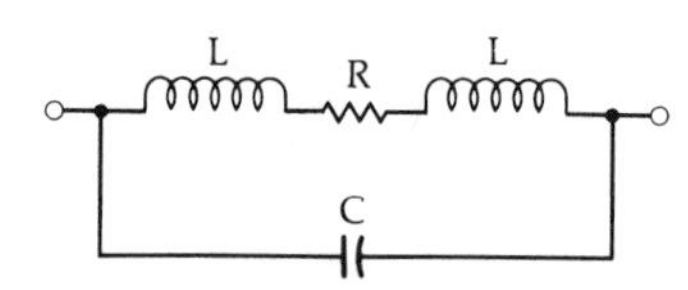

Fig. 1-2. Resistor equivalent circuit.

Resistors are used everywhere in circuits, as transistor bias networks, pads, and signal combiners. However, very rarely is there any thought given to how a resistor actually behaves once we depart from the world of direct current (dc). In some instances, such as in transistor biasing networks, the resistor will still perform its dc circuit function, but it may also disrupt the circuit's rf operating point.

Resistor Equivalent Circuit

The equivalent circuit of a resistor at radio frequencies is shown in Fig. 1-2. R is the resistor value itself, L is the lead inductance, and C is a combination of parasitic capacitances which varies from resistor to resistor depending on the resistor's structure. Carbon-composition resistors are notoriously poor high-frequency performers. A carbon-composition resistor consists of densely packed dielectric particulates or carbon granules. Between each pair of carbon granules is a very small parasitic capacitor. These parasitics, in aggregate, are not insignificant, however, and are the major component of the device's equivalent circuit.

Wirewound resistors have problems at radio frequencies too. As may be expected, these resistors tend to exhibit widely varying impedances over various frequencies. This is particularly true of the low resistance values in the frequency range of 10 MHz to 200 MHz. The inductor L, shown in the equivalent circuit of Fig. 1-2, is much larger for a wirewound resistor than for a carbon-composition resistor. Its value can be calculated using the single-layer air-core inductance approximation formula. This formula is discussed later in this chapter. Because wirewound resistors look like inductors, their impedances will first increase as the frequency increases. At some frequency (F_r), however, the inductance (L) will resonate with the shunt capaci-

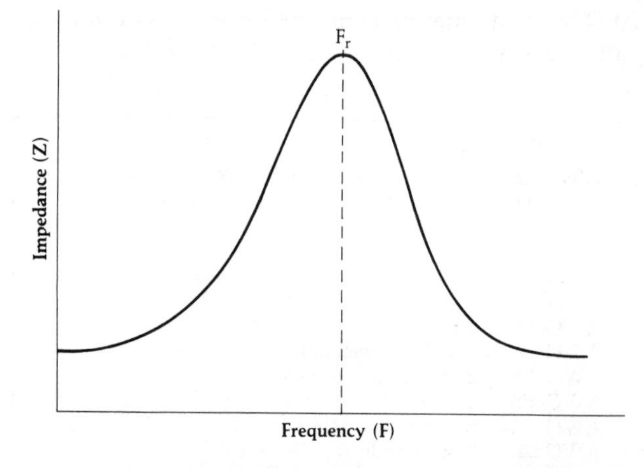

Fig. 1-3. Impedance characteristic of a wirewound resistor.

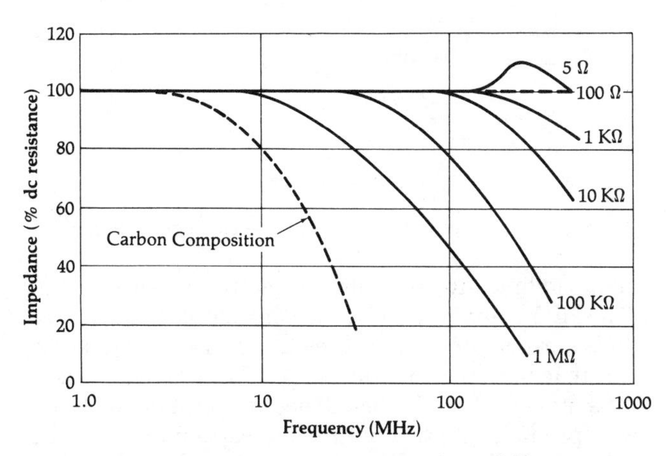

Fig. 1-4. Frequency characteristics of metal-film vs. carbon-composition resistors. (Adapted from *Handbook of Components for Electronics,* McGraw-Hill).

tance (C), producing an impedance peak. Any further increase in frequency will cause the resistor's impedance to decrease as shown in Fig. 1-3.

A metal-film resistor seems to exhibit the best characteristics over frequency. Its equivalent circuit is the same as the carbon-composition and wirewound resistor, but the values of the individual parasitic elements in the equivalent circuit decrease.

The impedance of a metal-film resistor tends to decrease with frequency above about 10 MHz, as shown in Fig. 1-4. This is due to the shunt capacitance in the equivalent circuit. At very high frequencies, and with low-value resistors (under 50 ohms), lead inductance and skin effect may become noticeable. The lead inductance produces a resonance peak, as shown for the 5-ohm resistance in Fig. 1-4, and skin effect decreases the slope of the curve as it falls off with frequency.

Many manufacturers will supply data on resistor behavior at radio frequencies but it can often be misleading. Once you understand the mechanisms involved in resistor behavior, however, it will not matter in what form the data is supplied. Example 1-3 illustrates that fact.

The recent trend in resistor technology has been to eliminate or greatly reduce the stray reactances associated with resistors. This has led to the development of thin-film chip resistors, such as those shown in Fig. 1-6. They are typically produced on alumina or beryllia substrates and offer very little parasitic reactance at frequencies from dc to 2 GHz.

Fig. 1-6. Thin-film chip resistors. (*Courtesy Piconics, Inc.*)

EXAMPLE 1-3

In Fig. 1-2, the lead lengths on the metal-film resistor are 1.27 cm (0.5 inch), and are made up of No. 14 wire. The total stray shunt capacitance (C) is 0.3 pF. If the resistor value is 10,000 ohms, what is its equivalent rf impedance at 200 MHz?

Solution

From Table 1-1, the diameter of No. 14 AWG wire is 64.1 mils (0.1628 cm). Therefore, using Equation 1-1:

$$L = 0.002(1.27)\left[2.3 \log\left(\frac{4(1.27)}{0.1628} - 0.75\right)\right]$$

$$= 8.7 \text{ nanohenries}$$

This presents an equivalent reactance at 200 MHz of:

$$X_L = \omega L$$

$$= 2\pi(200 \times 10^6)(8.7 \times 10^{-9})$$

$$= 10.93 \text{ ohms}$$

The capacitor (C) presents an equivalent reactance of:

$$X_c = \frac{1}{\omega C}$$

$$= \frac{1}{2\pi(200 \times 10^6)(0.3 \times 10^{-12})}$$

$$= 2653 \text{ ohms}$$

The combined equivalent circuit for this resistor, at 200 MHz, is shown in Fig. 1-5. From this sketch, we can see that, in this case, the lead inductance is insignificant when compared with the 10K series resistance and it may be

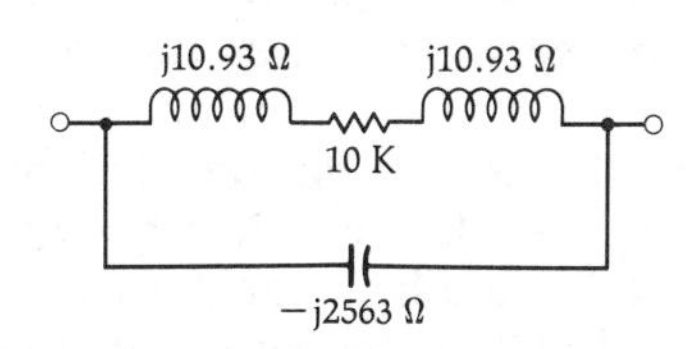

Fig. 1-5. Equivalent circuit values for Example 1-3.

neglected. The parasitic capacitance, on the other hand, cannot be neglected. What we now have, in effect, is a 2563-ohm reactance in parallel with a 10,000-ohm resistance. The magnitude of the combined impedance is:

$$Z = \frac{RX_c}{\sqrt{R^2 + X_c^2}}$$

$$= \frac{(10K)(2563)}{\sqrt{(10K)^2 + (2563)^2}}$$

$$= 1890.5 \text{ ohms}$$

Thus, our 10K resistor looks like 1890 ohms at 200 MHz.

CAPACITORS

Capacitors are used extensively in rf applications, such as bypassing, interstage coupling, and in resonant circuits and filters. It is important to remember, however, that not all capacitors lend themselves equally well to each of the above mentioned applications. The primary task of the rf circuit designer, with regard to capacitors, is to choose the best capacitor for his particular application. Cost effectiveness is usually a major factor in the selection process and, thus, many trade-offs occur. In this section, we'll take a look at the capacitor's equivalent circuit and we will examine a few of the various types of capacitors used at radio frequencies to see which are best suited for certain applications. But first, a little review.

Parallel-Plate Capacitor

A capacitor is any device which consists of two conducting surfaces separated by an insulating material or dielectric. The dielectric is usually ceramic, air, paper, mica, plastic, film, glass, or oil. The capacitance of a capacitor is that property which permits the storage of a charge when a potential difference exists between the conductors. Capacitance is measured in units of farads. A 1-farad capacitor's potential is raised by 1 volt when it receives a charge of 1 coulomb.

$$C = \frac{Q}{V}$$

where,

C = capacitance in farads,
Q = charge in coulombs,
V = voltage in volts.

However, the farad is much too impractical to work with, so smaller units were devised.

$$1 \text{ microfarad} = 1\ \mu F = 1 \times 10^{-6} \text{ farad}$$
$$1 \text{ picofarad} = 1\ pF = 1 \times 10^{-12} \text{ farad}$$

As stated previously, a capacitor in its fundamental form consists of two metal plates separated by a dielectric material of some sort. If we know the area (A) of each metal plate, the distance (d) between the plate (in inches), and the permittivity (ϵ) of the dielectric material in farads/meter (f/m), the capacitance of a parallel-plate capacitor can be found by:

$$C = \frac{0.2249\epsilon A}{d\epsilon_o} \text{ picofarads} \qquad \text{(Eq. 1-2)}$$

where,

ϵ_o = free-space permittivity = 8.854×10^{-12} f/m.

In Equation 1-2, the area (A) must be large with respect to the distance (d). The ratio of ϵ to ϵ_o is known as the dielectric constant (k) of the material. The dielectric constant is a number which provides a comparison of the given dielectric with air (see Fig. 1-7). The ratio of ϵ/ϵ_o for air is, of course, 1. If the dielectric constant of a material is greater than 1, its use in a capacitor as a dielectric will permit a greater amount of capacitance for the same dielectric thickness as air. Thus, if a material's dielectric constant is 3, it will produce a capacitor having three times the capacitance of one that has air as its dielectric. For a given value of capacitance, then, higher dielectric-constant materials will produce physically smaller capacitors. But, because the dielectric plays such a major role in determining the capacitance of a capacitor, it follows that the influence of a dielectric on capacitor operation, over frequency and temperature, is often important.

Dielectric	K
Air	1
Polystrene	2.5
Paper	4
Mica	5
Ceramic (low K)	10
Ceramic (high K)	100–10,000

Fig. 1-7. Dielectric constants of some common materials.

Real-World Capacitors

The usage of a capacitor is primarily dependent upon the characteristics of its dielectric. The dielectric's characteristics also determine the voltage levels and the temperature extremes at which the device may be used. Thus, any losses or imperfections in the dielectric have an enormous effect on circuit operation.

The equivalent circuit of a capacitor is shown in Fig. 1-8, where C equals the capacitance, R_s is the heat-dissipation loss expressed either as a power factor (PF) or as a dissipation factor (DF), R_p is the insulation resistance, and L is the inductance of the leads and plates. Some definitions are needed now.

Power Factor—In a *perfect* capacitor, the alternating current will lead the applied voltage by 90°. This phase angle (ϕ) will be smaller in a real capacitor due to the total series resistance ($R_s + R_p$) that is shown in the equivalent circuit. Thus,

$$PF = \cos \phi$$

The power factor is a function of temperature, frequency, and the dielectric material.

Insulation Resistance—This is a measure of the amount of dc current that flows through the dielectric of a capacitor with a voltage applied. No material is a perfect insulator; thus, some leakage current must flow. This current path is represented by R_p in the equivalent circuit and, typically, it has a value of 100,000 megohms or more.

Effective Series Resistance—Abbreviated ESR, this resistance is the combined equivalent of R_s and R_p, and is the ac resistance of a capacitor.

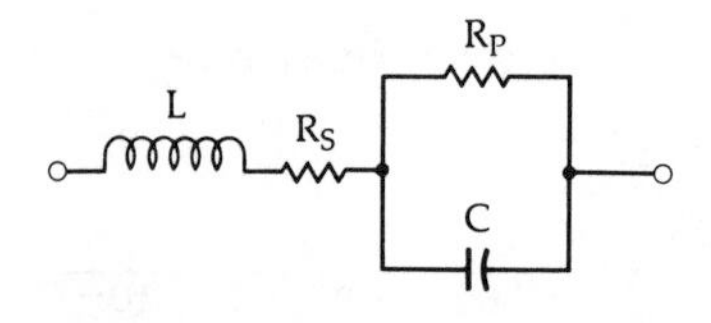

Fig. 1-8. Capacitor equivalent circuit.

$$ESR = \frac{PF}{\omega C}(1 \times 10^6)$$

where,

$\omega = 2\pi f$

Dissipation Factor—The DF is the ratio of ac resistance to the reactance of a capacitor and is given by the formula:

$$DF = \frac{ESR}{X_c} \times 100\%$$

Q—The Q of a circuit is the reciprocal of DF and is defined as the quality factor of a capacitor.

$$Q = \frac{1}{DF} = \frac{X_c}{ESR}$$

Thus, the larger the Q, the better the capacitor.

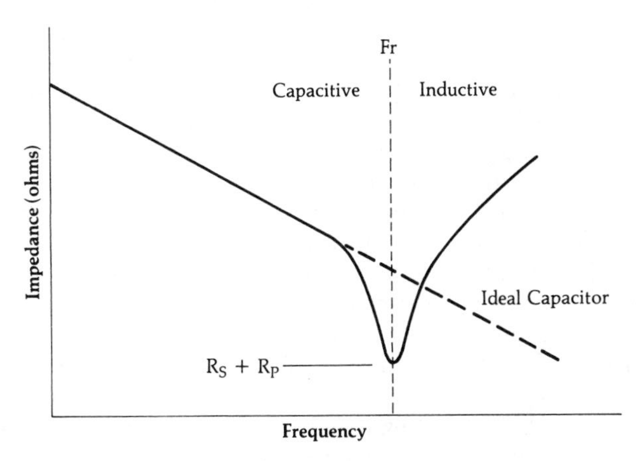

Fig. 1-9. Impedance characteristic vs. frequency.

The effect of these imperfections in the capacitor can be seen in the graph of Fig. 1-9. Here, the impedance characteristic of an ideal capacitor is plotted against that of a real-world capacitor. As shown, as the frequency of operation increases, the lead inductance becomes important. Finally, at F_r, the inductance becomes series resonant with the capacitor. Then, above F_r, the capacitor acts like an inductor. In general, larger-value capacitors tend to exhibit more internal inductance than smaller-value capacitors. Therefore, depending upon its internal structure, a 0.1-μF capacitor may not be as good as a 300-pF capacitor in a bypass application at 250 MHz. In other words, the classic formula for capacitive reactance, $X_c = \frac{1}{\omega C}$, might seem to indicate that larger-value capacitors have less reactance than smaller-value capacitors at a given frequency. At rf frequencies, however, the opposite may be true. At certain higher frequencies, a 0.1-μF capacitor might present a higher impedance to the signal than would a 330-pF capacitor. This is something that must be considered when designing circuits at frequencies above 100 MHz. Ideally, each component that is to be used in any vhf, or higher frequency, design should be examined on a network analyzer similar to the one shown in Fig. 1-10. This will allow the designer to know exactly what he is working with before it goes into the circuit.

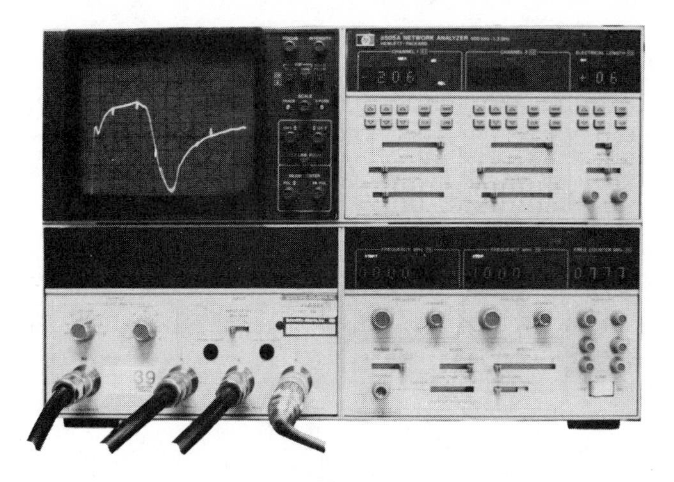

Fig. 1-10. Hewlett-Packard 8505A Network Analyzer.

Capacitor Types

There are many different dielectric materials used in the fabrication of capacitors, such as paper, plastic, ceramic, mica, polystyrene, polycarbonate, teflon, oil, glass, and air. Each material has its advantages and disadvantages. The rf designer is left with a myriad of capacitor types that he could use in any particular application and the ultimate decision to use a particular capacitor is often based on convenience rather than good sound judgement. In many applications, this approach simply cannot be tolerated. This is especially true in manufacturing environments where more than just one unit is to be built and where they must operate reliably over varying temperature extremes. It is often said in the engineering world that anyone can design something and make it work *once*, but it takes a good designer to develop a unit that can be produced in quantity and still operate as it should in different temperature environments.

Ceramic Capacitors—Ceramic dielectric capacitors vary widely in both dielectric constant (K = 5 to 10,000) and temperature characteristics. A good rule of thumb to use is: "The higher the K, the worse is its temperature characteristic." This is shown quite clearly in Fig. 1-11.

As illustrated, low-K ceramic capacitors tend to have linear temperature characteristics. These capacitors are generally manufactured using both magnesium titanate, which has a positive temperature coefficient (TC), and calcium titanate which has a negative TC. By combining the two materials in varying proportions, a range of controlled temperature coefficients can be generated. These capacitors are sometimes called temperature compensating capacitors, or NPO (negative positive zero) ceramics. They can have TCs that range anywhere from +150 to −4700 ppm/°C (parts-per-

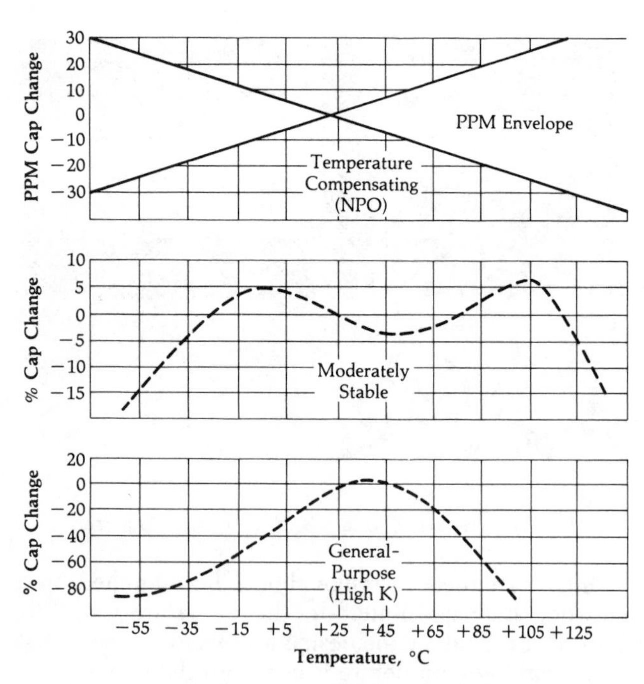

Fig. 1-11. Temperature characteristics for ceramic dielectric capacitors.

million-per-degree-Celsius) with tolerances as small as ±15 ppm/°C. Because of their excellent temperature stability, NPO ceramics are well suited for oscillator, resonant circuit, or filter applications.

Moderately stable ceramic capacitors (Fig. 1-11) typically vary ±15% of their rated capacitance over their temperature range. This variation is typically nonlinear, however, and care should be taken in their use in resonant circuits or filters where stability is important. These ceramics are generally used in switching circuits. Their main advantage is that they are generally smaller than the NPO ceramic capacitors and, of course, cost less.

High-K ceramic capacitors are typically termed general-purpose capacitors. Their temperature characteristics are very poor and their capacitance may vary as much as 80% over various temperature ranges (Fig. 1-11). They are commonly used only in bypass applications at radio frequencies.

There are ceramic capacitors available on the market which are specifically intended for rf applications. These capacitors are typically high-Q (low ESR) devices with flat ribbon leads or with no leads at all. The lead material is usually solid silver or silver plated and, thus, contains very low resistive losses. At vhf frequencies and above, these capacitors exhibit very low lead inductance due to the flat ribbon leads. These devices are, of course, more expensive and require special printed-circuit board areas for mounting. The capacitors that have no leads are called chip capacitors. These capacitors are typically used above 500 MHz where lead inductance cannot be tolerated. Chip capacitors and flat ribbon capacitors are shown in Fig. 1-12.

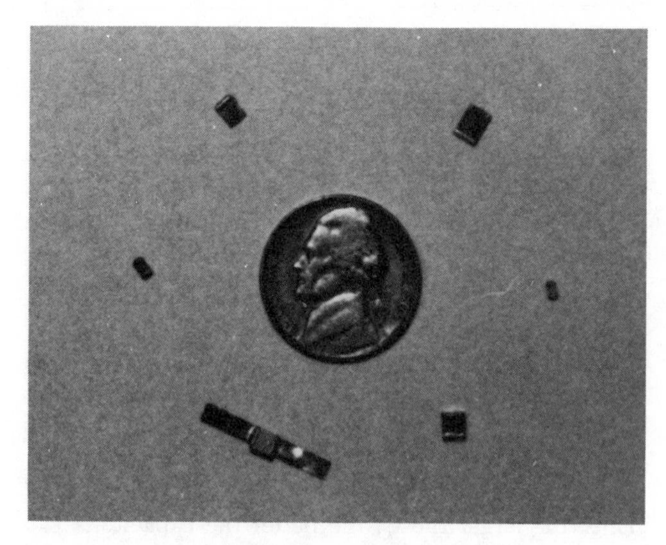

Fig. 1-12. Chip and ribbon capacitors.

Mica Capacitors—Mica capacitors typically have a dielectric constant of about 6, which indicates that for a particular capacitance value, mica capacitors are typically large. Their low K, however, also produces an extremely good temperature characteristic. Thus, mica capacitors are used extensively in resonant circuits and in filters where pc board area is of no concern.

Silvered mica capacitors are even more stable. Ordinary mica capacitors have plates of foil pressed against the mica dielectric. In silvered micas, the silver plates are applied by a process called *vacuum evaporation* which is a much more exacting process. This produces an even better stability with very tight and reproducible tolerances of typically +20 ppm/°C over a range −60 °C to +89 °C.

The problem with micas, however, is that they are becoming increasingly less cost effective than ceramic types. Therefore, if you have an application in which a mica capacitor would seem to work well, chances are you can find a less expensive NPO ceramic capacitor that will work just as well.

Metalized-Film Capacitors—"Metalized-film" is a

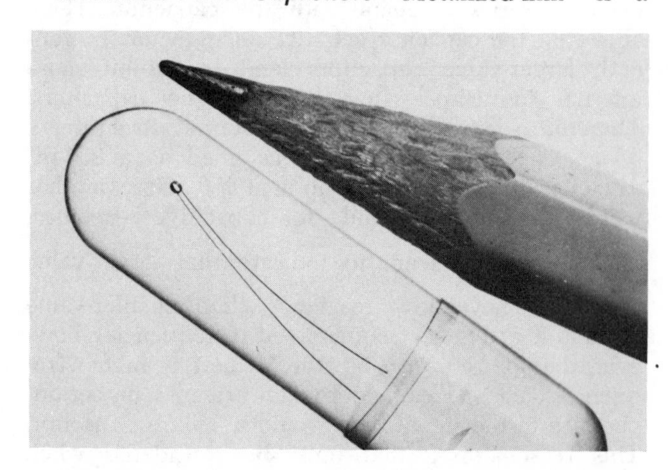

Fig. 1-13. A simple microwave air-core inductor. (*Courtesy Piconics, Inc.*)

broad category of capacitor encompassing most of the other capacitors listed previously and which we have not yet discussed. This includes teflon, polystyrene, polycarbonate, and paper dielectrics.

Metalized-film capacitors are used in a number of applications, including filtering, bypassing, and coupling. Most of the polycarbonate, polystyrene, and teflon styles are available in very tight (±2%) capacitance tolerances over their entire temperature range. Polystyrene, however, typically cannot be used over +85 °C as it is very temperature sensitive above this point. Most of the capacitors in this category are typically larger than the equivalent-value ceramic types and are used in applications where space is not a constraint.

INDUCTORS

An inductor is nothing more than a wire wound or coiled in such a manner as to increase the magnetic flux linkage between the turns of the coil (see Fig. 1-13). This increased flux linkage increases the wire's self-inductance (or just plain inductance) beyond that which it would otherwise have been. Inductors are used extensively in rf design in resonant circuits, filters, phase shift and delay networks, and as rf chokes used to prevent, or at least reduce, the flow of rf energy along a certain path.

Real-World Inductors

As we have discovered in previous sections of this chapter, there is no "perfect" component, and inductors are certainly no exception. As a matter of fact, of the components we have discussed, the inductor is probably the component most prone to very drastic changes over frequency.

Fig. 1-14 shows what an inductor really looks like

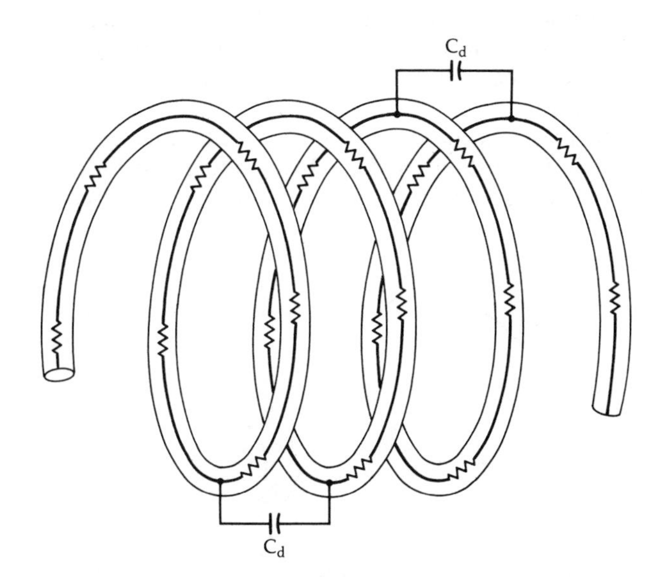

Fig. 1-14. Distributed capacitance and series resistance in an inductor.

at rf frequencies. As previously discussed, whenever we bring two conductors into close proximity but separated by a dielectric, and place a voltage differential between the two, we form a capacitor. Thus, if any wire resistance at all exists, a voltage drop (even though very minute) will occur between the windings, and small capacitors will be formed. This effect is shown in Fig. 1-14 and is called distributed capacitance (C_d). Then, in Fig. 1-15, the capacitance (C_d) is an aggregate of the individual parasitic distributed capacitances of the coil shown in Fig. 1-14.

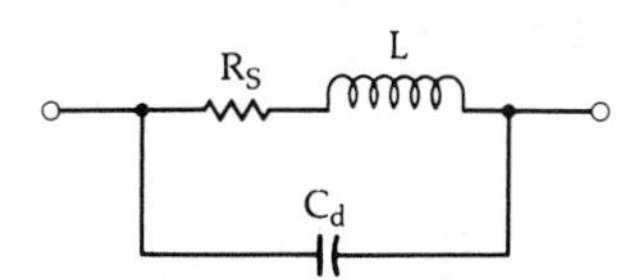

Fig. 1-15. Inductor equivalent circuit.

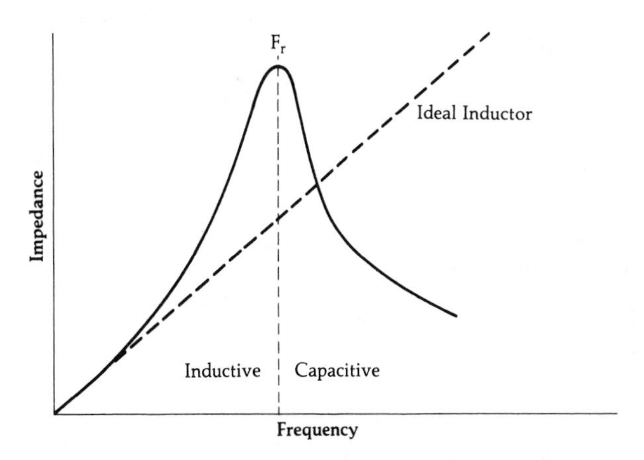

Fig. 1-16. Impedance characteristic vs. frequency for a practical and an ideal inductor.

The effect of C_d upon the reactance of an inductor is shown in Fig. 1-16. Initially, at lower frequencies, the inductor's reactance parallels that of an ideal inductor. Soon, however, its reactance departs from the ideal curve and increases at a much faster rate until it reaches a peak at the inductor's parallel resonant frequency (F_r). Above F_r, the inductor's reactance begins to decrease with frequency and, thus, the inductor begins to look like a capacitor. Theoretically, the resonance peak would occur at infinite reactance (see Example 1-4). However, due to the series resistance of the coil, some finite impedance is seen at resonance.

Recent advances in inductor technology have led to the development of microminiature fixed-chip inductors. One type is shown in Fig. 1-17. These inductors feature a ceramic substrate with gold-plated solderable wrap-around bottom connections. They come in values from 0.01 μH to 1.0 mH, with typical Qs that range from 40 to 60 at 200 MHz.

It was mentioned earlier that the series resistance of a coil is the mechanism that keeps the impedance of the coil finite at resonance. Another effect it has is

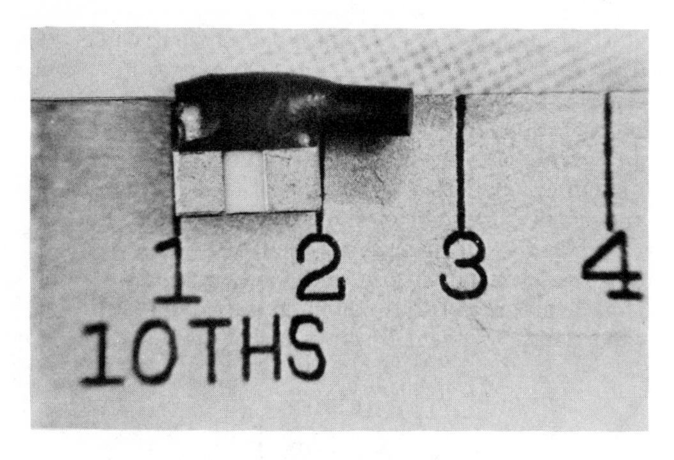

Fig. 1-17. Microminiature chip inductor.
(*Courtesy Piconics, Inc.*)

EXAMPLE 1-4

To show that the impedance of a lossless inductor at resonance is infinite, we can write the following:

$$Z = \frac{X_L X_C}{X_L + X_C} \quad \text{(Eq. 1-3)}$$

where,

Z = the impedance of the parallel circuit,
X_L = the inductive reactance ($j\omega L$),
X_C = the capacitive reactance $\left(\frac{1}{j\omega C}\right)$.

Therefore,

$$Z = \frac{j\omega L \left(\frac{1}{j\omega C}\right)}{j\omega L + \frac{1}{j\omega C}} \quad \text{(Eq. 1-4)}$$

Multiplying numerator and denominator by $j\omega C$, we get:

$$Z = \frac{j\omega L}{(j\omega L)(j\omega C) + 1}$$

$$= \frac{j\omega L}{j^2\omega^2 LC + 1} \quad \text{(Eq. 1-5)}$$

From algebra, $j^2 = -1$; then, rearranging:

$$Z = \frac{j\omega L}{1 - \omega^2 LC} \quad \text{(Eq. 1-6)}$$

If the term $\omega^2 LC$, in Equation 1-6, should ever become equal to 1, then the denominator will be equal to zero and impedance Z will become infinite. The frequency at which $\omega^2 LC$ becomes equal to 1 is:

$$\omega^2 LC = 1$$

$$LC = \frac{1}{\omega^2}$$

$$\sqrt{LC} = \frac{1}{\omega}$$

$$2\pi\sqrt{LC} = \frac{1}{f}$$

$$\frac{1}{2\pi\sqrt{LC}} = f \quad \text{(Eq. 1-7)}$$

which is the familiar equation for the resonant frequency of a tuned circuit.

to broaden the resonance peak of the impedance curve of the coil. This characteristic of resonant circuits is an important one and will be discussed in detail in Chapter 3.

The ratio of an inductor's reactance to its series resistance is often used as a measure of the quality of the inductor. The larger the ratio, the better is the inductor. This quality factor is referred to as the Q of the inductor.

$$Q = \frac{X}{R_s}$$

If the inductor were wound with a perfect conductor, its Q would be infinite and we would have a lossless inductor. Of course, there is no perfect conductor and, thus, an inductor always has some finite Q.

At low frequencies, the Q of an inductor is very good because the only resistance in the windings is the dc resistance of the wire—which is very small. But as the frequency increases, skin effect and winding capacitance begin to degrade the quality of the inductor. This is shown in the graph of Fig. 1-18. At low frequencies, Q will increase directly with frequency because its reactance is increasing and skin effect has not yet become noticeable. Soon, however, skin effect does become a factor. The Q still rises, but at a lesser rate, and we get a gradually decreasing slope in the curve. The flat portion of the curve in Fig. 1-18 occurs as the series resistance and the reactance are changing at the same rate. Above this point, the shunt capacitance and skin effect of the windings combine to decrease the Q of the inductor to zero at its resonant frequency.

Some methods of increasing the Q of an inductor and extending its useful frequency range are:

1. Use a larger diameter wire. This decreases the ac and dc resistance of the windings.
2. Spread the windings apart. Air has a lower dielectric constant than most insulators. Thus, an air gap between the windings decreases the interwinding capacitance.
3. Increase the permeability of the flux linkage path. This is most often done by winding the inductor around a magnetic-core material, such as iron or

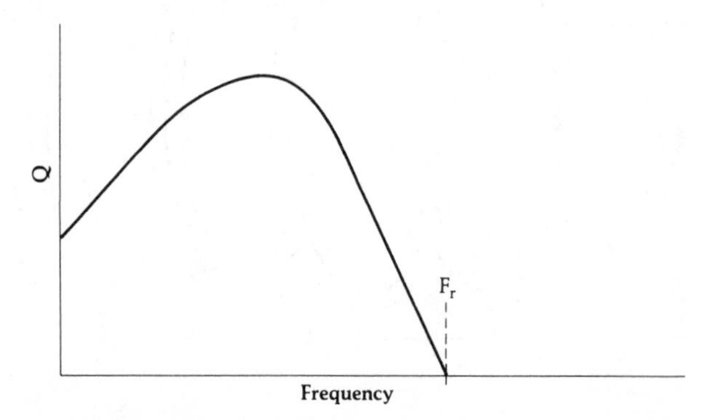

Fig. 1-18. The Q variation of an inductor vs. frequency.

ferrite. A coil made in this manner will also consist of fewer turns for a given inductance. This will be discussed in a later section of this chapter.

Single-Layer Air-Core Inductor Design

Every rf circuit designer needs to know how to design inductors. It may be tedious at times, but it's well worth the effort. The formula that is generally used to design single-layer air-core inductors is given in Equation 1-8 and diagrammed in Fig. 1-19.

$$L = \frac{0.394\, r^2 N^2}{9r + 10l} \qquad \text{(Eq. 1-8)}$$

where,

r = the coil radius in cm,
l = the coil length in cm,
L = the inductance in microhenries.

However, coil length l must be greater than 0.67r. This formula is accurate to within one percent. See Example 1-5.

Keep in mind that even though optimum Q is attained when the length of the coil (l) is equal to its diameter (2r), this is sometimes not practical and, in many cases, the length is much greater than the diameter. In Example 1-5, we calculated the need for 4.8 turns of wire in a length of 0.635 cm and decided that No. 18 AWG wire would fit. The only problem with this approach is that when the design is finished, we end up with a very tightly wound coil. This increases the distributed capacitance between the turns and, thus, lowers the useful frequency range of the inductor by lowering its resonant frequency. We could take either one of the following compromise solutions to this dilemma:

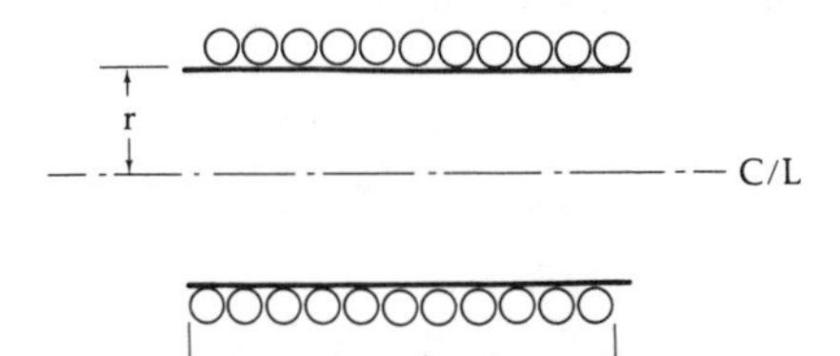

Fig. 1-19. Single-layer air-core inductor requirements.

EXAMPLE 1-5

Design a 100 nH (0.1 μH) air-core inductor on a ¼-inch (0.635 cm) coil form.

Solution

For optimum Q, the length of the coil should be equal to its diameter. Thus, l = 0.635 cm, r = 0.317 cm, and L = 0.1 μH.

Using Equation 1-8 and solving for N gives:

$$N = \sqrt{\frac{29L}{0.394r}}$$

where we have taken l = 2r, for optimum Q.

Substituting and solving:

$$N = \sqrt{\frac{29(0.1)}{(0.394)(0.317)}}$$
$$= 4.8 \text{ turns}$$

Thus, we need 4.8 turns of wire within a length of 0.635 cm. A look at Table 1-1 reveals that the largest diameter enamel-coated wire that will allow 4.8 turns in a length of 0.635 cm is No. 18 AWG wire which has a diameter of 42.4 mils (0.107 cm).

Table 1-1. AWG Wire Chart

Wire Size (AWG)	*Dia in Mils* (Bare)*	*Dia in Mils (Coated)*	*Ohms/ 1000 ft.*	*Area Circular Mils*
1	289.3		0.124	83690
2	257.6		0.156	66360
3	229.4		0.197	52620
4	204.3		0.249	41740
5	181.9		0.313	33090
6	162.0		0.395	26240
7	144.3		0.498	20820
8	128.5	131.6	0.628	16510
9	114.4	116.3	0.793	13090
10	101.9	104.2	0.999	10380
11	90.7	93.5	1.26	8230
12	80.8	83.3	1.59	6530
13	72.0	74.1	2.00	5180
14	64.1	66.7	2.52	4110
15	57.1	59.5	3.18	3260
16	50.8	52.9	4.02	2580
17	45.3	47.2	5.05	2050
18	40.3	42.4	6.39	1620
19	35.9	37.9	8.05	1290
20	32.0	34.0	10.1	1020
21	28.5	30.2	12.8	812
22	25.3	27.0	16.2	640
23	22.6	24.2	20.3	511
24	20.1	21.6	25.7	404
25	17.9	19.3	32.4	320
26	15.9	17.2	41.0	253
27	14.2	15.4	51.4	202
28	12.6	13.8	65.3	159
29	11.3	12.3	81.2	123
30	10.0	11.0	104.0	100
31	8.9	9.9	131	79.2
32	8.0	8.8	162	64.0
33	7.1	7.9	206	50.4
34	6.3	7.0	261	39.7
35	5.6	6.3	331	31.4
36	5.0	5.7	415	25.0
37	4.5	5.1	512	20.2
38	4.0	4.5	648	16.0
39	3.5	4.0	847	12.2
40	3.1	3.5	1080	9.61
41	2.8	3.1	1320	7.84
42	2.5	2.8	1660	6.25
43	2.2	2.5	2140	4.84
44	2.0	2.3	2590	4.00
45	1.76	1.9	3350	3.10
46	1.57	1.7	4210	2.46
47	1.40	1.6	5290	1.96
48	1.24	1.4	6750	1.54
49	1.11	1.3	8420	1.23
50	.99	1.1	10600	0.98

* 1 mil = 2.54×10^{-3} cm

1. Use the next smallest AWG wire size to wind the inductor while keeping the length (l) the same. This approach will allow a small air gap between windings and, thus, decrease the interwinding capacitance. It also, however, increases the resistance of the windings by decreasing the diameter of the conductor and, thus, it lowers the Q.
2. Extend the length of the inductor (while retaining the use of No. 18 AWG wire) just enough to leave a small air gap between the windings. This method will produce the same effect as Method No. 1. It reduces the Q somewhat but it decreases the interwinding capacitance considerably.

Magnetic-Core Materials

In many rf applications, where large values of inductance are needed in small areas, air-core inductors cannot be used because of their size. One method of decreasing the size of a coil while maintaining a given inductance is to decrease the number of turns while at the same time increasing its magnetic flux density. The flux density can be increased by decreasing the "reluctance" or magnetic resistance path that links the windings of the inductor. We do this by adding a magnetic-core material, such as iron or ferrite, to the inductor. The permeability (μ) of this material is much greater than that of air and, thus, the magnetic flux isn't as "reluctant" to flow between the windings. The net result of adding a high permeability core to an inductor is the gaining of the capability to wind a given inductance with fewer turns than what would be required for an air-core inductor. Thus, several advantages can be realized.

1. Smaller size—due to the fewer number of turns needed for a given inductance.
2. Increased Q—fewer turns means less wire resistance.
3. Variability—obtained by moving the magnetic core in and out of the windings.

There are some major problems that are introduced by the use of magnetic cores, however, and care must be taken to ensure that the core that is chosen is the right one for the job. Some of the problems are:

1. Each core tends to introduce its own losses. Thus, adding a magnetic core to an air-core inductor could possibly *decrease* the Q of the inductor, depending on the material used and the frequency of operation.
2. The permeability of all magnetic cores changes with frequency and usually decreases to a very small value at the upper end of their operating range. It eventually approaches the permeability of air and becomes "invisible" to the circuit.
3. The higher the permeability of the core, the more sensitive it is to temperature variation. Thus, over wide temperature ranges, the inductance of the coil may vary appreciably.
4. The permeability of the magnetic core changes with applied signal level. If too large an excitation is applied, saturation of the core will result.

These problems can be overcome if care is taken, in the design process, to choose cores wisely. Manufacturers now supply excellent literature on available sizes and types of cores, complete with their important characteristics.

TOROIDS

A toroid, very simply, is a ring or doughnut-shaped magnetic material that is widely used to wind rf inductors and transformers. Toroids are usually made of iron or ferrite. They come in various shapes and sizes (Fig. 1-20) with widely varying characteristics. When used as cores for inductors, they can typically yield very high Qs. They are self-shielding, compact, and best of all, easy to use.

The Q of a toroidal inductor is typically high because the toroid can be made with an extremely high permeability. As was discussed in an earlier section, high permeability cores allow the designer to construct an inductor with a given inductance (for example, 35 μH) with fewer turns than is possible with an air-core design. Fig. 1-21 indicates the potential savings obtained in number of turns of wire when coil design is changed from air-core to toroidal-core inductors. The air-core inductor, if wound for optimum

Fig. 1-20. Toroidal cores come in various shapes and sizes.

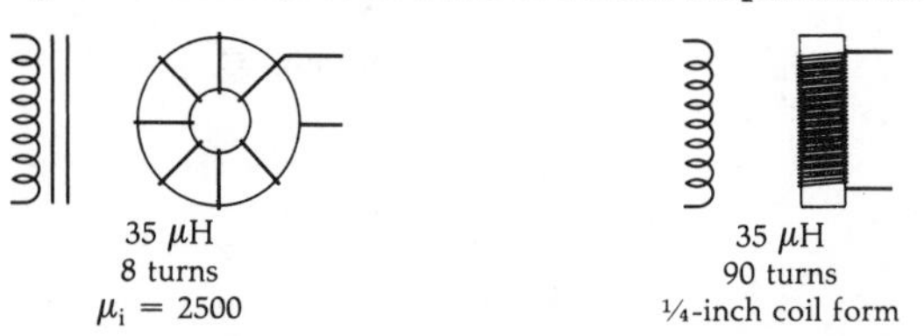

(A) *Toroid inductor.* (B) *Air-core inductor.*

Fig. 1-21. Turns comparison between inductors for the same inductance.

Q, would take 90 turns of a very small wire (in order to fit all turns within a ¼-inch length) to reach 35 μH; however, the toroidal inductor would only need 8 turns to reach the design goal. Obviously, this is an extreme case but it serves a useful purpose and illustrates the point. The toroidal core does require fewer turns for a given inductance than does an air-core design. Thus, there is less ac resistance and the Q *can be* increased dramatically.

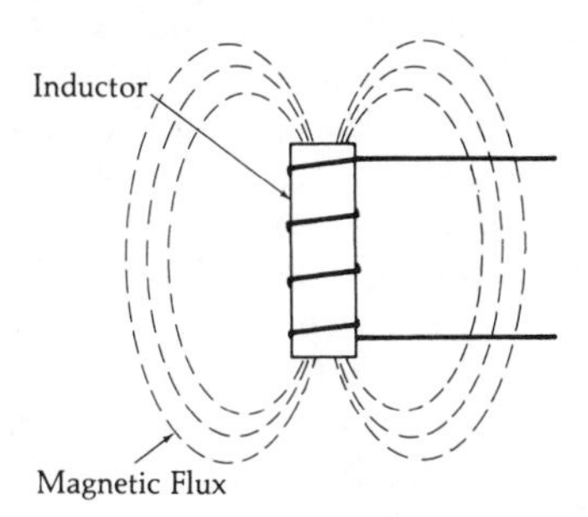

(A) *Typical inductor.*

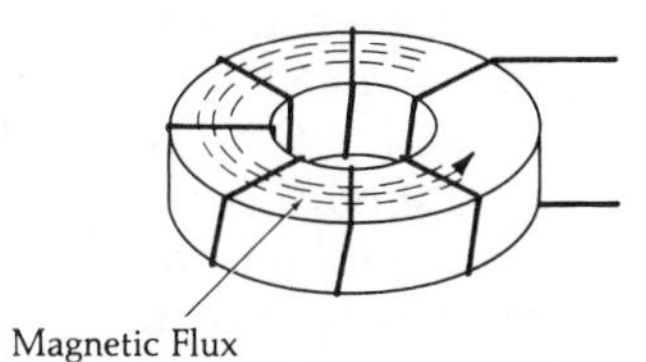

(B) *Toroidal inductor.*

Fig. 1-22. Shielding effect of a toroidal inductor.

The self-shielding properties of a toroid become evident when Fig. 1-22 is examined. In a typical air-core inductor, the magnetic-flux lines linking the turns of the inductor take the shape shown in Fig. 1-22A. The sketch clearly indicates that the air surrounding the inductor is definitely part of the magnetic-flux path. Thus, this inductor tends to radiate the rf signals flowing within. A toroid, on the other hand (Fig. 1-22B), completely contains the magnetic flux within the material itself; thus, no radiation occurs. In actual practice, of course, some radiation will occur but it is minimized. This characteristic of toroids eliminates the need for bulky shields surrounding the inductor. The shields not only tend to reduce available space, but they also reduce the Q of the inductor that they are shielding.

Core Characteristics

Earlier, we discussed, in general terms, the relative advantages and disadvantages of using magnetic cores. The following discussion of typical toroidal-core characteristics will aid you in specifying the core that you need for your particular application.

Fig. 1-23 is a typical magnetization curve for a magnetic core. The curve simply indicates the magnetic-flux density (B) that occurs in the inductor with a specific magnetic-field intensity (H) applied. As the magnetic-field intensity is increased from zero (by in-

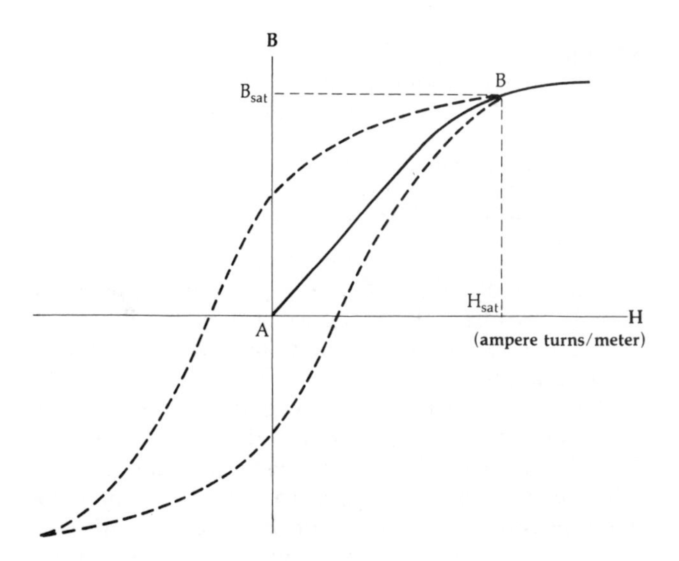

Fig. 1-23. Magnetization curve for a typical core.

creasing the applied signal voltage), the magnetic-flux density that links the turns of the inductor increases quite linearly. The ratio of the magnetic-flux density to the magnetic-field intensity is called the permeability of the material. This has already been mentioned on numerous occasions.

$$\mu = B/H \text{ (Webers/ampere-turn)} \quad \text{(Eq. 1-9)}$$

Thus, the permeability of a material is simply a measure of how well it transforms an electrical excitation into a magnetic flux. The better it is at this transformation, the higher is its permeability.

As mentioned previously, initially the magnetization curve is linear. It is during this linear portion of the curve that permeability is usually specified and, thus, it is sometimes called initial permeability (μ_i) in various core literature. As the electrical excitation increases, however, a point is reached at which the magnetic-flux intensity does not continue to increase at the same rate as the excitation and the slope of the curve begins to decrease. Any further increase in excitation may cause *saturation* to occur. H_{sat} is the excitation point above which no further increase in magnetic-flux density occurs (B_{sat}). The incremental permeability above this point is the same as air. Typically, in rf circuit applications, we keep the excitation small enough to maintain linear operation.

B_{sat} varies substantially from core to core, depending upon the size and shape of the material. Thus, it is necessary to read and understand the manufacturer's literature that describes the particular core you are using. Once B_{sat} is known for the core, it is a very simple matter to determine whether or not its use in a particular circuit application will cause it to saturate. The in-circuit operational flux density (B_{op}) of the core is given by the formula:

$$B_{op} = \frac{E \times 10^8}{(4.44) f N A_e} \quad \text{(Eq. 1-10)}$$

where,

B_{op} = the magnetic-flux density in gauss,
E = the maximum rms voltage across the inductor in volts,
f = the frequency in hertz,
N = the number of turns,
A_e = the effective cross-sectional area of the core in cm^2.

Thus, if the calculated B_{op} for a particular application is less than the published specification for B_{sat}, then the core will not saturate and its operation will be somewhat linear.

Another characteristic of magnetic cores that is very important to understand is that of internal loss. It has previously been mentioned that the careless addition of a magnetic core to an air-core inductor could possibly *reduce* the Q of the inductor. This concept might seem contrary to what we have studied so far, so let's examine it a bit more closely.

The equivalent circuit of an air-core inductor (Fig. 1-15) is reproduced in Fig. 1-24A for your convenience. The Q of this inductor is

$$Q = \frac{X_L}{R_s} \qquad \text{(Eq. 1-11)}$$

where,

$X_L = \omega L$,
R_s = the resistance of the windings.

If we add a magnetic core to the inductor, the equivalent circuit becomes like that shown in Fig. 1-24B. We have added resistance R_p to represent the losses which take place in the core itself. These losses are in the form of *hysteresis*. Hysteresis is the power lost in the core due to the realignment of the magnetic particles within the material with changes in excitation, and the eddy currents that flow in the core due to the voltages induced within. These two types of internal loss, which are inherent to some degree in every magnetic core and are thus unavoidable, combine to reduce the efficiency of the inductor and, thus, increase its loss. But what about the new Q for the magnetic-core inductor? This question isn't as easily answered. Remember, when a magnetic core is inserted into an existing inductor, the value of the inductance is increased. Therefore, at any given frequency, its reactance increases proportionally. The question that must be answered then, in order to determine the new Q of the inductor, is: By what factors did the inductance and loss increase? Obviously, if by adding a toroidal core, the inductance were increased by a factor of two and its total loss was also increased by a factor of two, the Q would remain unchanged. If, however, the total coil loss were increased to four times its previous value while only doubling the inductance, the Q of the inductor would be reduced by a factor of two.

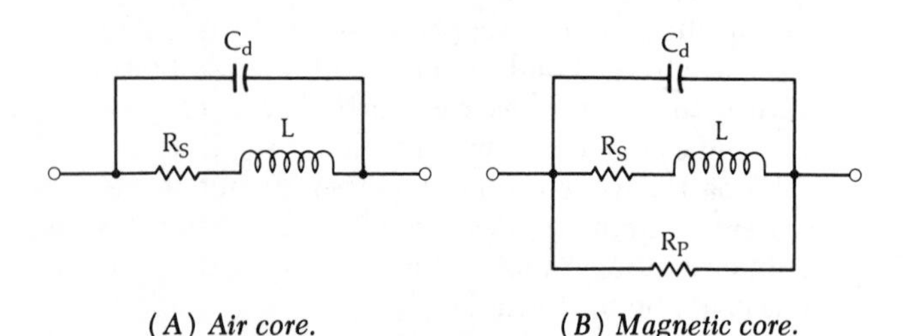

(*A*) *Air core.* (*B*) *Magnetic core.*

Fig. 1-24. Equivalent circuits for air-core and magnetic-core inductors.

Now, as if all of this isn't confusing enough, we must also keep in mind that the additional loss introduced by the core is not constant, but varies (usually increases) with frequency. Therefore, the designer must have a complete set of manufacturer's data sheets for every core he is working with.

Toroid manufacturers typically publish data sheets which contain all the information needed to design inductors and transformers with a particular core. (Some typical specification and data sheets are given in Figs. 1-25 and 1-26.) In most cases, however, each manufacturer presents the information in a unique manner and care must be taken in order to extract the information that is needed without error, and in a form that can be used in the ensuing design process. This is not always as simple as it sounds. Later in this chapter, we will use the data presented in Figs. 1-25 and 1-26 to design a couple of toroidal inductors so that we may see some of those differences. Table 1-2 lists some of the commonly used terms along with their symbols and units.

Powdered Iron Vs. Ferrite

In general, there are no hard and fast rules governing the use of ferrite cores versus powdered-iron cores in rf circuit-design applications. In many instances, given the same permeability and type, either core could be used without much change in performance of the actual circuit. There are, however, special applications in which one core might out-perform another, and it is those applications which we will address here.

Powdered-iron cores, for instance, can typically handle more rf power without saturation or damage than the same size ferrite core. For example, ferrite, if driven with a large amount of rf power, tends to retain its magnetism permanently. This ruins the core by changing its permeability permanently. Powdered iron, on the other hand, if overdriven will eventually return to its initial permeability (μ_i). Thus, in any application where high rf power levels are involved, iron cores might seem to be the best choice.

In general, powdered-iron cores tend to yield higher-Q inductors, at higher frequencies, than an equivalent size ferrite core. This is due to the inherent core characteristics of powdered iron which produce much less internal loss than ferrite cores. This characteristic of powdered iron makes it very useful in narrow-band or tuned-circuit applications. Table 1-3 lists a few of the common powdered-iron core materials along with their typical applications.

indiana general — BROAD BAND-RATED FERRAMIC COMPONENTS

7400 Series Toroids

Nom. μ_i 2500

- Values measured at 100 KHz, T = 25°C.
- Temperature Coefficient (TC) = 0 to +0.75% /°C max., -40 to +70°C.
- Disaccommodation (D) = 3.0% max., 10-100 min., 25°C.
- Hysteresis Core Constant (η_i) measured at 20 KHz to 30 gauss (3 milli Tesla).
- For mm dimensions and core constants, see page 30.

MECHANICAL SPECIFICATIONS

	PART NUMBER				TOL	UNITS
	BBR-7401	BBR-7402	BBR-7403	BBR-7404		
d_1	0.135	0.155	0.230	0.100	±0.005	in.
d_2	0.065	0.088	0.120	0.050	±0.005	in.
h	0.055	0.051	0.060	0.050	±0.005	in.

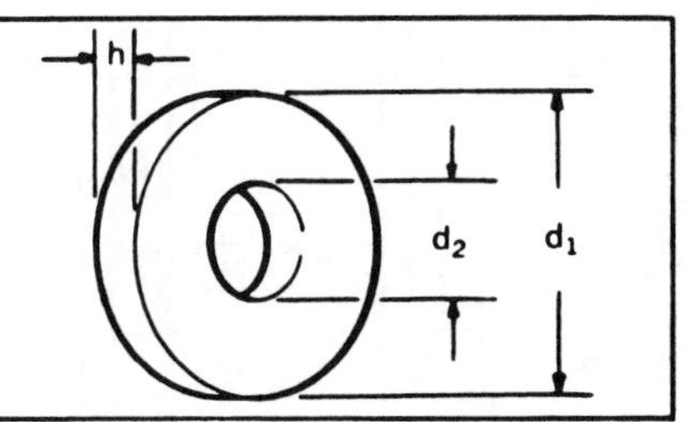

ELECTRICAL SPECIFICATIONS

	PART NUMBER				TOL	UNITS
	BBR-7401	BBR-7402	BBR-7403	BBR-7404		
A_L	510	365	495	440	±20%	nH/turn^2
X_p/N^2	0.320	0.229	0.310	0.276	±20%	ohm/turn^2
R_p/N^2	10.4	7.5	10.0	8.9	min.	ohm/turn^2
Q	54	54	54	54	min.	
V_{rms}	7.9	7.1	13.6	5.1	max.	mv
η_i	1,480	1,400	0,920	2,150	max.	VSA^{-2} H$^{-3/2}$

TYPICAL CHARACTERISTIC CURVES — Part Numbers 7401, 7402, 7403 and 7404

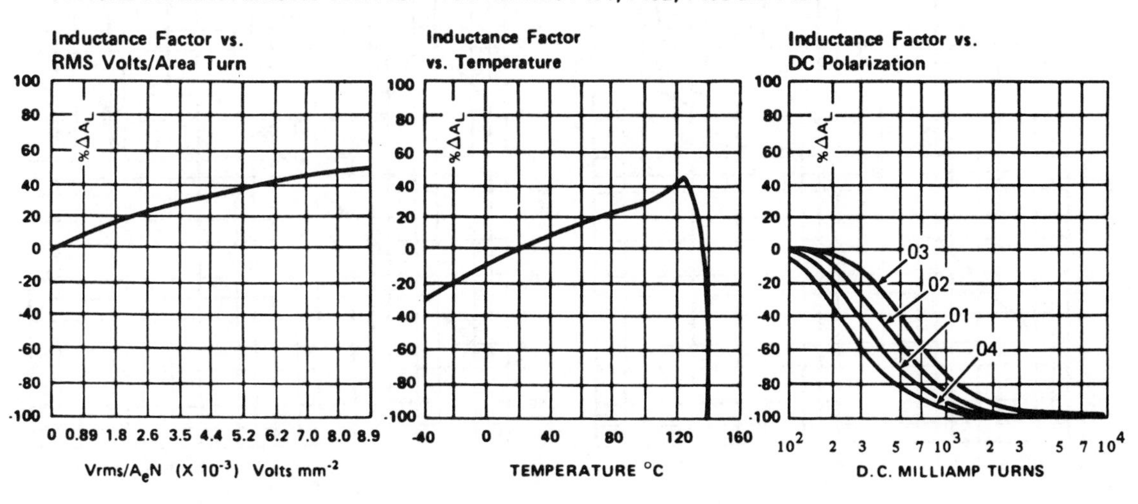

Cont. on next page

Fig. 1-25. Data sheet for ferrite toroidal cores. (*Courtesy Indiana General*)

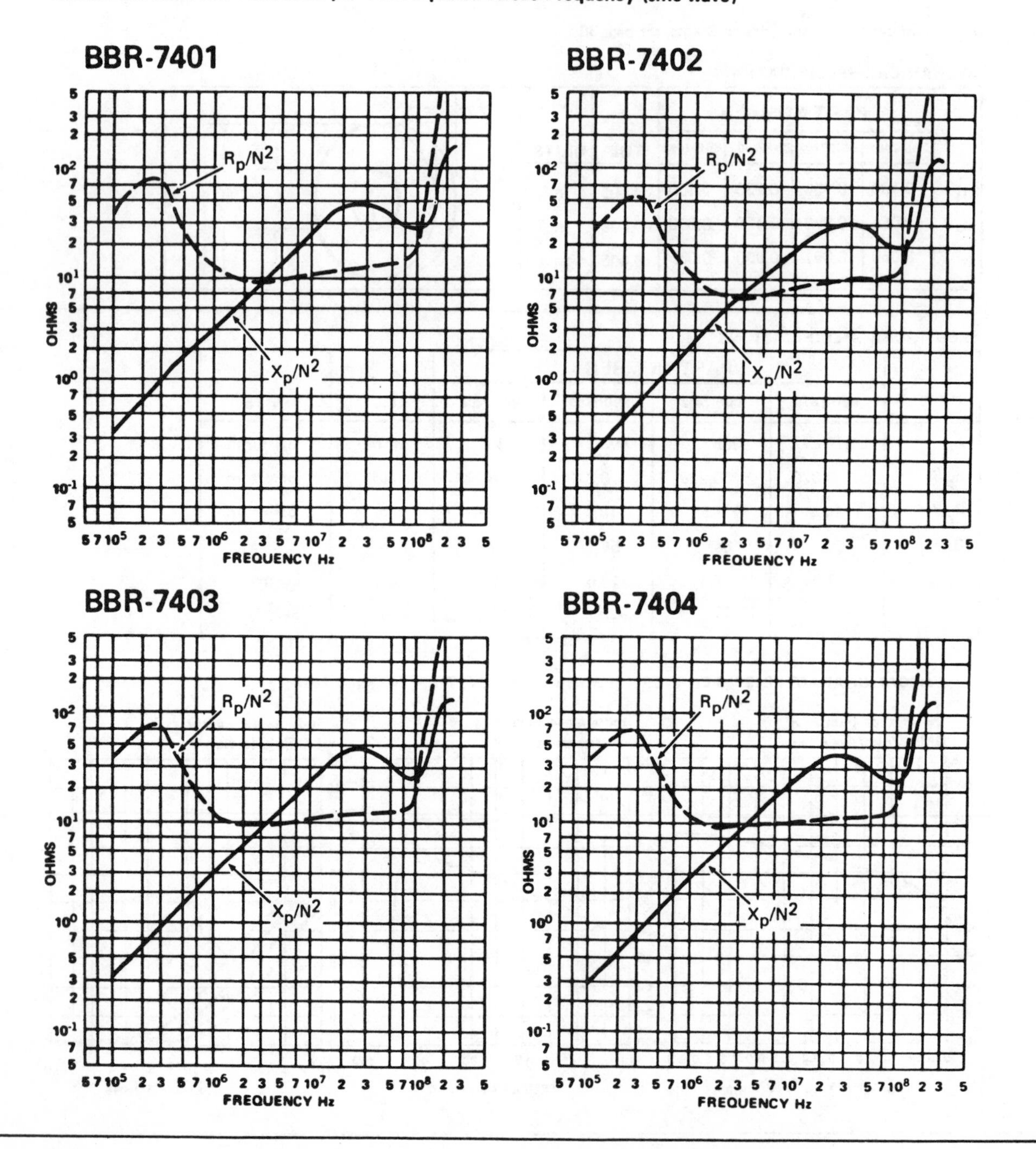

Fig. 1-25—cont. Data sheet for ferrite toroidal cores. (*Courtesy Indiana General*)

IRON-POWDER TOROIDAL CORES

Core size	PHYSICAL DIMENSIONS				
	Outer Diam. (in)	Inner Diam. (in)	Height (in)	Cross Sect. Area $(cm)^2$	Mean Length (cm)
T-225A - -	2.250	1.400	1.000	2.742	14.56
T-225 - -	2.250	1.400	.550	1.508	14.56
T-200 - -	2.000	1.250	.550	1.330	12.97
T-184 - -	1.840	.960	.710	2.040	11.12
T-157 - -	1.570	.950	.570	1.140	10.05
T-130 - -	1.300	.780	.437	.930	8.29
T-106 - -	1.060	.560	.437	.706	6.47
T- 94 - -	.942	.560	.312	.385	6.00
T- 80 - -	.795	.495	.250	.242	5.15
T- 68 - -	.690	.370	.190	.196	4.24
T- 50 - -	.500	.303	.190	.121	3.20
T- 44 - -	.440	.229	.159	.107	2.67
T- 37 - -	.375	.204	.128	.070	2.32
T- 30 - -	.307	.150	.128	.065	1.83
T- 25 - -	.255	.120	.096	.042	1.50
T- 20 - -	.200	.088	.070	.034	1.15
T- 16 - -	.160	.078	.060	.016	0.75
T- 12 - -	.120	.062	.050	.010	0.74

IRON - POWDER MATERIAL vs. FREQUENCY RANGE

Higher Q will be obtained in the upper portion of a materials frequency range when smaller cores are used. Likewise, in the lower portion of a materials frequency range, higher Q can be achieved when using the larger cores.

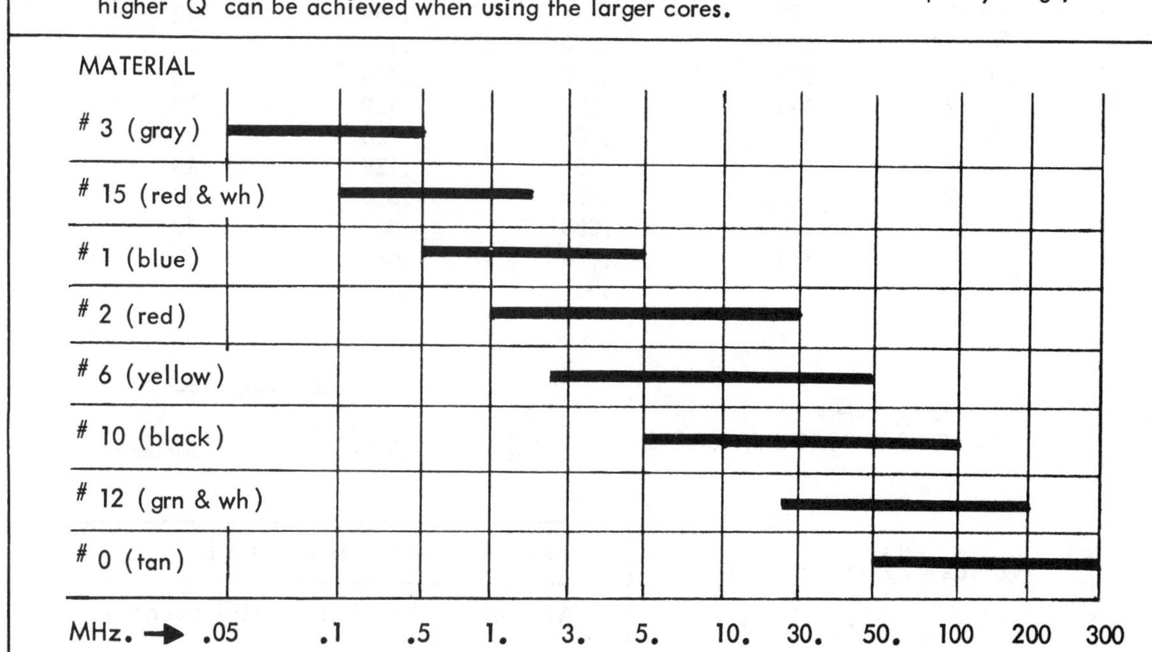

Cont. on next page

Fig. 1-26. Data sheet for powdered-iron toroidal cores. (*Courtesy Amidon Associates*)

IRON-POWDER TOROIDAL CORES

FOR RESONANT CIRCUITS

MATERIAL # 0 permeability 1 50 MHz to 300 MHz Tan

Core number	Outer diam. (in.)	Inner diam. (in.)	Height (in.)	A_L value uh / 100 t
T-130-0	1.300	.780	.437	15.0
T-106-0	1.060	.560	.437	19.2
T- 94-0	.942	.560	.312	10.6
T- 80-0	.795	.495	.250	8.5
T- 68-0	.690	.370	.190	7.5
T- 50-0	.500	.303	.190	6.4
T- 44-0	.440	.229	.159	6.5
T- 37-0	.375	.205	.128	4.9
T- 30-0	.307	.151	.128	6.0
T- 25-0	.255	.120	.096	4.5
T- 20-0	.200	.088	.067	3.5
T- 16-0	.160	.078	.060	3.0
T- 12-0	.125	.062	.050	3.0

MATERIAL # 12 permeability 3 20 MHz to 200 MHz Green & White

Core number	Outer diam. (in.)	Inner diam. (in.)	Height (in.)	A_L value uh / 100 t
T-80-12	.795	.495	.250	22
T-68-12	.690	.370	.190	21
T-50-12	.500	.300	.190	18
T-44-12	.440	.229	.159	18
T-37-12	.375	.205	.128	15
T-30-12	.307	.151	.128	16
T-25-12	.255	.120	.096	12
T-20-12	.200	.088	.067	10
T-16-12	.160	.078	.060	8
T-12-12	.125	.062	.050	7

Key to part numbers for :
IRON POWDER TOROIDAL CORES

T ------- 200 -------- 2
Toroid — Outer diameter — Material

$$\text{Number of turns} = 100\sqrt{\frac{\text{desired inductance (uh)}}{A_L \text{ value (uh per 100 turns)}}}$$

A_L values ± 5%

Cont. on next page

Fig. 1-26—cont. Data sheet for powdered-iron toroidal cores. (*Courtesy Amidon Associates*)

IRON-POWDER TOROIDAL CORES

FOR RESONANT CIRCUITS

MATERIAL # 10 permeability 6 10 MHz to 100 MHz Black

Core number	Outer diam. (in.)	Inner diam. (in.)	Height (in.)	A_L value uh / 100 t
T-94-10	.942	.560	.312	58
T-80-10	.795	.495	.250	32
T-68-10	.690	.370	.190	32
T-50-10	.500	.303	.190	31
T-44-10	.440	.229	.159	33
T-37-10	.375	.205	.128	25
T-30-10	.307	.151	.128	25
T-25-10	.255	.120	.096	19
T-20-10	.200	.088	.067	16
T-16-10	.160	.078	.060	13
T-12-10	.125	.062	.050	12

NUMBER OF TURNS vs. WIRE SIZE and CORE SIZE

Approximate number of turns of wire - single layer wound - single insulation

Core Size	wire size 40	38	36	34	32	30	28	26	24	22	20	18	16	14	12	10
T-12	47	37	29	21	15	11	8	5	4	2	1	1	1	0	0	0
T-16	63	49	38	29	21	16	11	8	5	3	3	1	1	1	0	0
T-20	72	56	43	33	25	18	14	9	6	5	4	3	1	1	1	0
T-25	101	79	62	48	37	28	21	15	11	7	5	4	3	1	1	1
T-30	129	101	79	62	48	37	28	21	15	11	7	5	4	3	1	1
T-37	177	140	110	87	67	53	41	31	23	17	12	9	7	5	3	1
T-44	199	157	124	97	76	60	46	35	27	20	15	10	7	6	5	3
T-50	265	210	166	131	103	81	63	49	37	28	21	16	11	8	6	5
T-68	325	257	205	162	127	101	79	61	47	36	28	21	15	11	9	7
T-80	438	347	276	219	172	137	108	84	66	51	39	30	23	17	12	8
T-94	496	393	313	248	195	156	123	96	75	58	45	35	27	20	14	10
T-106	496	393	313	248	195	156	123	96	75	58	45	35	27	20	14	10
T-130	693	550	439	348	275	220	173	137	107	83	66	51	40	30	23	17
T-157	846	672	536	426	336	270	213	168	132	104	82	64	50	38	29	22
T-184	846	672	536	426	336	270	213	168	132	104	82	64	50	38	29	22
T-200	1115	886	707	562	445	357	282	223	176	139	109	86	68	53	41	31
T-225	1250	993	793	631	499	400	317	250	198	156	123	98	77	60	46	36

Cont. on next page

Fig. 1-26—cont. Data sheet for powdered-iron toroidal cores. (*Courtesy Amidon Associates*)

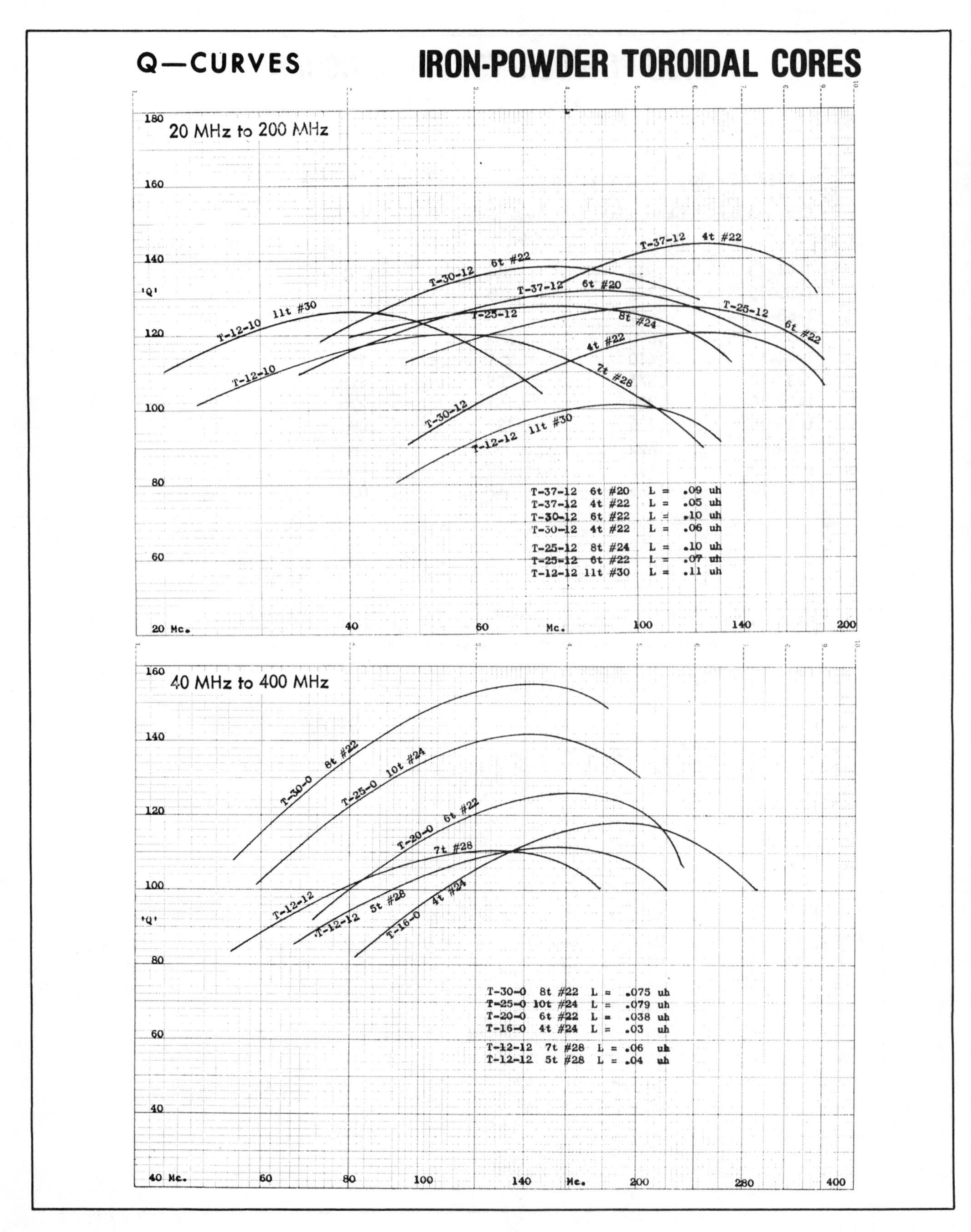

Cont. on next page

Fig. 1-26—cont. Data sheet for powdered-iron toroidal cores. (*Courtesy Amidon Associates*)

IRON-POWDER TOROIDAL CORES

TEMPERATURE COEFFICIENT CURVES

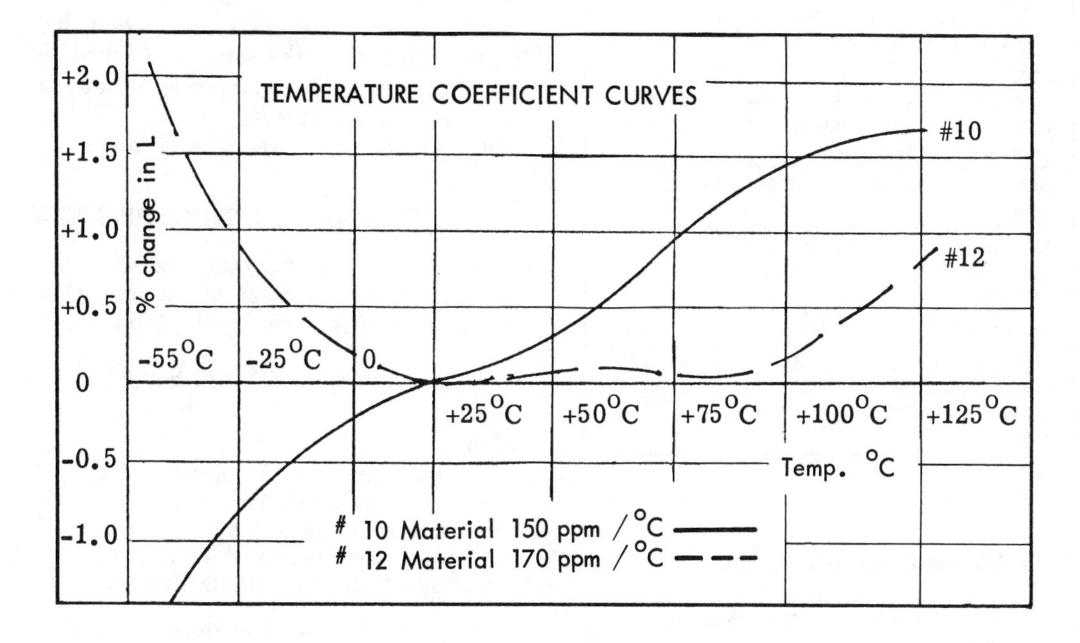

AMIDON Associates • 12033 OTSEGO STREET • NORTH HOLLYWOOD, CALIF. 91607

Fig. 1-26—cont. Data sheet for powdered-iron toroidal cores. (*Courtesy Amidon Associates*)

Table 1-2. Toroidal Core Symbols and Definitions

Symbol	*Description*	*Units*
A_c	Available cross-sectional area. The area (perpendicular to the direction of the wire) for winding turns on a particular core.	cm^2
A_e	Effective area of core. The cross-sectional area that an equivalent gapless core would have.	cm^2
A_L	Inductive index. This relates the inductance to the number of turns for a particular core.	$nH/turn^2$
B_{sat}	Saturation flux density of the core.	gauss
B_{op}	Operating flux density of the core. This is with an applied voltage.	gauss
l_e	Effective length of the flux path.	cm
μ_i	Initial permeability. This is the effective permeability of the core at low excitation in the linear region.	numeric

Table 1-3. Powdered-Iron Materials

Material	*Application/Classification*
Carbonyl C	A medium-Q powdered-iron material at 150 kHz. A high-cost material for am tuning applications and low-frequency if transformers.
Carbonyl E	The most widely used of all powdered-iron materials. Offers high-Q and medium permeability in the 1 MHz to 30 MHz frequency range. A medium-cost material for use in if transformers, antenna coils, and general-purpose designs.
Carbonyl J	A high-Q powdered-iron material at 40 to 100 MHz, with a medium permeability. A high-cost material for fm and tv applications.
Carbonyl SF	Similar to carbonyl E, but with a better Q up through 50 MHz. Costs more than carbonyl E.
Carbonyl TH	A powdered-iron material with a higher Q than carbonyl E up to 30 MHz, but less than carbonyl SF. Higher cost than carbonyl E.
Carbonyl W	The highest cost powdered-iron material. Offers a high Q to 100 MHz, with medium permeability.
Carbonyl HP	Excellent stability and a good Q for lower frequency operation—to 50 kHz. A powdered-iron material.
Carbonyl GS6	For commercial broadcast frequencies. Offers good stability and a high Q.
IRN-8	A synthetic oxide hydrogen-reduced material with a good Q from 50 to 150 MHz. Medium priced for use in fm and tv applications.

At very low frequencies, or in broad-band circuits which span the spectrum from vlf up through vhf, ferrite seems to be the general choice. This is true because, for a given core size, ferrite cores have a much higher permeability. The higher permeability is needed at the low end of the frequency range where, for a given inductance, fewer windings would be needed with the ferrite core. This brings up another point. Since ferrite cores, in general, have a higher permeability than the same size powdered-iron core, a coil of a given inductance can usually be wound on a much smaller ferrite core and with fewer turns. Thus, we can save circuit board area.

TOROIDAL INDUCTOR DESIGN

For a toroidal inductor operating on the linear (nonsaturating) portion of its magnetization curve, its inductance is given by the following formula:

$$L = \frac{0.4\pi N^2 \mu_i A_c \times 10^{-2}}{l_e} \qquad \text{(Eq. 1-12)}$$

where,

L = the inductance in microhenries,
N = the number of turns,
μ_i = initial permeability,
A_c = the cross-sectional area of the core in cm^2,
l_e = the effective length of the core in cm.

In order to make calculations easier, most manufacturers have combined μ_i, A_c, l_e, and other constants for a given core into a single quantity called the *inductance index*, A_L. The inductance index relates the inductance to the number of turns for a particular core. This simplification reduces Equation 1-12 to:

$$L = N^2 A_L \text{ nanohenries} \qquad \text{(Eq. 1-13)}$$

where,

L = the inductance in nanohenries,
N = the number of turns,
A_L = the inductance index in nanohenries/turn^2

Thus, the number of turns to be wound on a given core for a specific inductance is given by:

$$N = \sqrt{\frac{L}{A_L}} \qquad \text{(Eq. 1-14)}$$

This is shown in Example 1-6.

The Q of the inductor cannot be *calculated* with the information given in Fig. 1-25. If we look at the X_p/N^2, R_p/N^2 vs. Frequency curves given for the BBR-7403, however, we can make a calculated guess. At low frequencies (100 kHz), the Q of the coil would be approximately 54, where,

$$Q = \frac{R_p/N^2}{X_p/N^2} = \frac{R_p}{X_p} \qquad \text{(Eq. 1-16)}$$

As the frequency increases, resistance R_p decreases

EXAMPLE 1-6

Using the data given in Fig. 1-25, design a toroidal inductor with an inductance of 50 μH. What is the largest AWG wire that we could possibly use while still maintaining a single-layer winding? What is the inductor's Q at 100 MHz?

Solution

There are numerous possibilities in this particular design since no constraints were placed on us. Fig. 1-25 is a data sheet for the Indiana General 7400 Series of ferrite toroidal cores. This type of core would normally be used in broadband or low-Q transformer applications rather than in narrow-band tuned circuits. This exercise will reveal why.

The mechanical specifications for this series of cores indicate a fairly typical size for toroids used in small-signal rf circuit design. The largest core for this series is just under a quarter of an inch in diameter. Since no size constraints were placed on us in the problem statement, we will use the BBR-7403 which has an outside diameter of 0.0230 inch. This will allow us to use a larger diameter wire to wind the inductor.

The published value for A_L for the given core is 495 nH/turn². Using Equation 1-14, the number of turns required for this core is:

$$N = \sqrt{\frac{50{,}000 \text{ nH}}{495 \text{ nH/turn}^2}}$$

$$= 10 \text{ turns}$$

Note that the inductance of 50 μH was replaced with its equivalent of 50,000 nH. The next step is to determine the largest diameter wire that can be used to wind the transformer while still maintaining a single-layer winding. In some cases, the data supplied by the manufacturer will include this type of winding information. Thus, in those cases, the designer need only look in a table to determine the maximum wire size that can be used. In our case, this information was not given, so a simple calculation must be made. Fig. 1-27 illustrates the geometry of the problem. It is obvious from the diagram that the inner radius (r_1) of

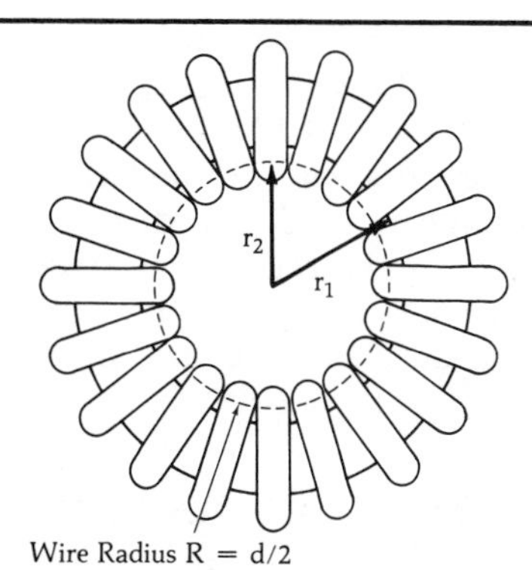

Fig. 1-27. Toroid coil winding geometry.

the toroid is the limiting factor in determining the maximum number of turns for a given wire diameter. The exact maximum diameter wire for a given number of turns can be found by:

$$d = \frac{2\pi r_1}{N + \pi} \qquad \text{(Eq. 1-15)}$$

where,

d = the diameter of the wire in inches,
r_1 = the inner radius of the core in inches,
N = the number of turns.

For this example, we obtain the value of r_1 from Fig. 1-25 (d_2 = 0.120 inch).

$$d = \frac{2\pi \frac{0.120}{2}}{10 + \pi}$$

$$= 28.69 \times 10^{-3} \text{ inches}$$

$$= 28.69 \text{ mils}$$

As a practical rule of thumb, however, taking into account the insulation thickness variation among manufacturers, it is best to add a "fudge factor" and take 90% of the calculated value, or 25.82 mils. Thus, the largest diameter wire used would be the next size below 25.82 mils, which is AWG No. 22 wire.

EXAMPLE 1-7

Using the information provided in the data sheet of Fig. 1-26, design a high-Q ($Q > 80$), 300 nH, toroidal inductor for use at 100 MHz. Due to pc board space available, the toroid may not be any larger than 0.3 inch in diameter.

Solution

Fig. 1-26 is an excerpt from an Amidon Associates iron-powder toroidal-core data sheet. The recommended operating frequencies for various materials are shown in the Iron-Powder Material vs. Frequency Range graph. Either material No. 12 or material No. 10 seems to be well suited for operation at 100 MHz. Elsewhere on the data sheet, material No. 12 is listed as IRN-8. (IRN-8 is described in Table 1-3.) Material No. 10 is not described, so choose material No. 12.

Then, under a heading of Iron-Powder Toroidal Cores, the data sheet lists the physical dimensions of the toroids along with the value of A_L for each. Note, however, that this particular company chooses to specify A_L in μH/100 turns rather than μH/100 turns². The conversion factor between their value of A_L and A_L in nH/turn² is to divide their value of A_L by 10. Thus, the T-80-12 core with an A_L of 22 μH/100 turns is equal to 2.2 nH/turn².

Next, the data sheet lists a set of Q-curves for the cores listed in the preceding charts. Note that all of the curves shown indicate Qs that are greater than 80 at 100 MHz.

Choose the largest core available that will fit in the allotted pc board area. The core you should have chosen is the number T-25-12, with an outer diameter of 0.255 inch.

$$A_L = 12 \ \mu\text{H}/100 \text{ t}$$

$$= 1.2 \text{ nH/turn}^2$$

Therefore, using Equation 1-14, the number of turns required is

$$N = \sqrt{\frac{L}{A_L}}$$

$$= \sqrt{\frac{300}{1.2}}$$

$$= 15.81$$

$$= 16 \text{ turns}$$

Finally, the chart of Number of Turns vs. Wire Size and Core Size on the data sheet clearly indicates that, for a T-25 size core, the largest size wire we can use to wind this particular toroid is No. 28 AWG wire.

while reactance X_p increases. At about 3 MHz, X_p equals R_p and the Q becomes unity. The Q then falls below unity until about 100 MHz where resistance R_p begins to increase dramatically and causes the Q to again pass through unity. Thus, due to losses in the core itself, the Q of the coil at 100 MHz is probably very close to 1. Since the Q is so low, this coil would not be a very good choice for use in a narrow-band tuned circuit. See Example 1-7.

PRACTICAL WINDING HINTS

Fig. 1-28 depicts the correct method for winding a toroid. Using the technique of Fig. 1-28A, the interwinding capacitance is minimized, a good portion of the available winding area is utilized, and the resonant frequency of the inductor is increased, thus extending the useful frequency range of the device. Note that by using the methods shown in Figs. 1-28B and 1-28C, both lead capacitance and interwinding capacitance will affect the toroid.

30°—40°

(A) *Correct.*

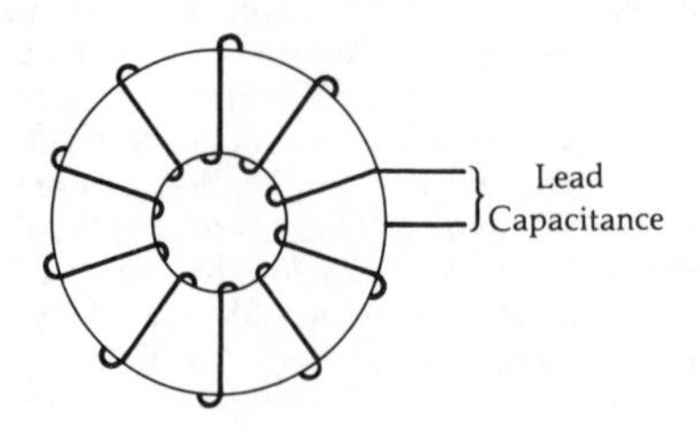

(B) *Incorrect.*

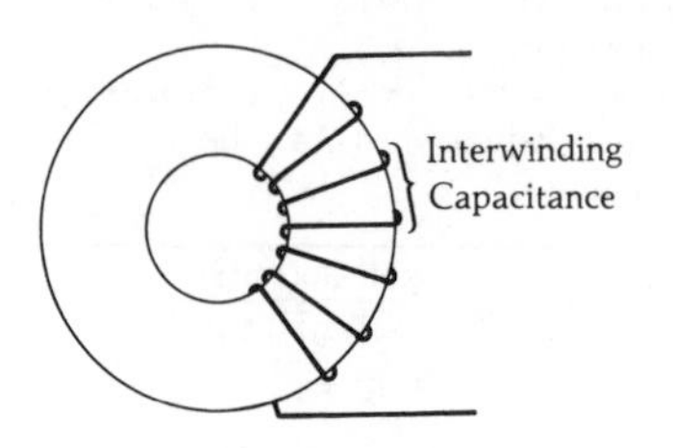

(C) *Incorrect.*

Fig. 1-28. Practical winding hints.

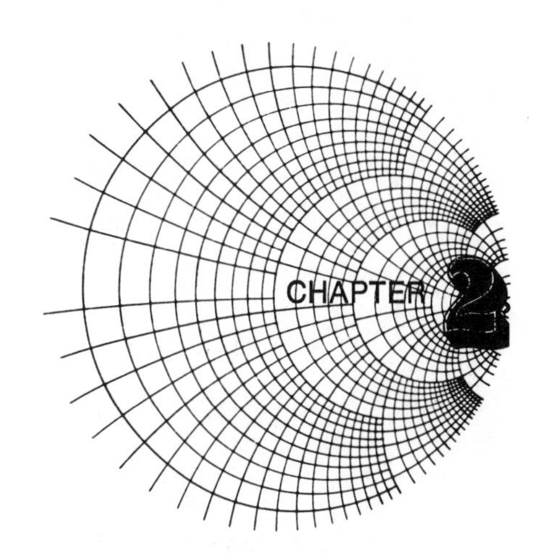

RESONANT CIRCUITS

In this chapter, we will explore the parallel resonant circuit and its characteristics at radio frequencies. We will examine the concept of loaded-Q and how it relates to source and load impedances. We will also see the effects of component losses and how they affect circuit operation. Finally, we will investigate some methods of coupling resonant circuits to increase their selectivity.

SOME DEFINITIONS

The resonant circuit is certainly nothing new in rf circuitry. It is used in practically every transmitter, receiver, or piece of test equipment in existence, to selectively pass a certain frequency or group of frequencies from a source to a load while attenuating all other frequencies outside of this passband. The perfect resonant-circuit *passband* would appear as shown in Fig. 2-1. Here we have a perfect rectangular-shaped passband with infinite attenuation above and below the frequency band of interest, while allowing the desired signal to pass undisturbed. The realization of this filter is, of course, impossible due to the physical characteristics of the components that make up a filter. As we learned in Chapter 1, there is no perfect component and, thus, there can be no perfect filter. If we understand the mechanics of resonant circuits, however, we can certainly tailor an imperfect circuit to suit our needs just perfectly.

Fig. 2-2 is a diagram of what a practical filter response might resemble. Appropriate definitions are presented below:

1. *Bandwidth*—The bandwidth of any resonant circuit is most commonly defined as being the difference between the upper and lower frequency $(f_2 - f_1)$ of the circuit at which its amplitude response is 3 dB below the passband response. It is often called the half-power bandwidth.
2. *Q*—The ratio of the center frequency of the resonant circuit to its bandwidth is defined as the circuit Q.

$$Q = \frac{f_c}{f_2 - f_1} \qquad \text{(Eq. 2-1)}$$

   This Q should not be confused with component Q which was defined in Chapter 1. Component Q does have an effect on circuit Q, but the reverse is not true. Circuit Q is a measure of the selectivity of a resonant circuit. The higher its Q, the narrower its bandwidth, the higher is the selectivity of a resonant circuit.
3. *Shape Factor*—The shape factor of a resonant circuit is typically defined as being the ratio of the 60-dB bandwidth to the 3-dB bandwidth of the resonant circuit. Thus, if the 60-dB bandwidth $(f_4 - f_3)$ were 3 MHz and the 3-dB bandwidth $(f_2 - f_1)$ were 1.5 MHz, then the shape factor would be:

$$SF = \frac{3\ \text{MHz}}{1.5\ \text{MHz}} = 2$$

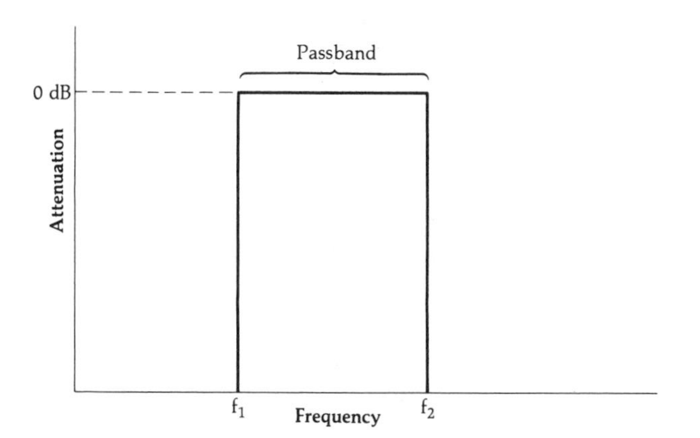

Fig. 2-1. The perfect filter response.

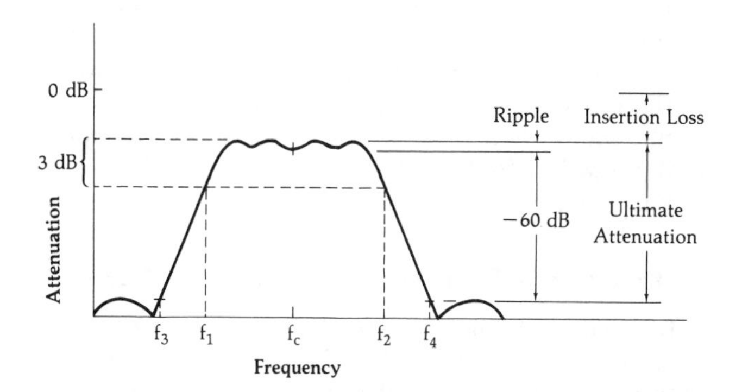

Fig. 2-2. A practical filter response.

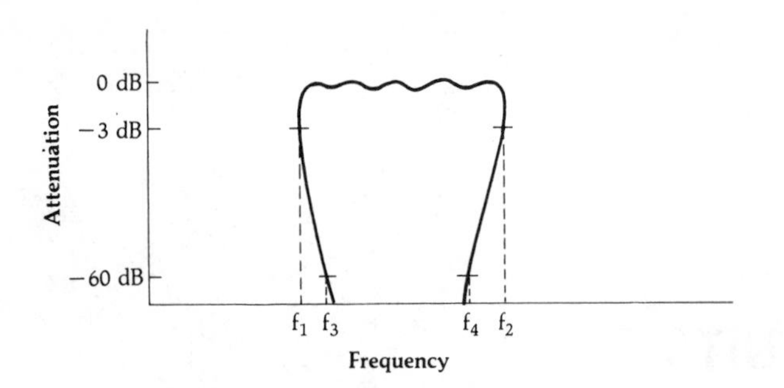

Fig. 2-3. An impossible shape factor.

Shape factor is simply a degree of measure of the steepness of the skirts. The smaller the number, the steeper are the response skirts. Notice that our perfect filter in Fig. 2-1 has a shape factor of 1, which is the ultimate. The passband for a filter with a shape factor smaller than 1 would have to look similar to the one shown in Fig. 2-3. Obviously, this is a physical impossibility.

4. *Ultimate Attenuation*—Ultimate attenuation, as the name implies, is the final *minimum* attenuation that the resonant circuit presents outside of the specified passband. A perfect resonant circuit would provide infinite attenuation outside of its passband. However, due to component imperfections, infinite attenuation is infinitely impossible to get. Keep in mind also, that if the circuit presents response peaks outside of the passband, as shown in Fig. 2-2, then this, of course, detracts from the ultimate attenuation specification of that resonant circuit.
5. *Insertion Loss*—Whenever a component or group of components is inserted between a generator and its load, some of the signal from the generator is absorbed in those components due to their inherent resistive losses. Thus, not as much of the transmitted signal is transferred to the load as when the load is connected directly to the generator. (I am assuming here that no impedance matching function is being performed.) The attenuation that results is called *insertion loss* and it is a very important characteristic of resonant circuits. It is usually expressed in decibels (dB).
6. *Ripple*—Ripple is a measure of the flatness of the passband of a resonant circuit and it is also expressed in decibels. Physically, it is measured in the response characteristics as the difference between the maximum attenuation *in the passband* and the minimum attenuation in the passband. In Chapter 3, we will actually design filters for a specific passband ripple.

RESONANCE (LOSSLESS COMPONENTS)

In Chapter 1, the concept of resonance was briefly mentioned when we studied the parasitics associated with individual component elements. We will now examine the subject of resonance in detail. We will determine what causes resonance to occur and how we can use it to our best advantage.

The voltage division rule (illustrated in Fig. 2-4) states that whenever a shunt element of impedance Z_p is placed across the output of a generator with an internal resistance R_s, the maximum output voltage available from this circuit is

$$V_{out} = \frac{Z_p}{R_s + Z_p}(V_{in}) \quad \text{(Eq. 2-2)}$$

Thus, V_{out} will always be less than V_{in}. If Z_p is a frequency-dependent impedance, such as a capacitive or inductive reactance, then V_{out} will also be frequency dependent and the ratio of V_{out} to V_{in}, which is the gain (or, in this case, loss) of the circuit, will also be frequency dependent. Let's take, for example, a 25-pF capacitor as the shunt element (Fig. 2-5A) and plot the function of V_{out}/V_{in} in dB versus frequency, where we have:

$$\frac{V_{out}}{V_{in}} = 20 \log_{10} \frac{X_C}{R_s + X_C} \quad \text{(Eq. 2-3)}$$

where,

$\frac{V_{out}}{V_{in}}$ = the loss in dB,
R_s = the source resistance,
X_C = the reactance of the capacitor.

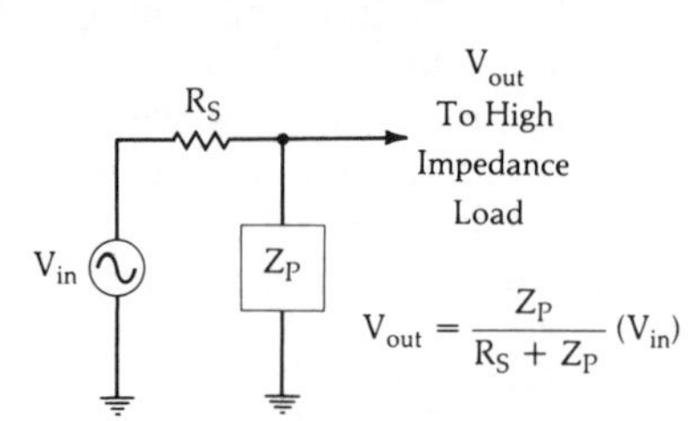

Fig. 2-4. Voltage division rule.

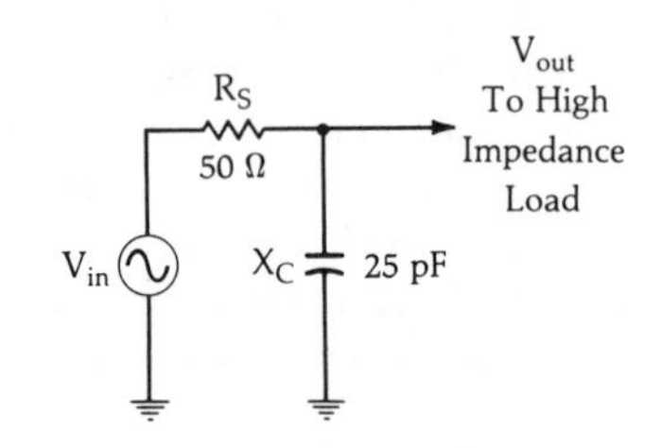

(*A*) *Simple circuit.*

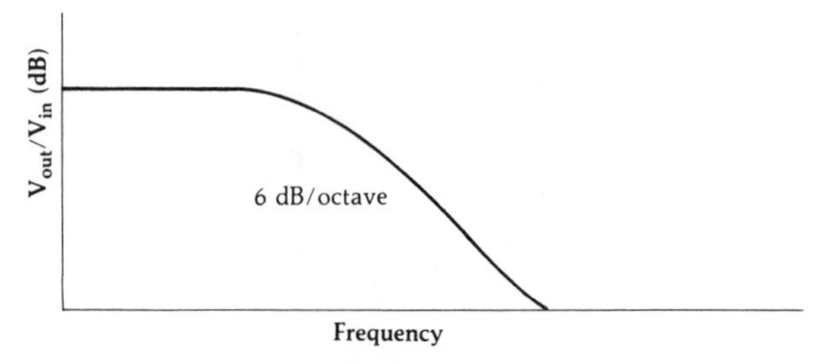

(*B*) *Response curve.*

Fig. 2-5. Frequency response of a simple RC low-pass filter.

Book Mark

Sams books cover a wide range of technical topics. We are always interested in hearing from our readers regarding their informational needs. Please complete this questionnaire and return it to us with your suggestions. We appreciate your comments.

1. Which brand and model of computer do you use?

☐Apple ______

☐Commodore ______

☐IBM ______

☐Other (please specify) ______

______

2. Where do you use your computer?

☐ Home ☐Work

3. Are you planning to buy a new computer?

☐Yes ☐No

If yes, what brand are you planning to buy? ______

4. Please specify the brand/ type of software, operating systems or languages you use.

☐Word Processing ______

☐Spreadsheets ______

☐Data Base Management ______

☐Integrated Software ______

☐Operating Systems ______

☐Computer Languages ______

5. Are you interested in any of the following electronics or technical topics?

☐Amateur radio

☐Antennas and propagation

☐Artificial intelligence/ expert systems

☐Audio

☐Data communications/ telecommunications

☐Electronic projects

☐Instrumentation and measurements

☐Lasers

☐Power engineering

☐Robotics

☐Satellite receivers

6. Are you interested in servicing and repair of any of the following (please specify)?

☐VCRs ______

☐Compact disc players ______

☐Microwave ovens ______

☐Television ______

☐Computers ______

☐Automotive electronics ______

☐Mobile telephones ______

☐Other ______

7. How many computer or electronics books did you buy in the last year?

☐One or two ☐Three or four

☐Five or six ☐More than six

8. What is the average price you paid per book?

☐Less than $10 ☐$10-$15

☐$16-$20 ☐$21-$25 ☐$26+

9. What is your occupation?

☐Manager

☐Engineer

☐Technician

☐Programmer/analyst

☐Student

☐Other ______

10. Please specify your educational level.

☐High school

☐Technical school

☐College graduate

☐Postgraduate

11. Are there specific books you would like to see us publish? ______

______

______

Comments ______

______

______

Name ______

Address ______

City ______

State/Zip ______

21868

SAMS™ Book Mark

Book Mark SAMS™

NO POSTAGE
NECESSARY
IF MAILED
IN THE
UNITED STATES

BUSINESS REPLY CARD

FIRST CLASS PERMIT NO. 1076 INDIANAPOLIS, IND.

POSTAGE WILL BE PAID BY ADDRESSEE

HOWARD W. SAMS & CO.
ATTN: Public Relations Department
P.O. BOX 7092
Indianapolis, IN 46206

and, where,

$$X_C = \frac{1}{j\omega C}.$$

The plot of this equation is shown in the graph of Fig. 2-5B. Notice that the loss of this circuit increases as the frequency increases; thus, we have formed a simple *low-pass filter.* Notice, also, that the attenuation slope eventually settles down to the rate of 6 dB for every octave (doubling) increase in frequency. This is due to the single reactive element in the circuit. As we will see later, this attenuation slope will increase an additional 6 dB for each ***significant*** reactive element that we insert into the circuit.

If we now delete the capacitor from the circuit and insert a 0.05-μH inductor in its place, we obtain the circuit of Fig. 2-6A and the plot of Fig. 2-6B, where we are plotting:

$$\frac{V_{out}}{V_{in}} = 20 \log_{10} \frac{X_L}{R_s + X_L} \qquad \text{(Eq. 2-4)}$$

where,

$\frac{V_{out}}{V_{in}}$ = the loss in dB,
R_s = the source resistance,
X_L = the reactance of the coil.

and, where,

$$X_L = j\omega L.$$

Here, we have formed a simple ***high-pass filter*** with a final attenuation slope of 6 dB per octave.

Thus, through simple calculations involving the basic voltage division formula (Equation 2-2), we were able to plot the frequency response of two separate and opposite reactive components. But what happens if we place both the inductor and capacitor

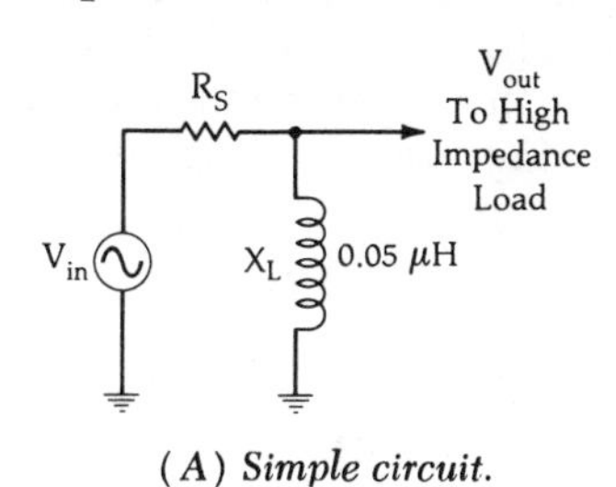

(A) Simple circuit.

(B) Response curve.

Fig. 2-6. Simple high-pass filter.

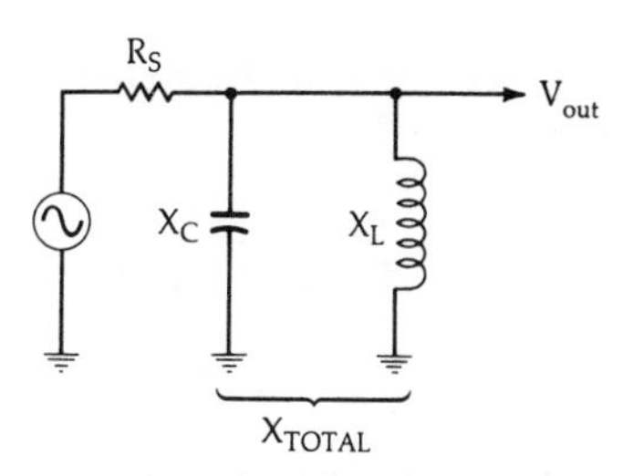

Fig. 2-7. Resonant circuit with two reactive components.

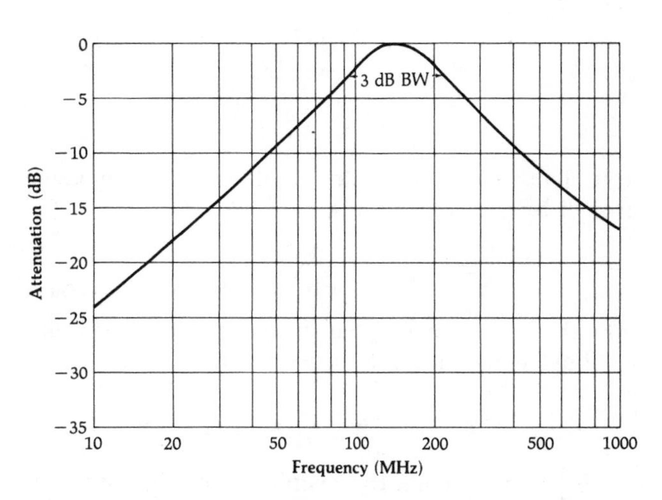

Fig. 2-8. Frequency response of an LC resonant circuit.

across the generator simultaneously? Actually, this case is no more difficult to analyze than the previous two circuits. In fact, at any frequency, we can simply apply the basic voltage division rule as before. The only difference here is that we now have two reactive components to deal with instead of one and these components are in parallel (Fig. 2-7). If we make the calculation for all frequencies of interest, we will obtain the plot shown in Fig. 2-8. The mathematics behind this calculation are as follows:

$$V_{out} = \frac{X_{total}}{R_s + X_{total}}(V_{in}) \qquad \text{(Eq. 2-5)}$$

where,

$$X_{total} = \frac{X_C X_L}{X_C + X_L}.$$

and, where,

$$X_C = \frac{1}{j\omega C},$$

$$X_L = j\omega L.$$

Therefore, we have:

$$X_{total} = \frac{\frac{1}{j\omega C}(j\omega L)}{\frac{1}{j\omega C} + j\omega L} = \frac{\frac{L}{C}}{\frac{1}{j\omega C} + j\omega L}$$

Multiply the numerator and the denominator by $j\omega C$. (Remember that $j^2 = -1$.)

$$X_{total} = \frac{j\omega L}{1 + (j\omega L)(j\omega C)}$$
$$= \frac{j\omega L}{1 - \omega^2 LC}$$

Thus, substituting and transposing in Equation 2-5, we have:

$$\frac{V_{out}}{V_{in}} = \frac{\dfrac{j\omega L}{1 - \omega^2 LC}}{R_s + \dfrac{j\omega L}{1 - \omega^2 LC}}$$

Multiplying the numerator and the denominator through by $1 - \omega^2 LC$ yields:

$$\frac{V_{out}}{V_{in}} = \frac{j\omega L}{(R_s - \omega^2 R_s LC) + j\omega L}$$

Thus, the loss at any frequency may be calculated from the above equation or, if needed, in dB.

$$\frac{V_{out}}{V_{in}} = 20 \log_{10} \left| \frac{j\omega L}{R_s - \omega^2 R_s LC + j\omega L} \right|$$

where $| \ |$ represents the magnitude of the quantity within the brackets.

Notice, in Fig. 2-8, that as we near the resonant frequency of the tuned circuit, the slope of the resonance curve increases to 12 dB/octave. This is due to the fact that we now have two *significant* reactances present and each one is changing at the rate of 6 dB/octave and sloping in opposite directions. As we move away from resonance in either direction, however, the curve again settles to a 6-dB/octave slope because, again, only one reactance becomes significant. The other reactance presents a very high impedance to the circuit at these frequencies and the circuit behaves as if the reactance were no longer there.

LOADED Q

The Q of a resonant circuit was defined earlier to be equal to the ratio of the center frequency of the circuit to its 3-dB bandwidth (Equation 2-1). This "circuit Q," as it was called, is often given the label *loaded* Q because it describes the passband characteristics of the resonant circuit under actual in-circuit or *loaded* conditions. The loaded Q of a resonant circuit is dependent upon three main factors. (These are illustrated in Fig. 2-9.)

1. The source resistance (R_s).
2. The load resistance (R_L).
3. The component Q as defined in Chapter 1.

Effect of R_s and R_L on the Loaded Q

Let's discuss briefly the role that source and load impedances play in determining the loaded Q of a resonant circuit. This role is probably best illustrated

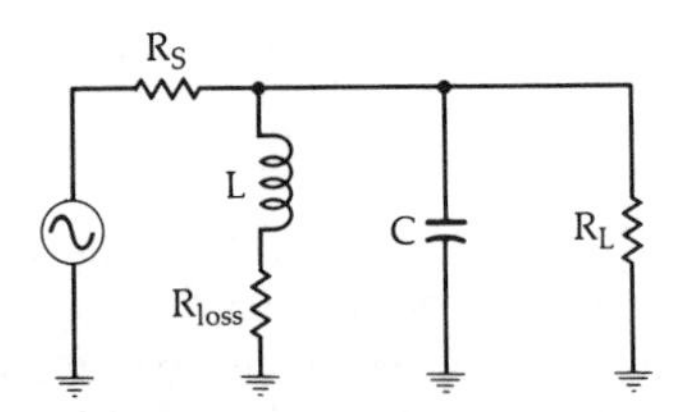

Fig. 2-9. Circuit for loaded-Q calculations.

through an example. In Fig. 2-8, we plotted a resonance curve for a circuit consisting of a 50-ohm source, a 0.05-μH lossless inductor, and a 25-pF lossless capacitor. The loaded Q of this circuit, as defined by Equation 2-1 and determined from the graph, is approximately 1.1. Obviously, this is not a very narrow-band or high-Q design. But now, let's replace the 50-ohm source with a 1000-ohm source and again plot our results using the equation derived in Fig. 2-7 (Equation 2-5). This new plot is shown in Fig. 2-10. (The resonance curve for the 50-ohm source circuit is shown with dashed lines for comparison purposes.) Notice that the Q, or selectivity of the resonant circuit, has been increased dramatically to about 22. Thus, by raising the source impedance, we have increased the Q of our resonant circuit.

Neither of these plots addresses the effect of a load impedance on the resonance curve. If an external load of some sort were attached to the resonant circuit, as shown in Fig. 2-11A, the effect would be to broaden or "de-Q" the response curve to a degree that depends on the value of the load resistance. The equivalent circuit, for resonance calculations, is shown in Fig. 2-11B. The resonant circuit sees an equivalent resistance of R_s in parallel with R_L as its true load. This total external resistance is, by definition, smaller in value than either R_s or R_L, and the loaded Q must decrease. If we put this observation in equation form, it becomes (assuming lossless components):

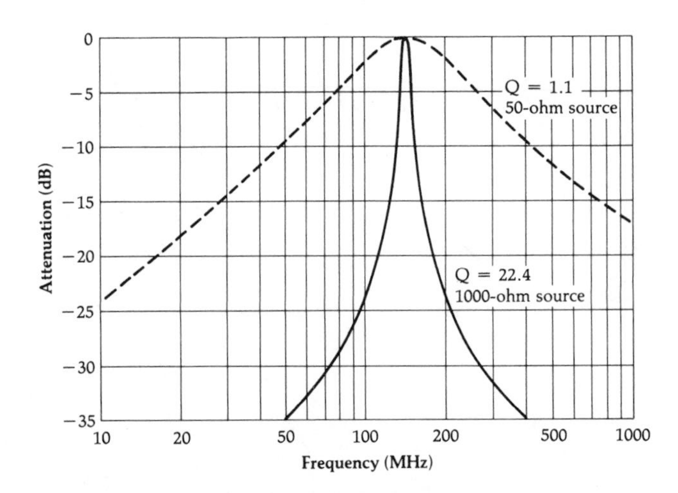

Fig. 2-10. The effect of R_s and R_L on loaded Q.

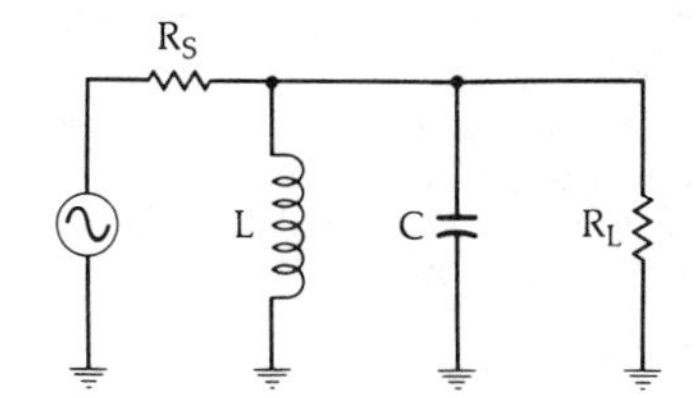

(A) *Resonant circuit with an external load.*

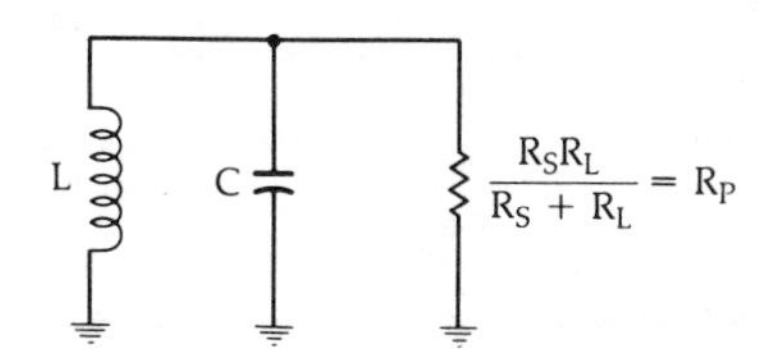

(B) *Equivalent circuit for Q calculations.*

Fig. 2-11. The equivalent parallel impedance across a resonant circuit.

$$Q = \frac{R_p}{X_p} \qquad \text{(Eq. 2-6)}$$

where,

R_p = the equivalent parallel resistance of R_s and R_L,
X_p = either the inductive or capacitive reactance. (They are equal at resonance.)

Equation 2-6 illustrates that a decrease in R_p will decrease the Q of the resonant circuit and an increase in R_p will increase the circuit Q, and it also illustrates another very important point. The same effect can be obtained by keeping R_p constant and varying X_p. Thus, for a given source and load impedance, the optimum Q of a resonant circuit is obtained when the inductor is a small value and the capacitor is a large value. Therefore, in either case, X_p is decreased. This effect is shown using the circuits in Fig. 2-12 and the characteristics curves in Fig. 2-13.

The circuit designer, therefore, has two approaches he can follow in designing a resonant circuit with a particular Q (Example 2-1).

1. He can select an optimum value of source and load impedance.
2. He can select component values of L and C which optimize Q.

Often there is no real choice in the matter because, in many instances, the source and load are defined and we have no control over them. When this occurs, X_p

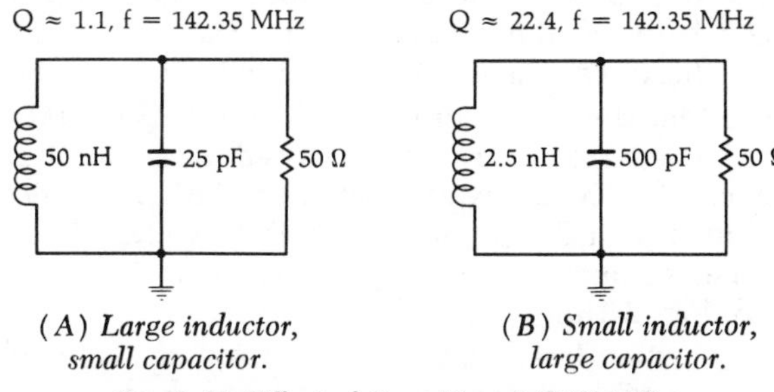

(A) *Large inductor, small capacitor.*

(B) *Small inductor, large capacitor.*

Fig. 2-12. Effect of Q vs. X_p at 142.35 MHz.

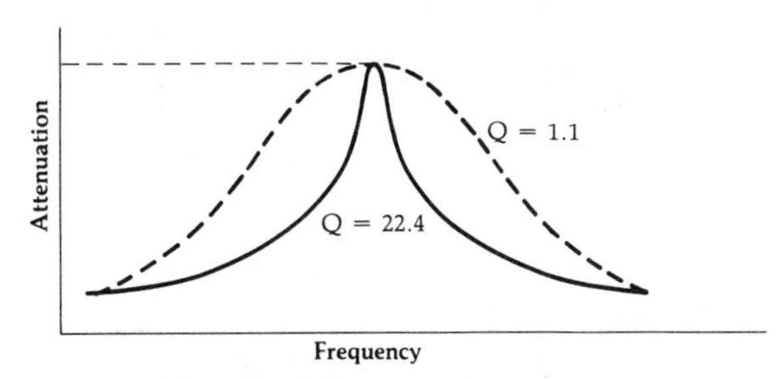

Fig. 2-13. Plot of loaded-Q curves for circuits in Fig. 2-12.

is automatically defined for a given Q and we usually end up with component values that are impractical at best. Later in this chapter, we will study some methods of eliminating this problem.

EXAMPLE 2-1

Design a resonant circuit to operate between a source resistance of 150 ohms and a load resistance of 1000 ohms. The loaded Q must be equal to 20 at the resonant frequency of 50 MHz. Assume lossless components and no impedance matching.

Solution

The effective parallel resistance across the resonant circuit is 150 ohms in parallel with 1000 ohms, or

$$R_p = 130 \text{ ohms}$$

Thus, using Equation 2-6:

$$X_p = \frac{R_p}{Q} = \frac{130}{20} = 6.5 \text{ ohms}$$

and,

$$X_p = \omega L = \frac{1}{\omega C}$$

Therefore, L = 20.7 nH, and C = 489.7 pF.

The Effect of Component Q on Loaded Q

Thus far in this chapter, we have assumed that the components used in the resonant circuits are lossless and, thus, produce no degradation in loaded Q. In reality, however, such is not the case and the individual component Q's must be taken into account. In a lossless resonant circuit, the impedance seen across the circuit's terminals at resonance is infinite. In a practical circuit, however, due to component losses, there exists some finite equivalent parallel resistance. This is illustrated in Fig. 2-14. The resistance (R_p) and its associated shunt reactance (X_p) can be found from the following transformation equations:

$$R_p = (Q^2 + 1) R_s \qquad \text{(Eq. 2-7)}$$

where,

R_p = the equivalent parallel resistance,
R_s = the series resistance of the component,

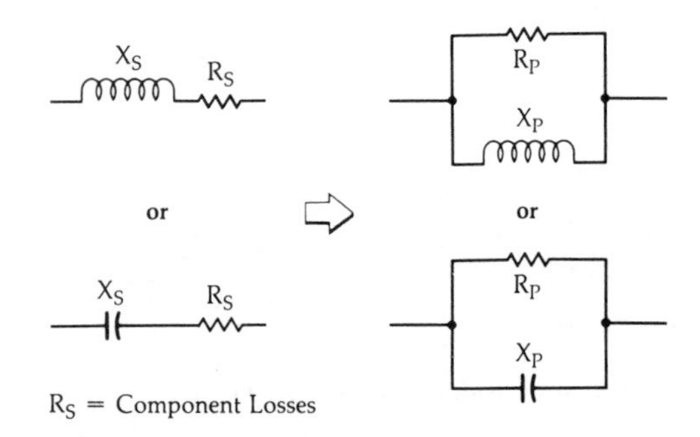

Fig. 2-14. A series-to-parallel transformation.

$Q = Q_s$ which equals Q_p which equals the Q of the component.

and,

$$X_p = \frac{R_p}{Q_p} \qquad \text{(Eq. 2-8)}$$

If the Q of the component is greater than 10, then,

$$R_p \approx Q^2 R_s \qquad \text{(Eq. 2-9)}$$

and,

$$X_p \approx X_s \qquad \text{(Eq. 2-10)}$$

These transformations are valid at only one frequency because they involve the component reactance which is frequency dependent (Example 2-2).

Example 2-2 vividly illustrates the potential drastic effects that can occur if poor-quality (low Q) components are used in highly selective resonant circuit designs. The net result of this action is that we effectively place a low-value shunt resistor directly across the circuit. As was shown earlier, any low-value resistance that shunts a resonant circuit drastically reduces its loaded Q and, thus, increases its bandwidth.

In most cases, we only need to involve the Q of the inductor in loaded-Q calculations. The Q of most capacitors is quite high over their useful frequency range, and the equivalent shunt resistance they present to the circuit is also quite high and can usually be neglected. Care must be taken, however, to ensure that this is indeed the case.

INSERTION LOSS

Insertion loss (defined earlier in this chapter) is another direct effect of component Q. If inductors and capacitors were perfect and contained no internal resistive losses, then insertion loss for LC resonant circuits and filters would not exist. This is, of course, not the case and, as it turns out, insertion loss is a very critical parameter in the specification of any resonant circuit.

Fig. 2-16 illustrates the effect of inserting a resonant circuit between a source and its load. In Fig. 2-16A, the source is connected directly to the load. Using the voltage division rule, we find that:

$$V_1 = 0.5\, V_{in}$$

EXAMPLE 2-2

Given a 50-nanohenry coil as shown in Fig. 2-15A, compute its Q at 100 MHz. Then, transform the series circuit of Fig. 2-15A into the equivalent parallel inductance and resistance circuit of Fig. 2-15B.

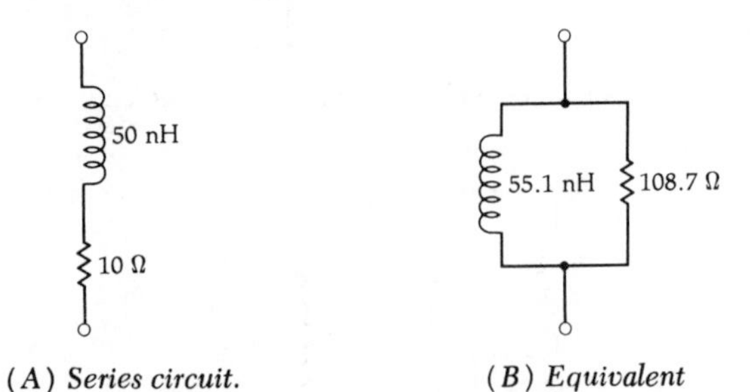

(A) Series circuit. *(B) Equivalent parallel circuit.*

Fig. 2-15. Example of a series-to-parallel transformation.

Solution

The Q of this coil at 100 MHz is, from Chapter 1,

$$Q = \frac{X_s}{R_s} = \frac{2\pi(100 \times 10^6)(50 \times 10^{-9})}{10} = 3.14$$

Then, since the Q is less than 10, use Equation 2-7 to find R_p.

$$R_p = (Q^2 + 1)R_s = [(3.14)^2 + 1]\, 10 = 108.7 \text{ ohms}$$

Next, we find X_p using Equation 2-8:

$$X_p = \frac{R_p}{Q_p} = \frac{108.7}{3.14} = 34.62$$

Thus, the parallel inductance becomes:

$$L_p = \frac{X_p}{\omega} = \frac{34.62}{2\pi(100 \times 10^6)} = 55.1 \text{ nH}$$

These values are shown in the equivalent circuit of Fig. 2-15B.

Fig. 2-16B shows that a resonant circuit has been placed between the source and the load. Then, Fig. 2-16C illustrates the equivalent circuit at resonance. Notice that the use of an inductor with a Q of 100 at the resonant frequency creates an effective shunt resistance of 4500 ohms at resonance. This resistance, combined with R_L, produces an 0.9-dB voltage loss at V_1 when compared to the equivalent point in the circuit of Fig. 2-16A.

An insertion loss of 0.9 dB doesn't sound like much, but it can add up very quickly if we cascade several

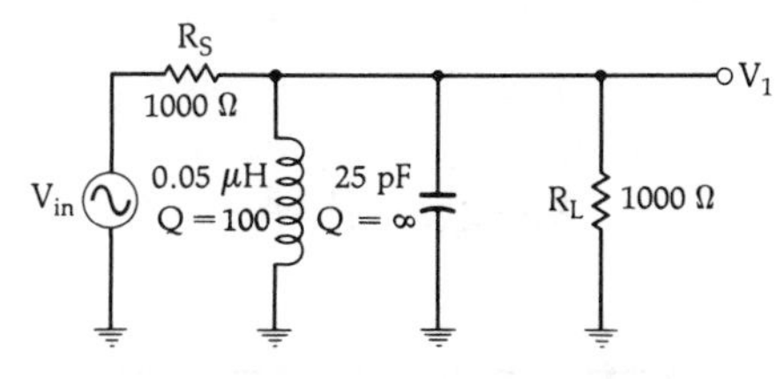

(A) Source connected directly to the load.

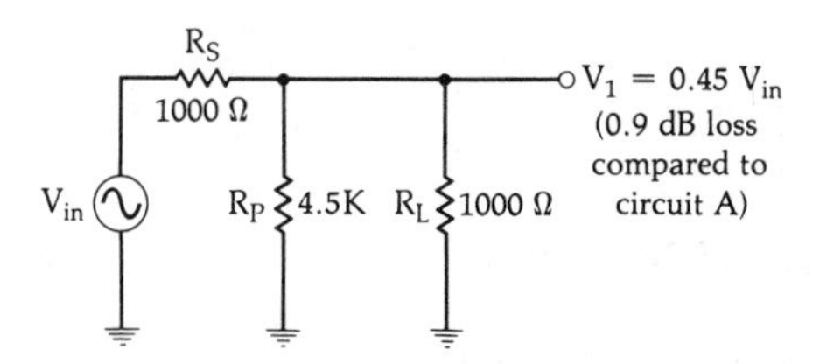

(B) Insertion of a resonant circuit.

(C) Equivalent circuit at resonance.

Fig. 2-16. The effect of component Q on insertion loss.

resonant circuits. We will see some very good examples of this later in Chapter 3. For now, examine the problem given in Example 2-3.

IMPEDANCE TRANSFORMATION

As we have seen in earlier sections of this chapter, low values of source and load impedance tend to load a given resonant circuit down and, thus, tend to decrease its loaded Q and increase its bandwidth. This makes it very difficult to design a simple LC high-Q resonant circuit for use between two very low values of source and load resistance. In fact, even if we were able to come up with a design on paper, it most likely would be impossible to build due to the extremely small (or negative) inductor values that would be required.

One method of getting around this potential design problem is to make use of one of the impedance transforming circuits shown in Fig. 2-18. These handy circuits fool the resonant circuit into seeing a source or load resistance that is much larger than what is actually present. For example, an impedance transformer could present an impedance (R_s') of 500 ohms to the resonant circuit, when in reality there is an impedance (R_s) of 50 ohms. Consequently, by utilizing these transformers, both the Q of the resonant tank and its selectivity can be increased. In many cases, these methods can make a previously unworkable problem workable again, complete with realistic values for the coils and capacitors involved.

The design equations for each of the transformers are presented in the following equations and are useful for designs that need loaded Q's that are greater than 10 (Example 2-4). For the tapped-C transformer (Fig. 2-18A), we use the formula:

$$R_s' = R_s\left(1 + \frac{C_1}{C_2}\right)^2 \qquad \text{(Eq. 2-13)}$$

The equivalent capacitance (C_T) that will resonate with the inductor is equal to C_1 in series with C_2, or:

$$C_T = \frac{C_1C_2}{C_1 + C_2} \qquad \text{(Eq. 2-14)}$$

For the tapped-L network of Fig. 2-18B, we use the following formula:

$$R_s' = R_s\left(\frac{n}{n_1}\right)^2 \qquad \text{(Eq. 2-15)}$$

As an exercise, you might want to rework Example 2-4 without the aid of an impedance transformer. You will find that the inductor value which results is much more difficult to obtain and control physically because it is so small.

COUPLING OF RESONANT CIRCUITS

In many applications where steep passband skirts and small shape factors are needed, a single resonant circuit might not be sufficient. In situations such as this, individual resonant circuits are often coupled together to produce more attenuation at certain frequencies than would normally be available with a single resonator. The coupling mechanism that is used is generally chosen specifically for each application as each type of coupling has its own peculiar characteristics that must be dealt with. The most common forms of coupling are: capacitive, inductive, transformer (mutual), and active (transistor).

Capacitive Coupling

Capacitive coupling is probably the most frequently used method of linking two or more resonant circuits. This is true mainly due to the simplicity of the arrangement but another reason is that it is relatively inexpensive. Fig. 2-19 indicates the circuit arrangement for a two-resonator capacitively coupled filter.

The value of the capacitor that is used to couple each resonator cannot be just chosen at random, as Fig. 2-20 indicates. If capacitor C_{12} of Fig. 2-19 is too large, too much coupling occurs and the frequency response broadens drastically with two response peaks in the filter's passband. If capacitor C_{12} is too small, not enough signal energy is passed from one resonant circuit to the other and the insertion loss can increase to an unacceptable level. The compromise solution to these two extremes is the point of *critical coupling*, where we obtain a reasonable bandwidth and the lowest possible insertion loss and, consequently, a maximum transfer of signal power. There are instances in

EXAMPLE 2-3

Design a simple parallel resonant circuit to provide a 3-dB bandwidth of 10 MHz at a center frequency of 100 MHz. The source and load impedances are each 1000 ohms. Assume the capacitor to be lossless. The Q of the inductor (that is available to us) is 85. What is the insertion loss of the network?

Solution

From Equation 2-1, the required loaded Q of the resonant circuit is:

$$Q = \frac{f_c}{f_2 - f_1}$$

$$= \frac{100\text{ MHz}}{10\text{ MHz}}$$

$$= 10$$

To find the inductor and capacitor values needed to complete the design, it is necessary that we know the equivalent shunt resistance and reactance of the components at resonance. Thus, from Equation 2-8:

$$X_p = \frac{R_p}{Q_p}$$

where,
X_p = the reactance of the inductor and capacitor at resonance,
R_p = the equivalent shunt resistance of the inductor,
Q_p = the Q of the inductor.

Thus,

$$R_p = (85)X_p \qquad \text{(Eq. 2-11)}$$

The loaded Q of the resonant circuit is equal to:

$$Q = \frac{R_{total}}{X_p}$$

$$10 = \frac{R_{total}}{X_p}$$

where,
R_{total} = the shunt resistance, which equals $R_p \parallel R_s \parallel R_L$.

Therefore, we have:

$$10 = \frac{\dfrac{R_p(500)}{R_p + 500}}{X_p} \qquad \text{(Eq. 2-12)}$$

We now have two equations and two unknowns (X_p, R_p). If we substitute Equation 2-11 into Equation 2-12 and solve for X_p, we get:

$$X_p = 44.1 \text{ ohms}$$

Plugging this value back into Equation 2-11 gives:

$$R_p = 3.75\text{K}$$

Thus, our component values must be

$$L = \frac{X_p}{\omega} = 70 \text{ nH}$$

$$C = \frac{1}{\omega X_p} = 36 \text{ pF}$$

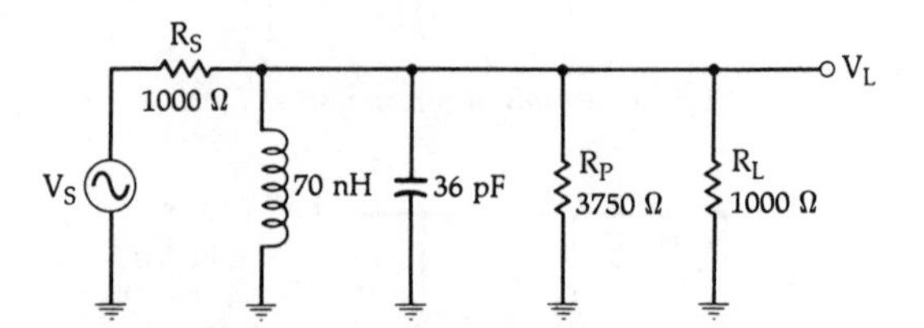

Fig. 2-17. Resonant circuit design for Example 2-3.

The final circuit is shown in Fig. 2-17.

The insertion-loss calculation, at center frequency, is now very straightforward and can be found by applying the voltage division rule as follows. Resistance R_p in parallel with resistance R_L is equal to 789.5 ohms. The voltage at V_L is, therefore,

$$V_L = \frac{789.5}{789.5 + 1000}(V_s)$$

$$= .44\ V_s$$

The voltage at V_L, without the resonant circuit in place, is equal to 0.5 V_s due to the 1000-ohm load. Thus, we have:

$$\text{Insertion Loss} = 20 \log_{10} \frac{0.44\ V_s}{0.5\ V_s}$$

$$= 1.1 \text{ dB}$$

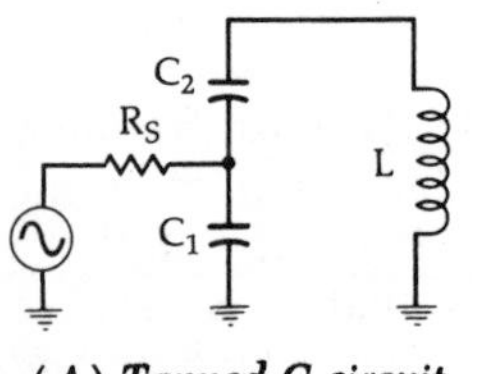

(A) *Tapped-C circuit.*

(B) *Tapped-L circuit.*

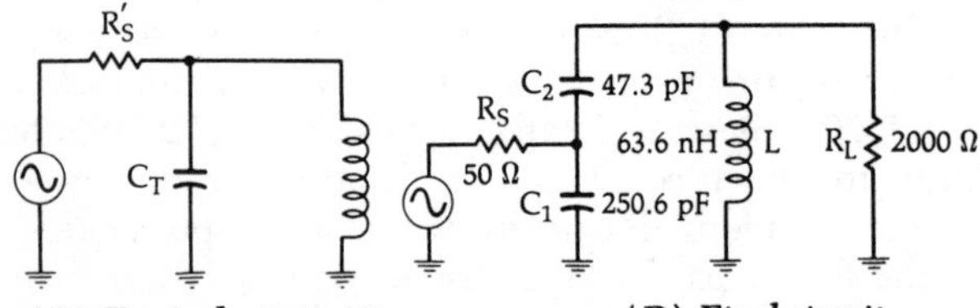

(C) *Equivalent circuit.* (D) *Final circuit.*

Fig. 2-18. Two methods used to perform an impedance transformation.

which overcoupling or undercoupling might serve a useful purpose in a design, such as in tailoring a specific frequency response that a critically coupled filter cannot provide. But these applications are generally left to the multiple resonator filter. The multiple resonator filter is covered in Chapter 3. In this section, we will only concern ourselves with critical coupling as it pertains to resonant circuit design.

The loaded Q of a critically coupled two-resonator circuit is approximately equal to 0.707 times the loaded Q of one of its resonators. Therefore, the 3-dB bandwidth of a two-resonator circuit is actually wider than that of one of its resonators. This might seem contrary to what we have studied so far, but remember, the

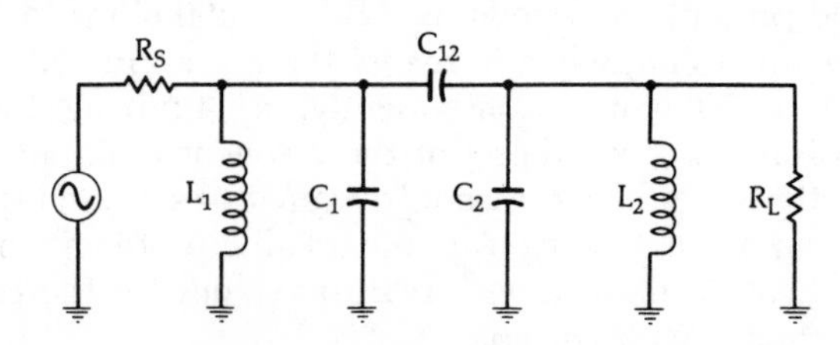

Fig. 2-19. Capacitive coupling.

EXAMPLE 2-4

Design a resonant circuit with a loaded Q of 20 at a center frequency of 100 MHz that will operate between a source resistance of 50 ohms and a load resistance of 2000 ohms. Use the tapped-C approach and assume that inductor Q is 100 at 100 MHz.

Solution

We will use the tapped-C transformer to step the source resistance up to 2000 ohms to match the load resistance for optimum power transfer. (Impedance matching will be covered in detail in Chapter 4.) Thus,

$$R_s' = 2000 \text{ ohms}$$

and from Equation 2-13, we have:

$$\frac{C_1}{C_2} = \sqrt{\frac{R_s'}{R_s}} - 1$$
$$= 5.3$$

or,

$$C_1 = 5.3C_2 \qquad \text{(Eq. 2-16)}$$

Proceeding as we did in Example 2-3, we know that for the inductor:

$$Q_p = \frac{R_p}{X_p} = 100$$

Therefore,

$$R_p = 100\, X_p \qquad \text{(Eq. 2-17)}$$

We also know that the loaded Q of the resonant circuit is equal to:

$$Q = \frac{R_{total}}{X_p}$$

where,

R_{total} = the total equivalent shunt resistance,
$= R_s' \,||\, R_p \,||\, R_L$
$= 1000 \,||\, R_p$

and, where we have taken R_s' and R_L to each be 2000 ohms, in parallel. Hence, the loaded Q is

$$Q = \frac{1000R_p}{(1000 + R_p)X_p} \qquad \text{(Eq. 2-18)}$$

Substituting Equation 2-17 (and the value of the desired loaded Q) into Equation 2-18, and solving for X_p, yields:

$$X_p = 40 \text{ ohms}$$

And, substituting this result back into Equation 2-17 gives

$$R_p = 4000 \text{ ohms}$$

and,

$$L = \frac{X_p}{\omega}$$
$$= 63.6 \text{ nH}$$
$$C_T = \frac{1}{X_p\omega}$$
$$= 39.78 \text{ pF}$$

We now know what the total capacitance must be to resonate with the inductor. We also know from Equation 2-16 that C_1 is 5.3 times larger than C_2. Thus, if we substitute Equation 2-16 into Equation 2-14, and solve the equations simultaneously, we get:

$$C_2 = 47.3 \text{ pF}$$
$$C_1 = 250.6 \text{ pF}$$

The final circuit is shown in Fig. 2-18D.

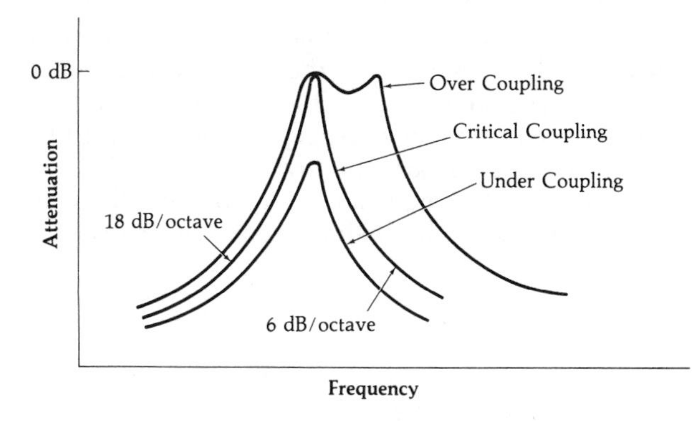

Fig. 2-20. The effects of various values of capacitive coupling on passband response.

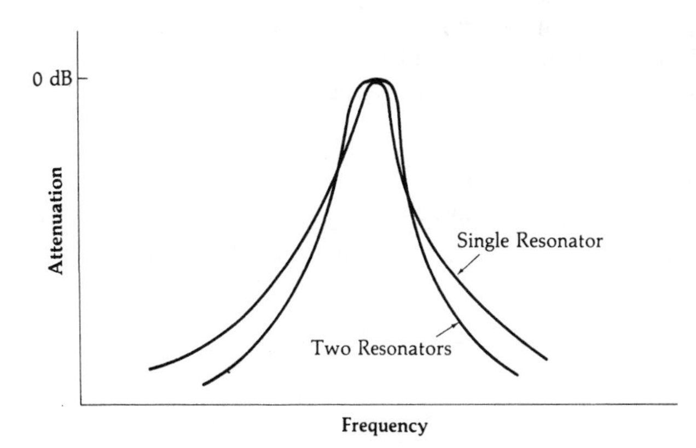

Fig. 2-21. Selectivity of single- and two-resonator designs.

main purpose of the two-resonator passively coupled filter is not to provide a narrower 3-dB bandwidth, but to increase the steepness of the stopband skirts and, thus, to reach an ultimate attenuation much faster than a single resonator could. This characteristic is shown in Fig. 2-21. Notice that the shape factor has decreased for the two-resonator design. Perhaps one way to get an intuitive feel for how this occurs is to consider that each resonator is itself a load for the other resonator, and each decreases the loaded Q of the other. But as we move away from the passband and into the stopband, the response tends to fall much more quickly due to the *combined* response of each resonator.

The value of the capacitor used to couple two identical resonant circuits is given by

$$C_{12} = \frac{C}{Q} \qquad \text{(Eq. 2-19)}$$

where,

C_{12} = the coupling capacitance.
C = the resonant circuit capacitance,
Q = the loaded Q of a single resonator.

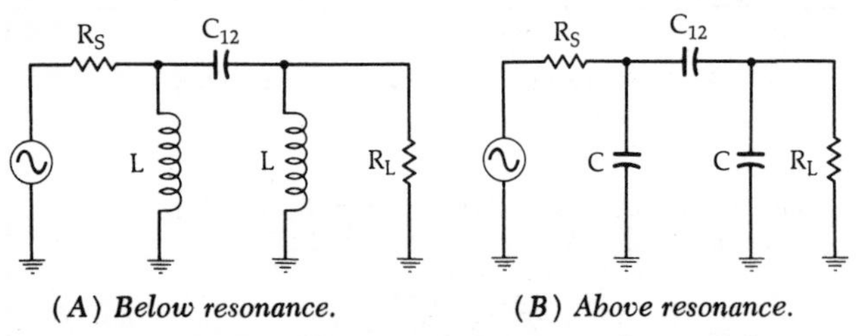

(A) Below resonance. *(B) Above resonance.*

Fig. 2-22. Equivalent circuit of capacitively coupled resonant circuits.

One other important characteristic of a capacitively coupled resonant circuit can be seen if we take another look at Fig. 2-20. Notice that even for the critically coupled case, the response curve is not symmetric around the center frequency but is skewed somewhat. The lower frequency portion of the response plummets down at the rate of 18 dB per octave while the upper slope decreases at only 6 dB per octave. This can be explained if we take a look at the equivalent circuit both above and below resonance. Below resonance, we have the circuit of Fig. 2-22A. The reactance of the two resonant-circuit capacitors (Fig. 2-19) has increased, and the reactance of the two inductors has decreased to the point that only the inductor is seen as a shunt element and the capacitors can be ignored. This leaves three reactive components and each contributes 6 dB per octave to the response.

On the high side of resonance, the equivalent circuit approaches the configuration of Fig. 2-22B. Here the inductive reactance has increased above the capacitive reactance to the point where the inductive reactance can be ignored as a shunt element. We now have an arrangement of three capacitors that effectively looks like a single shunt capacitor and yields a slope of 6 dB per octave.

Inductive Coupling

Two types of inductively coupled resonant circuits are shown in Fig. 2-23. One type (Fig. 2-23A) uses a series inductor or coil to transfer energy from the first resonator to the next, and the other type (Fig. 2-23B) uses transformer coupling for the same purpose. In either case, the frequency response curves will resemble those of Fig. 2-24 depending on the amount of coupling. If we compare Fig. 2-24A with Fig. 2-20, we see that the two are actually mirror images of each other. The response of the inductively coupled resonator is skewed toward the higher end of the frequency spectrum, while the capacitively coupled response is skewed toward the low frequency side. An examination of the equivalent circuit reveals why. Fig. 2-25A indicates that below resonance, the capaci-

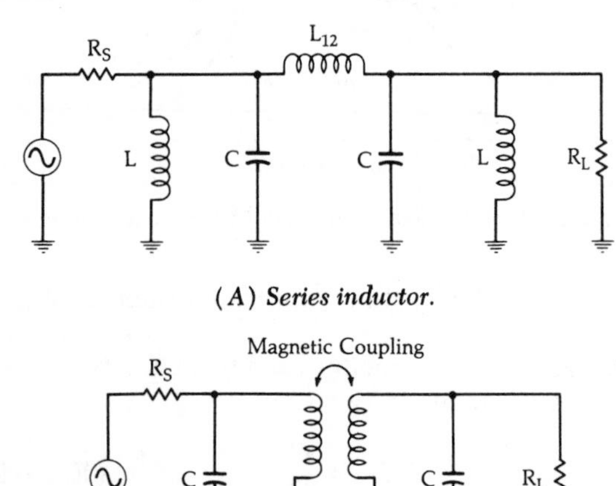

(A) Series inductor.

(B) Transformer.

Fig. 2-23. Inductive coupling.

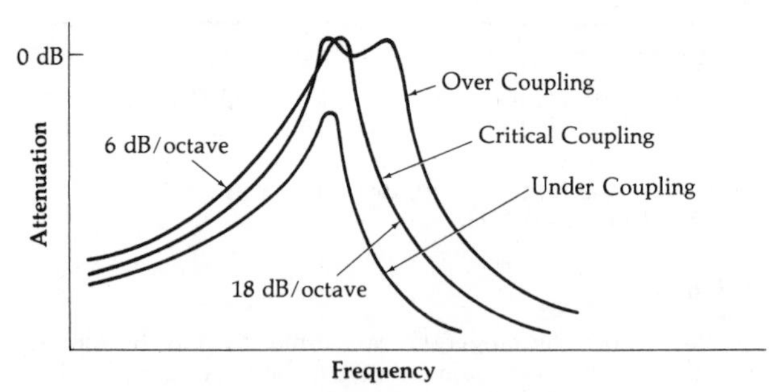

(A) Inductive coupling.

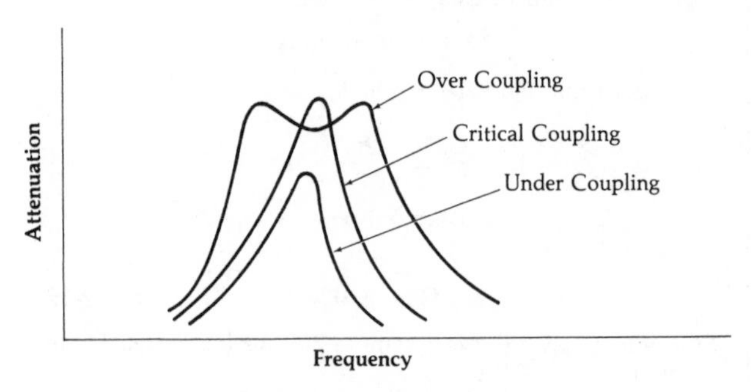

(B) Transformer coupling.

Fig. 2-24. The effects of various values of inductive coupling on passband response.

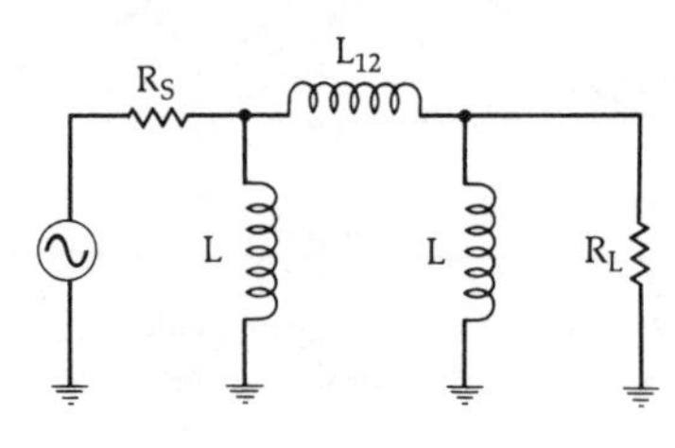

(A) Below resonance.

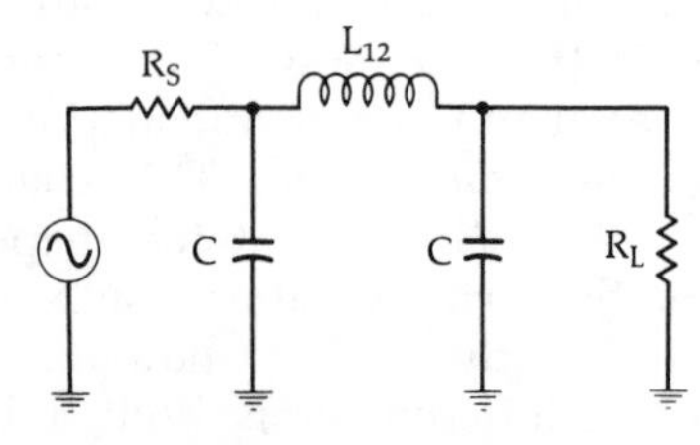

(B) Above resonance.

Fig. 2-25. Equivalent circuit of inductively coupled resonant circuits.

tors drop out of the equivalent circuit very quickly because their reactance becomes much greater than the shunt inductive reactance. This leaves an arrangement of three inductors which can be thought of as a single tapped inductor and which produces a 6-dB per octave rolloff. Above resonance, the shunt inductors can be ignored for the same reasons, and you have the circuit of Fig. 2-25B. We now have three effective elements in the equivalent circuit with each contributing 6 dB per octave to the response for a combined slope of 18 dB per octave.

The mirror-image characteristic of inductively and capacitively coupled resonant circuits is a very useful concept. This is especially true in applications that require symmetrical response curves. For example, suppose that a capacitively coupled design exhibited too much skew for your application. One very simple way to correct the problem would be to add a "top-L" coupled section to the existing network. The top-L coupling would attempt to skew the response in the opposite direction and would, therefore, tend to counteract any skew caused by the capacitive coupling. The net result is a more symmetric response shape.

The value of the inductor used to couple two identical resonant circuits can be found by

$$L_{12} = QL \qquad \text{(Eq. 2-20)}$$

where,

L_{12} = the inductance of the coupling inductor,
Q = the loaded Q of a single resonator,
L = the resonant circuit inductance.

A little manipulation of Equations 2-19 and 2-20 will reveal a very interesting point. The reactance of C_{12} calculated with Equation 2-19 will equal the reactance of L_{12} calculated with Equation 2-20 for the same operating Q and resonant frequency. The designer now has the option of changing any "top-C" coupled resonator to a top-L design simply by replacing the coupling capacitor with an inductor of equal reactance at the resonant frequency. When this is done, the degree of coupling, Q, and resonant frequency of the design will remain unchanged while the slope of the stopband skirts will flip-flop from one side to the other. For obvious reasons, top-L coupled designs work best in applications where the primary objective is a certain ultimate attenuation that must be met above the passband. Likewise, top-C designs are best for meeting ultimate attenuation specifications below the passband.

Transformer coupling does not lend itself well to an exact design procedure because there are so many factors which influence the degree of coupling. The geometry of the coils, the spacing between them, the core materials used, and the shielding, all have a pronounced effect on the degree of coupling attained in any design.

Probably the best way to design your own transformer is to use the old trial-and-error method. But do it in an orderly fashion and be consistent. It's a very sad day when one forgets how he got from point A to point B, especially if point B is an improvement in the design. Remember:

1. Decreasing the spacing between the primary and secondary increases the coupling.
2. Increasing the permeability of the magnetic path increases the coupling.
3. Shielding a transformer decreases its loaded Q and has the effect of increasing the coupling.

Begin the design by setting the loaded Q of each resonator to about twice what will be needed in the actual design. Then, slowly decrease the spacing between the primary and secondary until the response broadens to the loaded Q that is actually needed. If that response can't be met, try changing the geometry of the windings or the permeability of the magnetic path. Then, vary the spacing again. Use this as an iterative process to zero-in on the response that is needed. Granted, this is not an exact process, but it works and, if documented, can be reproduced.

There are literally thousands of commercially available transformers on the market that just might suit your needs perfectly. So before the trial-and-error method is put into practice, try a little research—it just might save a lot of time and money.

Active Coupling

It is possible to achieve very narrow 3-dB bandwidths in cascaded resonant circuits through the use of active coupling. Active coupling, for this purpose, is defined as being either a transistor or vacuum tube which, at least theoretically, allows signal flow in only one direction (Fig. 2-26). If each of the tuned circuits is the same and if each has the same loaded Q, the total loaded Q of the cascaded circuit is approximately equal to

$$Q_{total} = \frac{Q}{\sqrt{2^{1/n} - 1}} \qquad \text{(Eq. 2-21)}$$

where,

Q_{total} = the total Q of the cascaded circuit,
Q = the Q of each individual resonant circuit,
n = the number of resonant circuits.

The first step in any design procedure must be to

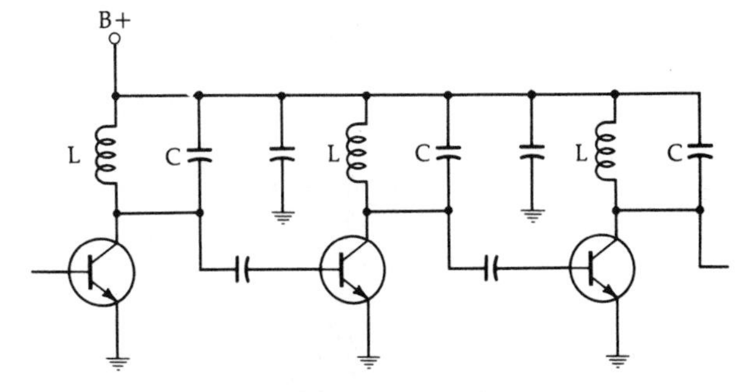

Fig. 2-26. Active coupling.

relate the required Q_{total} of the network back to the individual loaded Q of each resonator. This is done by rearranging Equation 2-21 to solve for Q. As an example, with $n = 4$ resonators, and given that Q_{total} of the cascaded circuit must be 50, Equation 2-21 tells us that the Q of the individual resonator need only be about 22—a fairly simple and realizable design task.

Active coupling is obviously more expensive than passive coupling due to the added cost of each active device. But, in some applications, there is no real trade-off involved because passive coupling just might not yield the required loaded Q. Example 2-5 illustrates some of the factors you must deal with.

This chapter was meant to provide an insight into how resonant circuits actually perform their function as well as to provide you with the capability for designing one to operate at a certain value of loaded Q. In Chapter 3, we will carry our study one step further to include low-pass, high-pass, and bandpass filters of various shapes and sizes.

EXAMPLE 2-5

Design a top-L coupled two-resonator tuned circuit to meet the following requirements:

1. Center frequency = 75 MHz
2. 3-dB bandwidth = 3.75 MHz
3. Source resistance = 100 ohms
4. Load resistance = 1000 ohms

Assume that inductors are available that have an unloaded Q of 85 at the frequency of interest. Finally, use a tapped-C transformer to present an effective source resistance (R_s') of 1000 ohms to the filter.

Solution

The solution to this design problem is not a very difficult one, but it does involve quite a few separate and distinct calculations which might tend to make you lose sight of our goal. For this reason, we will walk through the solution in a very orderly fashion with a complete explanation of each calculation.

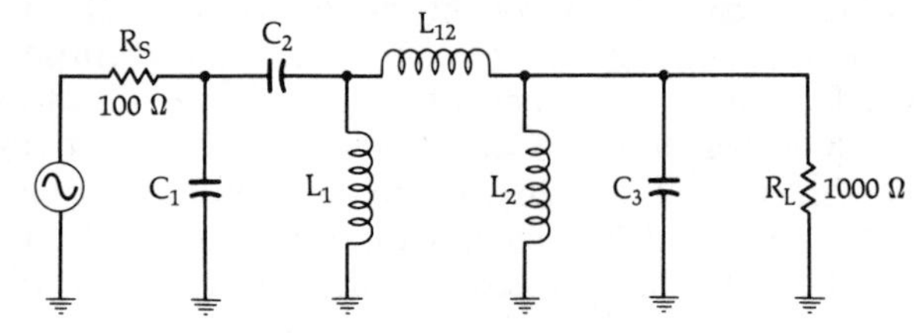

Fig. 2-27. Circuit for Example 2-5.

The circuit we are designing is shown in Fig. 2-27. Let's begin with a few definitions.

Q_{total} = the loaded Q of the entire circuit
Q_p = the Q of the inductor
Q_R = the loaded Q of each resonator

From our discussion on coupling and its effects on bandwidth, we know that

$$Q_R = \frac{Q_{total}}{0.707}$$

and,

$$Q_{total} = \frac{f_c}{B} = \frac{75\text{ MHz}}{3.75\text{ MHz}} = 20$$

so,

$$Q_R = \frac{20}{0.707} = 28.3$$

Thus, to provide a total loaded Q of 20, it is necessary that the loaded Q of each resonator be 28.3. For the inductor,

$$Q_p = \frac{R_p}{X_p} = 85$$

or,

$$R_p = 85\,X_p \qquad \text{(Eq. 2-22)}$$

The loaded Q of each resonant circuit is

$$Q_R = \frac{R_{total}}{X_p} \qquad \text{(Eq. 2-23)}$$

where,

R_{total} = the total equivalent shunt resistance for each resonator and
$= R_s' \,||\, R_p$
$= R_L \,||\, R_p$

since both circuits are identical. Remember, we have already taken into account the loading effect that each resonant circuit has on the other through the factor 0.707, which was used at the beginning of the example. Now, we have:

$$R_{total} = \frac{R_s'R_p}{R_s' + R_p}$$

Substituting into Equation 2-23:

$$Q_R = \frac{R_s'R_p}{(R_s' + R_p)X_p}$$

and,

$$X_p = \frac{R_s'R_p}{(R_s' + R_p)Q_R} = \frac{1000R_p}{(1000 + R_p)28.3} \qquad \text{(Eq. 2-24)}$$

We can now substitute Equation 2-22 into Equation 2-24 and solve for X_p.

$$X_p = \frac{(1000)(85\,X_p)}{(1000 + 85\,X_p)28.3} = 23.57\text{ ohms}$$

and,

$$R_p = 85\,X_p = 2003\text{ ohms}$$

To find the component values

$$L_1 = L_2 = \frac{X_p}{\omega} = 50\text{ nH}$$

Continued on next page

EXAMPLE 2-5—cont

and,

$$C_3 = \frac{1}{\omega X_p}$$
$$= 90 \text{ pF}$$

Now all that remains is to design the tapped-C transformer and the coupling inductor. From Equation 2-12:

$$R_s' = R_s\left(1 + \frac{C_1}{C_2}\right)^2$$

or,

$$\frac{C_1}{C_2} = \sqrt{\frac{R_s'}{R_s}} - 1$$
$$= 2.16$$

and,

$$C_1 = 2.16C_2 \qquad \text{(Eq. 2-25)}$$

We know that the total capacitance that must be used to resonate with the inductor is 90 pF and

$$C_{total} = \frac{C_1C_2}{C_1 + C_2} \qquad \text{(Eq. 2-26)}$$

Substituting Equation 2-25 into Equation 2-26 and taking C_{total} to be 90 pF yields:

$$90 \text{ pF} = \frac{2.16C_2^2}{3.16C_2}$$

and,

$$C_2 = 132 \text{ pF}$$
$$C_1 = 285 \text{ pF}$$

To solve for the coupling inductance from Equation 2-20:

$$L_{12} = Q_BL$$
$$= (28.3)(50 \text{ nH})$$
$$= 1.415 \ \mu\text{H}$$

The design is now complete. Notice that the tapped-C transformer is actually serving a dual purpose. It provides a dc block between the source and load in addition to its transformation properties.

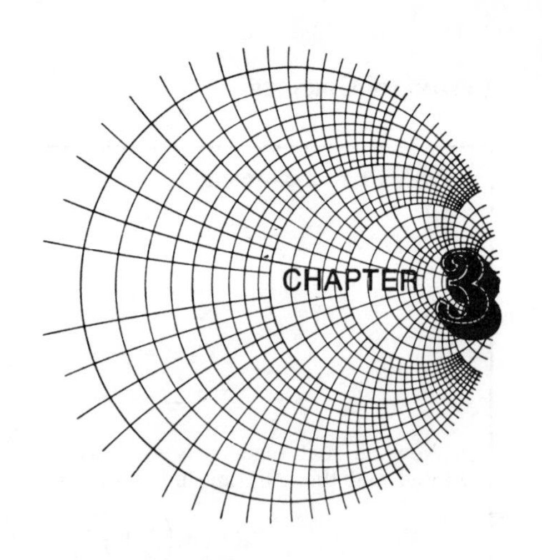

FILTER DESIGN

Filters occur so frequently in the instrumentation and communications industries that no book covering the field of rf circuit design could be complete without at least one chapter devoted to the subject. Indeed, entire books have been written on the art of filter design alone, so this single chapter cannot possibly cover all aspects of all types of filters. But it will familiarize you with the characteristics of four of the most commonly used filters and will enable you to design very quickly and easily a filter that will meet, or exceed, most of the common filter requirements that you will encounter.

We will cover Butterworth, Chebyshev, and Bessel filters in all of their common configurations: *low-pass, high-pass, bandpass,* and *bandstop*. We will learn how to take advantage of the attenuation characteristics unique to each type of filter. Finally, we will learn how to design some very powerful filters in as little as 5 minutes by merely looking through a catalog to choose a design to suit your needs.

BACKGROUND

In Chapter 2, the concept of resonance was explored and we determined the effects that component value changes had on resonant circuit operation. You should now be somewhat familiar with the methods that are used in analyzing passive resonant circuits to find quantities, such as loaded Q, insertion loss, and bandwidth. You should also be capable of designing one- or two-resonator circuits for any loaded Q desired (or, at least, determine why you cannot). Quite a few of the filter applications that you will encounter, however, cannot be satisfied with the simple bandpass arrangement given in Chapter 2. There are occasions when, instead of passing a certain band of frequencies while rejecting frequencies above and below (bandpass), we would like to attenuate a small band of frequencies while passing all others. This type of filter is called, appropriately enough, a *bandstop filter*. Still other requirements call for a low-pass or high-pass response. The characteristic curves for these responses are shown in Fig. 3-1. The low-pass filter will allow all signals below a certain cutoff frequency to pass while attenuating all others. A high-pass filter's response is the mirror-image of the low-pass response and attenuates all signals below a certain cutoff frequency while allowing those above cutoff to pass. These types of response simply cannot be handled very well with the two-resonator bandpass designs of Chapter 2.

In this chapter, we will use the low-pass filter as our workhorse, as all other responses will be derived from it. So let's take a quick look at a simple low-pass filter and examine its characteristics. Fig. 3-2 is an example of a very simple *two-pole,* or *second-order* low-pass filter. The order of a filter is determined by the slope of the attenuation curve it presents in the stopband. A second-order filter is one whose rolloff is a function of the frequency squared, or 12 dB per octave. A third-order filter causes a rolloff that is proportional to frequency cubed, or 18 dB per octave. Thus, the order of a filter can be equated with the number of significant reactive elements that it presents to the source as the signal deviates from the passband.

The circuit of Fig. 3-2 can be analyzed in much the same manner as was done in Chapter 2. For instance, an examination of the effects of loaded Q on the response would yield the family of curves shown in Fig. 3-3. Surprisingly, even this circuit configuration can cause a peak in the response. This is due to the fact that at some frequency, the inductor and capacitor will become resonant and, thus, peak the response if the loaded Q is high enough. The resonant frequency can be determined from

$$F_r = \frac{1}{2\pi\sqrt{LC}} \qquad \text{(Eq. 3-1)}$$

For low values of loaded Q, however, no response peak will be noticed.

The loaded Q of this filter is dependent upon the individual Q's of the series leg and the shunt leg where, assuming perfect components,

$$Q_1 = \frac{X_L}{R_s} \qquad \text{(Eq. 3-2)}$$

and,

$$Q_2 = \frac{R_L}{X_c} \qquad \text{(Eq. 3-3)}$$

and the total Q is:

$$Q_{total} = \frac{Q_1 Q_2}{Q_1 + Q_2} \qquad \text{(Eq. 3-4)}$$

If the total Q of the circuit is greater than about 0.5, then for optimum transfer of power from the source to the load, Q_1 should equal Q_2. In this case, at the peak frequency, the response will approach 0-dB insertion loss. If the total Q of the network is less than about 0.5, there will be no peak in the response and, for optimum transfer of power, R_s should equal R_L. The peaking of the filter's response is commonly called ripple (defined in Chapter 2) and can vary consider-

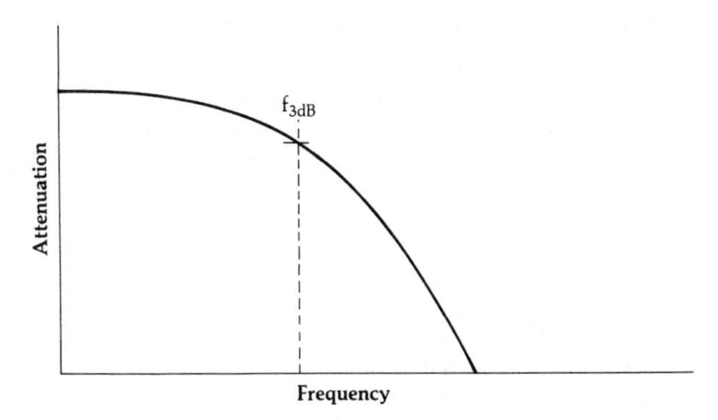

(A) *Low-pass.*

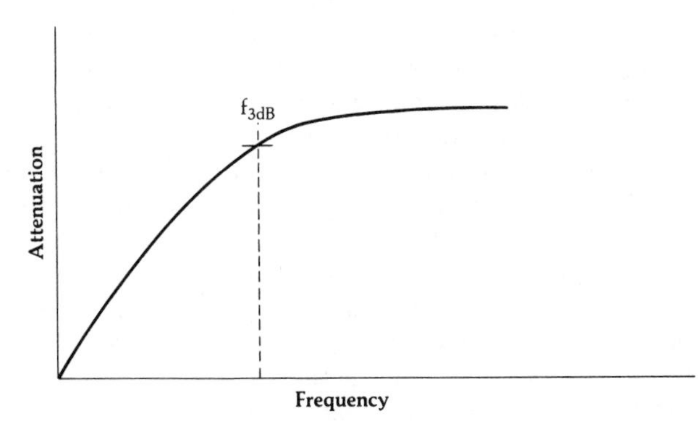

(B) *High-pass.*

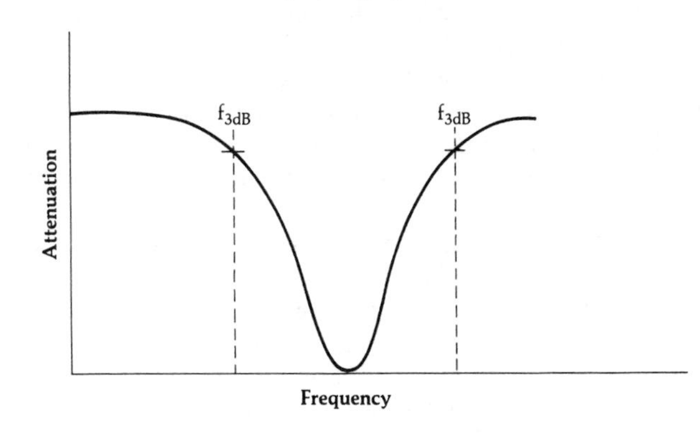

(C) *Bandstop.*

Fig. 3-1. Typical filter response curves.

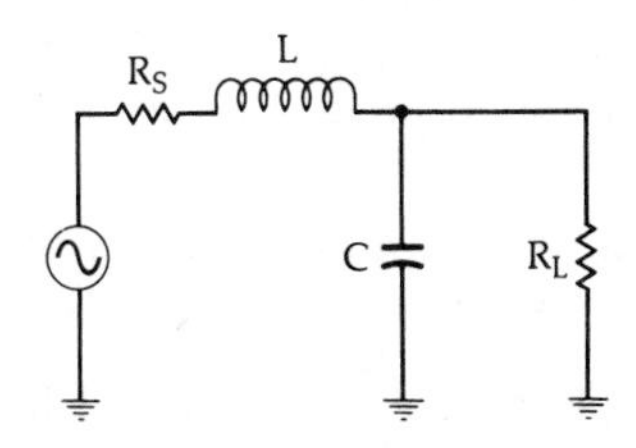

Fig. 3-2. A simple low-pass filter.

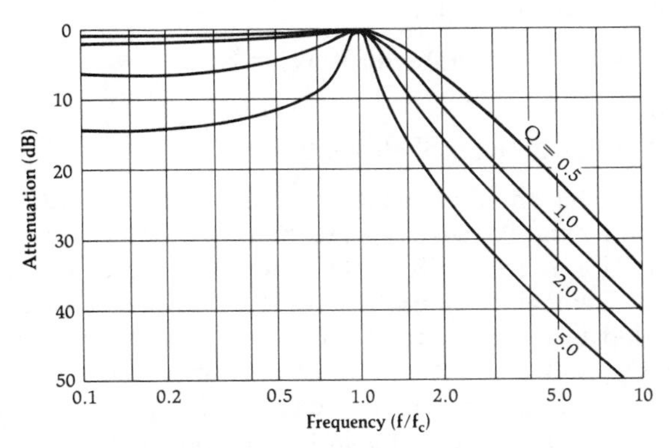

Fig. 3-3. Typical two-pole filter response curves.

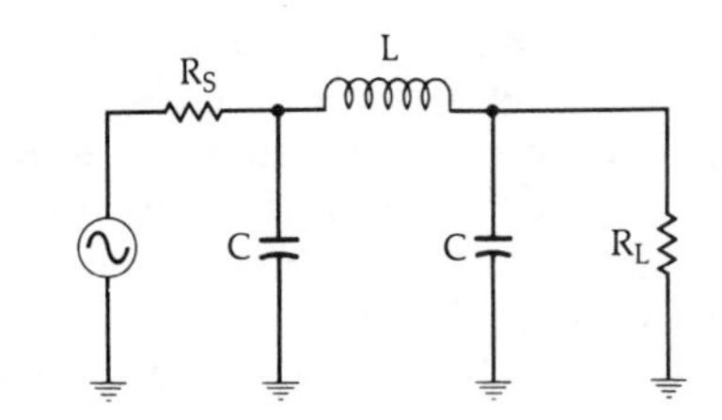

Fig. 3-4. Three-element low-pass filter.

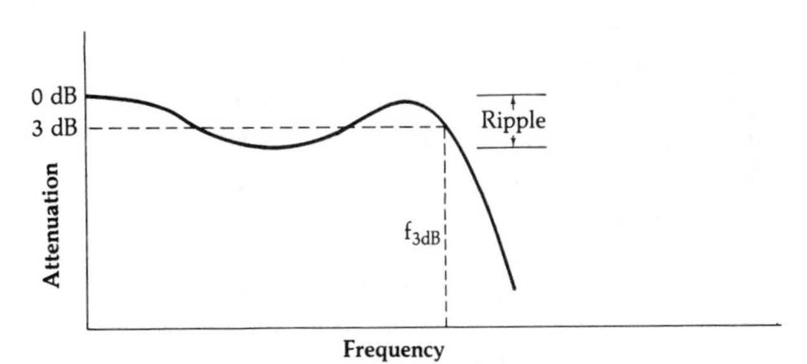

Fig. 3-5. Typical response of a three-element low-pass filter.

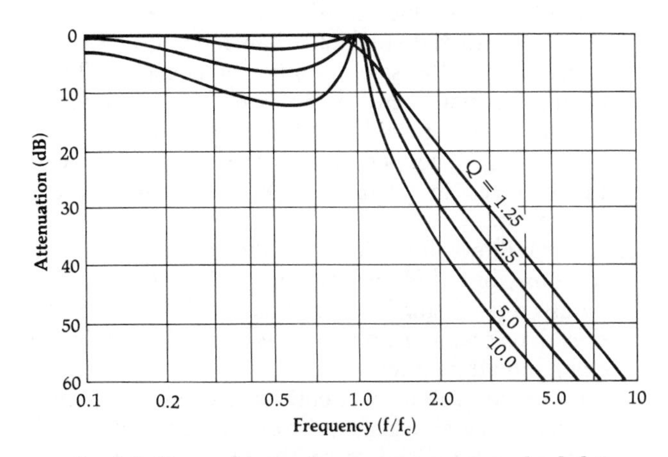

Fig. 3-6. Curves showing frequency response vs. loaded Q for three-element low-pass filters.

ably from one filter design to the next depending on the application. As shown, the two-element filter exhibits only one response peak at the edge of the passband.

It can be shown that the number of peaks within the passband is directly related to the number of elements in the filter by:

$$\text{Number of Peaks} = N - 1$$

where,

N = the number of elements.

Thus, the three-element low-pass filter of Fig. 3-4 should exhibit two response peaks as shown in Fig. 3-5. This is true only if the loaded Q is greater than one. Typical response curves for various values of loaded Q for the circuit given in Fig. 3-4 are shown in Fig. 3-6. For all odd-order networks, the response at dc and at the upper edge of the passband approaches 0 dB with dips in the response between the two frequencies. All even-order networks will produce an insertion loss at dc equal to the amount of passband ripple in dB. Keep in mind, however, that either of these two networks, if designed for low values of loaded Q, can be made to exhibit little or no passband ripple. But, as you can see from Figs. 3-3 and 3-6, the elimination of passband ripple can be made only at the expense of bandwidth. The smaller the ripple that is allowed, the wider the bandwidth becomes and, therefore, selectivity suffers. Optimum flatness in the passband occurs when the loaded Q of the three-element circuit is equal to one (1). Any value of loaded Q that is less than one will cause the response to roll off noticeably even at very low frequencies, within the defined passband. Thus, not only is the selectivity poorer but the passband insertion loss is too. In an application where there is not much signal to begin with, an even further decrease in signal strength could be disastrous.

Now that we have taken a quick look at two representative low-pass filters and their associated responses, let's discuss filters in general:

1. High-Q filters tend to exhibit a far greater initial slope toward the stopband than their low-Q counterparts with the same number of elements. Thus, at any frequency in the stopband, the attenuation will be greater for a high-Q filter than for one with a lower Q. The penalty for this improvement is the increase in passband ripple that must occur as a result.
2. Low-Q filters tend to have the flattest passband response but their initial attenuation slope at the band edge is small. Thus, the penalty for the reduced passband ripple is a decrease in the *initial* stopband attenuation.
3. As with the resonant circuits discussed in Chapter 2, the source and load resistors loading a filter will have a profound effect on the Q of the filter and, therefore, on the passband ripple and shape factor of the filter. If a filter is inserted between two resistance values for which it was not designed, the performance will suffer to an extent, depending upon the degree of error in the terminating impedance values.
4. The *final* attenuation slope of the response is dependent upon *the order of the network.* The order of the network is equal to the number of reactive elements *in the low-pass filter.* Thus, a second-order network (2 elements) falls off at a final attenuation slope of 12 dB per octave, a third-order network (3 elements) at the rate of 18 dB per octave, and so on, with the addition of 6 dB per octave per element.

MODERN FILTER DESIGN

Modern filter design has evolved through the years from a subject known only to specialists in the field (because of the advanced mathematics involved) to a practical well-organized catalog of ready-to-use circuits available to anyone with a knowledge of eighth grade level math. In fact, an average individual with absolutely no prior practical filter design experience should be able to sit down, read this chapter, and within 30 minutes be able to design a practical high-pass, low-pass, bandpass, or bandstop filter to his specifications. It sounds simple and it is—once a few basic rules are memorized.

The approach we will take in all of the designs in this chapter will be to make use of the myriad of normalized *low-pass prototypes* that are now available to the designer. The actual design procedure is, therefore, nothing more than determining your requirements and, then, finding a filter in a catalog which satisfies these requirements. Each normalized element value is then scaled to the frequency and impedance you desire and, then, transformed to the type of response (bandpass, high-pass, bandstop) that you wish. With practice, the procedure becomes very simple and soon you will be defining and designing filters.

The concept of normalization may at first seem foreign to the person who is a newcomer to the field of filter design, and the idea of transforming a low-pass filter into one that will give one of the other three types of responses might seem absurd. The best advice I can give (to anyone not familiar with these practices and who might feel a bit skeptical at this point) is to press on. The only way to truly realize the beauty and simplicity of this approach is to try a few actual designs. Once you try a few, you will be hooked, and any other approach to filter design will suddenly seem tedious and unnecessarily complicated.

NORMALIZATION AND THE LOW-PASS PROTOTYPE

In order to offer a catalog of useful filter circuits to the electronic filter designer, it became necessary to standardize the presentation of the material. Obviously, in practice, it would be extremely difficult to compare the performance and evaluate the usefulness of two filter networks if they were operating under two totally different sets of circumstances. Similarly, the presentation of any comparative design information for filters, if not standardized, would be totally useless. This concept of standardization or *normalization*, then, is merely a tool used by filter experts to present all filter design and performance information in a manner useful to circuit designers. Normalization assures the designer of the capability of comparing the performance of any two filter types when given the same operating conditions.

All of the catalogued filters in this chapter are low-pass filters normalized for a cutoff frequency of one radian per second (0.159 Hz) and for source and load resistors of one ohm. A characteristic response of such a filter is shown in Fig. 3-7. The circuit used to generate this response is called the *low-pass prototype*.

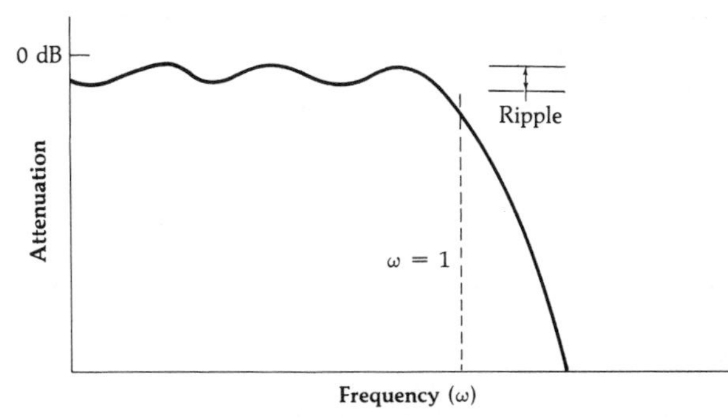

Fig. 3-7. Normalized low-pass response.

Obviously, the design of a filter with such a low cutoff frequency would require component values much larger than those we are accustomed to working with; capacitor values would be in farads rather than microfarads and picofarads, and the inductor values would be in henries rather than in microhenries and nanohenries. But once we choose a suitable low-pass prototype from the catalog, we can change the impedance level and cutoff frequency of the filter to any value we wish through a simple process called *scaling*. The net result of this process is a practical filter design with realizable component values.

FILTER TYPES

Many of the filters used today bear the names of the men who developed them. In this section, we will take a look at three such filters and examine their attenuation characteristics. Their relative merits will be discussed and their low-pass prototypes presented. The three filter types discussed will include the Butterworth, Chebyshev, and Bessel responses.

The Butterworth Response

The Butterworth filter is a medium-Q filter that is used in designs which require the amplitude response of the filter to be as flat as possible. The Butterworth response is the flattest passband response available and contains no ripple. The typical response of such a filter might look like that of Fig. 3-8.

Since the Butterworth response is only a medium-Q filter, its initial attenuation steepness is not as good as some filters but it is better than others. This characteristic often causes the Butterworth response to be called a middle-of-the-road design.

The attenuation of a Butterworth filter is given by

$$A_{dB} = 10 \log\left[1 + \left(\frac{\omega}{\omega_c}\right)^{2n}\right] \qquad \text{(Eq. 3-5)}$$

where,

ω = the frequency at which the attenuation is desired,

ω_c = the cutoff frequency ($\omega_{3\,dB}$) of the filter,

n = the number of elements in the filter.

If Equation 3-5 is evaluated at various frequencies for various numbers of elements, a family of curves is generated which will give a very good graphical representation of the attenuation provided by any order of filter at any frequency. This information is illustrated in Fig. 3-9. Thus, from Fig. 3-9, a 5-element (fifth order) Butterworth filter will provide an attenuation of approximately 30 dB at a frequency equal to

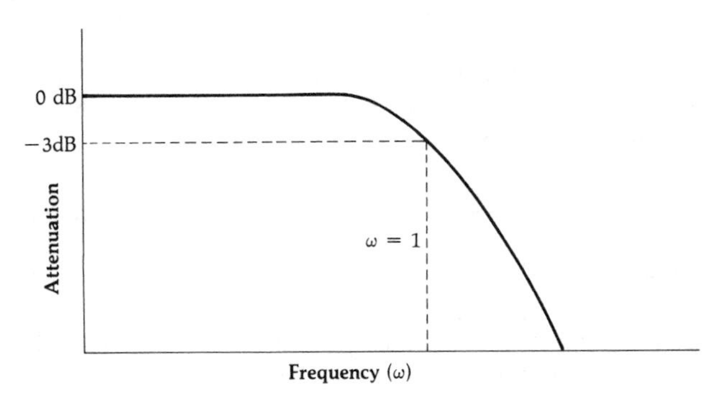

Fig. 3-8. The Butterworth response.

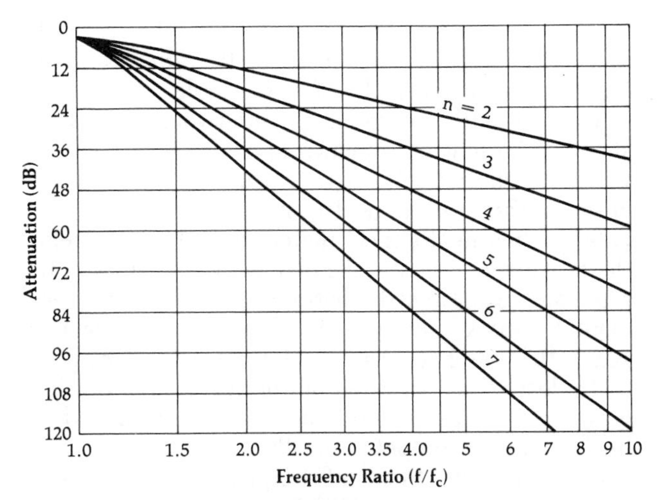

Fig. 3-9. Attenuation characteristics for Butterworth filters.

twice the cutoff frequency of the filter. Notice here that the frequency axis is normalized to ω/ω_c and the graph begins at the cutoff (−3 dB) point. This graph is extremely useful as it provides you with a method of determining, at a glance, the order of a filter needed to meet a given attenuation specification. A brief example should illustrate this point (Example 3-1).

EXAMPLE 3-1

How many elements are required to design a Butterworth filter with a cutoff frequency of 50 MHz, if the filter must provide at least 50 dB of attenuation at 150 MHz?

Solution

The first step in the solution is to find the ratio of $\omega/\omega_c = f/f_c$.

$$\frac{f}{f_c} = \frac{150 \text{ MHz}}{50 \text{ MHz}} = 3$$

Thus, at 3 times the cutoff frequency, the response must be down by at least 50 dB. Referring to Fig. 3-9, it is seen very quickly that a minimum of 6 elements is required to meet this design goal. At an f/f_c of 3, a 6-element design would provide approximately 57 dB of attenuation, while a 5-element design would provide only about 47 dB, which is not quite good enough.

The element values for a normalized Butterworth low-pass filter operating between equal 1-ohm terminations (source and load) can be found by

$$A_k = 2 \sin \frac{(2k-1)\pi}{2n}, \quad k = 1,2, \ldots n \qquad \text{(Eq. 3-6)}$$

where,

n is the number of elements,

A_k is the k-th reactance in the ladder and may be either an inductor or capacitor.

The term $(2k-1)\pi/2n$ is in radians. We can use Equation 3-6 to generate our first entry into the catalog of low-pass prototypes shown in Table 3-1. The placement of each component of the filter is shown immediately above and below the table.

The rules for interpreting Butterworth tables are simple. The schematic shown above the table is used whenever the ratio R_S/R_L is calculated as the design criteria. The table is read from the top down. Alternately, when R_L/R_S is calculated, the schematic below the table is used. Then, the element designators in the table are read from the bottom up. Thus, a four-element low-pass prototype could appear as shown in Fig. 3-10. Note here that the element values not given in Table 3-1 are simply left out of the prototype ladder network. The 1-ohm load resistor is then placed directly across the output of the filter.

Remember that the cutoff frequency of each filter is 1 radian per second, or 0.159 Hz. Each capacitor value given is in farads, and each inductor value is in henries. The network will later be scaled to the impedance and frequency that is desired through a simple multiplication and division process. The component values will then appear much more realistic.

Occasionally, we have the need to design a filter that will operate between two unequal terminations as shown in Fig. 3-11. In this case, the circuit is normal-

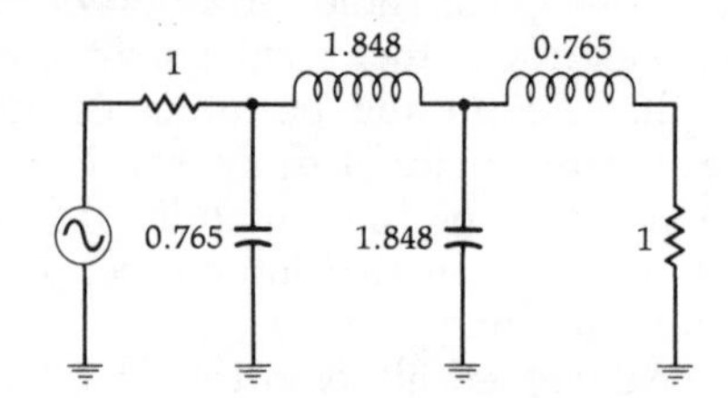

Fig. 3-10. A four-element Butterworth low-pass prototype circuit.

Table 3-1. Butterworth Equal Termination Low-Pass Prototype Element Values ($R_S = R_L$)

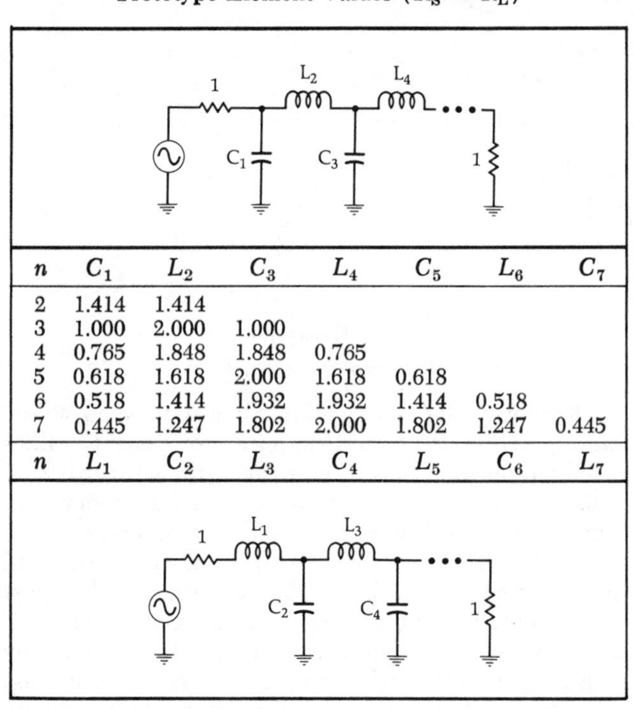

n	C_1	L_2	C_3	L_4	C_5	L_6	C_7
2	1.414	1.414					
3	1.000	2.000	1.000				
4	0.765	1.848	1.848	0.765			
5	0.618	1.618	2.000	1.618	0.618		
6	0.518	1.414	1.932	1.932	1.414	0.518	
7	0.445	1.247	1.802	2.000	1.802	1.247	0.445
n	L_1	C_2	L_3	C_4	L_5	C_6	L_7

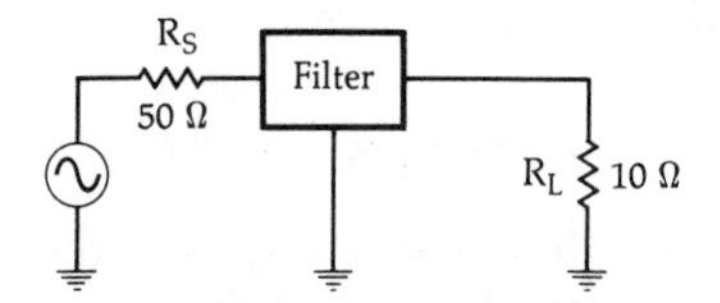

Fig. 3-11. Unequal terminations.

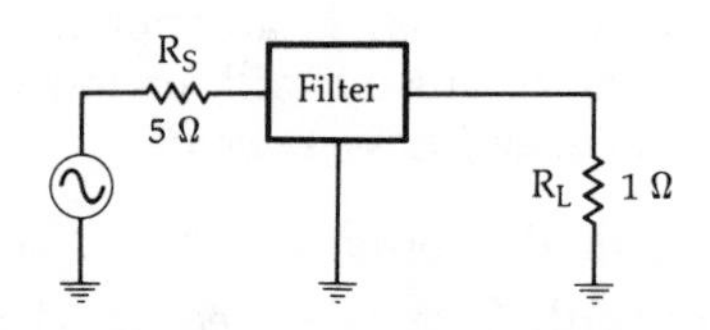

Fig. 3-12. Normalized unequal terminations.

ized for a load resistance of 1 ohm, while taking what we get for the source resistance. Dividing both the load and source resistor by 10 will yield a load resistance of 1 ohm and a source resistance of 5 ohms as shown in Fig. 3-12. We can use the normalized terminating resistors to help us find a low-pass prototype circuit.

Table 3-2 is a list of Butterworth low-pass prototype values for various ratios of source to load impedance (R_S/R_L). The schematic shown above the table is used when R_S/R_L is calculated, and the element values are read down from the top of the table.

Table 3-2A. Butterworth Low-Pass Prototype Element Values

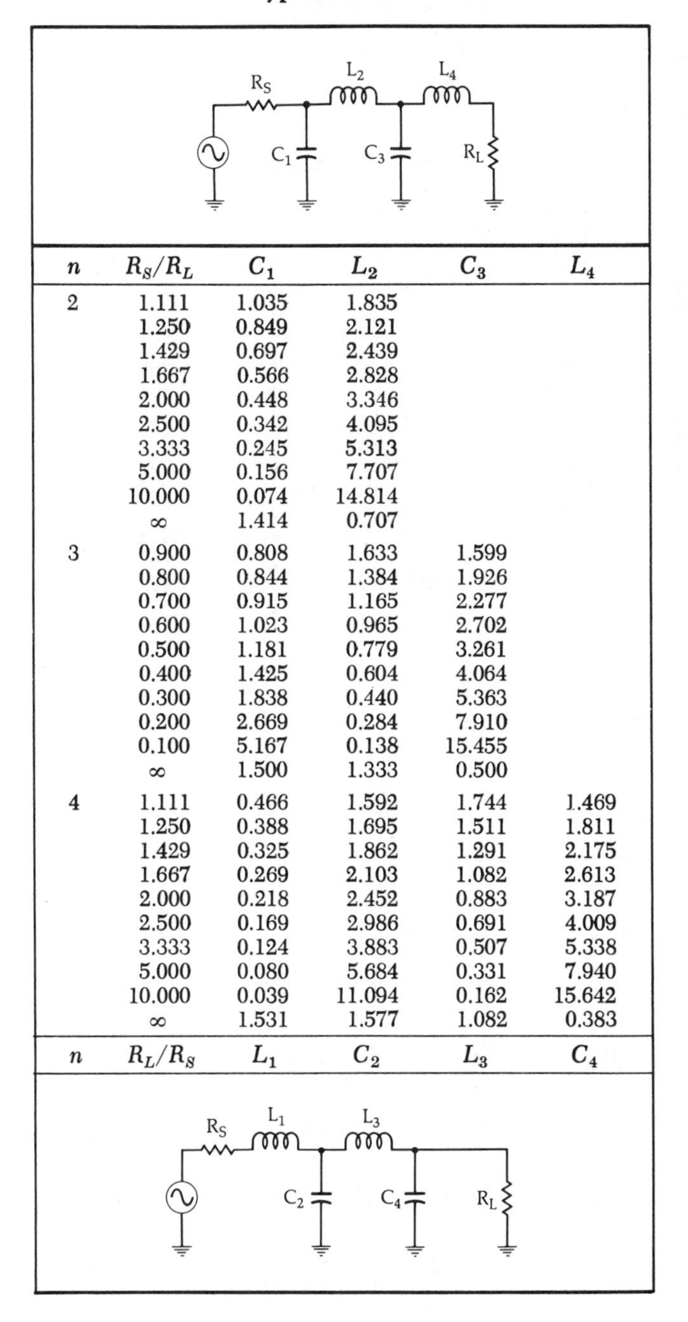

n	R_S/R_L	C_1	L_2	C_3	L_4
2	1.111	1.035	1.835		
	1.250	0.849	2.121		
	1.429	0.697	2.439		
	1.667	0.566	2.828		
	2.000	0.448	3.346		
	2.500	0.342	4.095		
	3.333	0.245	5.313		
	5.000	0.156	7.707		
	10.000	0.074	14.814		
	∞	1.414	0.707		
3	0.900	0.808	1.633	1.599	
	0.800	0.844	1.384	1.926	
	0.700	0.915	1.165	2.277	
	0.600	1.023	0.965	2.702	
	0.500	1.181	0.779	3.261	
	0.400	1.425	0.604	4.064	
	0.300	1.838	0.440	5.363	
	0.200	2.669	0.284	7.910	
	0.100	5.167	0.138	15.455	
	∞	1.500	1.333	0.500	
4	1.111	0.466	1.592	1.744	1.469
	1.250	0.388	1.695	1.511	1.811
	1.429	0.325	1.862	1.291	2.175
	1.667	0.269	2.103	1.082	2.613
	2.000	0.218	2.452	0.883	3.187
	2.500	0.169	2.986	0.691	4.009
	3.333	0.124	3.883	0.507	5.338
	5.000	0.080	5.684	0.331	7.940
	10.000	0.039	11.094	0.162	15.642
	∞	1.531	1.577	1.082	0.383
n	R_L/R_S	L_1	C_2	L_3	C_4

Alternately, when R_L/R_S is calculated, the schematic below the table is used while reading up from the bottom of the table to get the element values (Example 3-2).

EXAMPLE 3-2

Find the low-pass prototype value for an $n = 4$ Butterworth filter with unequal terminations: $R_S = 50$ ohms, $R_L = 100$ ohms.

Solution

Normalizing the two terminations for $R_L = 1$ ohm will yield a value of $R_S = 0.5$. Reading down from the top of Table 3-2, for an $n = 4$ low-pass prototype value, we see that there is no $R_S/R_L = 0.5$ ratio listed. Our second choice, then, is to take the value of $R_L/R_S = 2$, and read up from the bottom of the table while using the schematic below the table as the form for the low-pass prototype values. This approach results in the low-pass prototype circuit of Fig. 3-13.

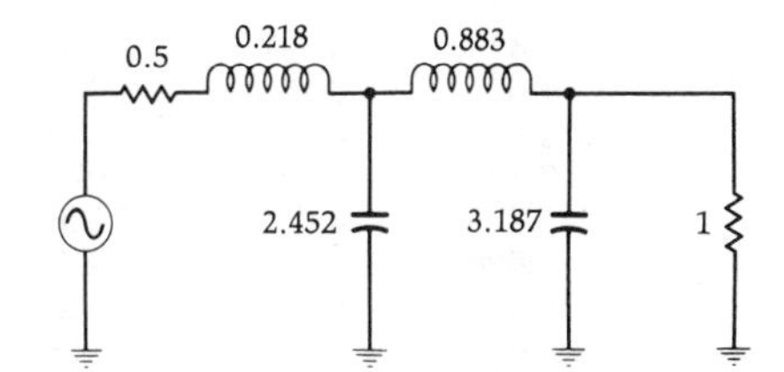

Fig. 3-13. Low-pass prototype circuit for Example 3-2.

Obviously, all possible ratios of source to load resistance could not possibly fit on a chart of this size. This, of course, leaves the potential problem of not being able to find the ratio that you need for a particular design task. The solution to this dilemma is to simply choose a ratio which most closely matches the ratio you need to complete the design. For ratios of 100:1 or so, the best results are obtained if you assume this value to be so high for practical purposes as to be infinite. Since, in these instances, you are only approximating the ratio of source to load resistance, the filter derived will only approximate the response that was originally intended. This is usually not too much of a problem.

The Chebyshev Response

The Chebyshev filter is a high-Q filter that is used when: (1) a steeper initial descent into the stopband is required, and (2) the passband response is no longer required to be flat. With this type of requirement, ripple can be allowed in the passband. As more ripple is introduced, the initial slope at the beginning of the stopband is increased and produces a more rectangular attenuation curve when compared to the rounded Butterworth response. This comparison is made in Fig. 3-14. Both curves are for $n = 3$ filters. The Chebyshev response shown has 3 dB of passband ripple and produces a 10 dB improvement in stopband attenuation over the Butterworth filter.

Table 3-2B. Butterworth Low-Pass Prototype Element Values

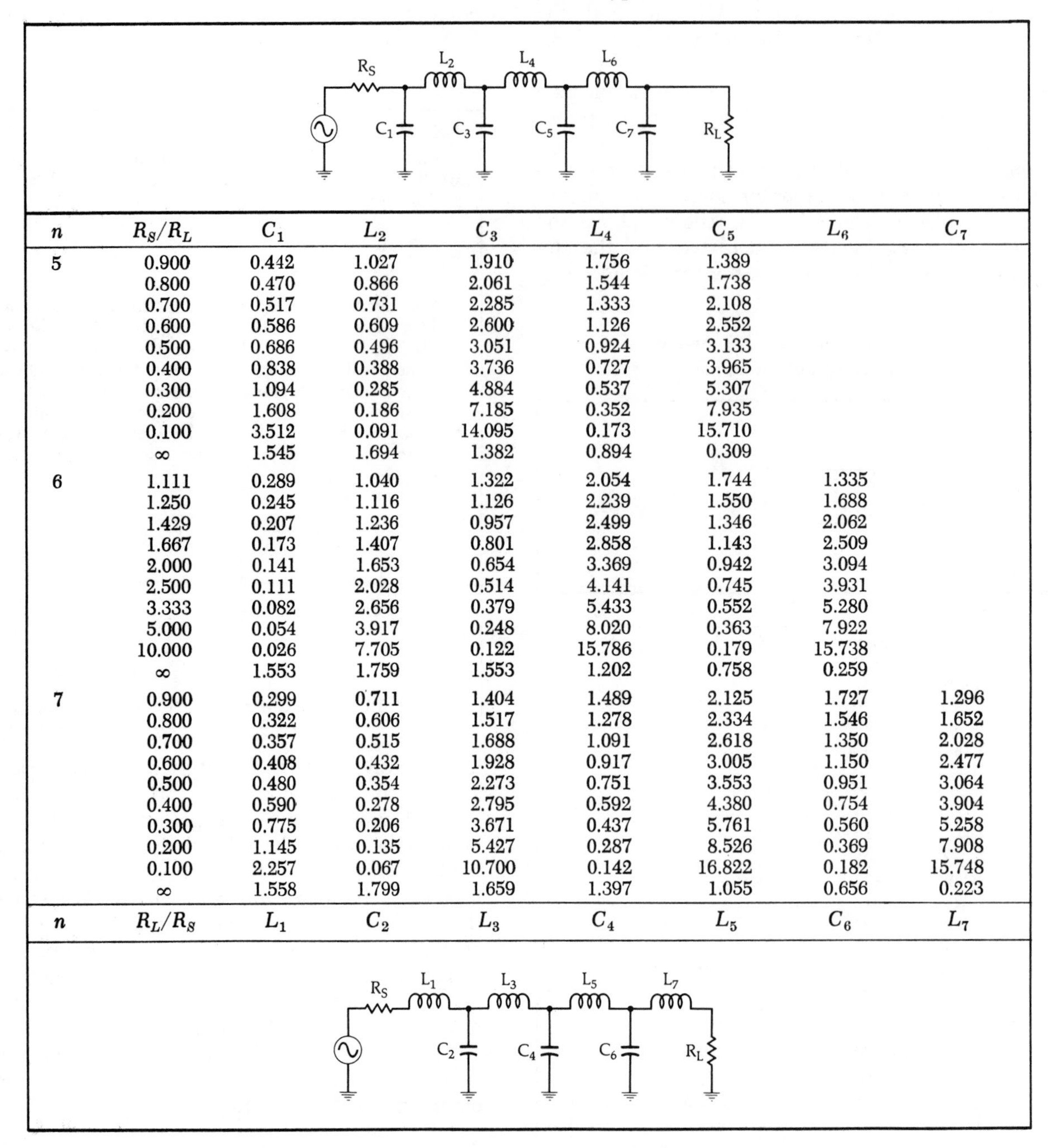

n	R_S/R_L	C_1	L_2	C_3	L_4	C_5	L_6	C_7
5	0.900	0.442	1.027	1.910	1.756	1.389		
	0.800	0.470	0.866	2.061	1.544	1.738		
	0.700	0.517	0.731	2.285	1.333	2.108		
	0.600	0.586	0.609	2.600	1.126	2.552		
	0.500	0.686	0.496	3.051	0.924	3.133		
	0.400	0.838	0.388	3.736	0.727	3.965		
	0.300	1.094	0.285	4.884	0.537	5.307		
	0.200	1.608	0.186	7.185	0.352	7.935		
	0.100	3.512	0.091	14.095	0.173	15.710		
	∞	1.545	1.694	1.382	0.894	0.309		
6	1.111	0.289	1.040	1.322	2.054	1.744	1.335	
	1.250	0.245	1.116	1.126	2.239	1.550	1.688	
	1.429	0.207	1.236	0.957	2.499	1.346	2.062	
	1.667	0.173	1.407	0.801	2.858	1.143	2.509	
	2.000	0.141	1.653	0.654	3.369	0.942	3.094	
	2.500	0.111	2.028	0.514	4.141	0.745	3.931	
	3.333	0.082	2.656	0.379	5.433	0.552	5.280	
	5.000	0.054	3.917	0.248	8.020	0.363	7.922	
	10.000	0.026	7.705	0.122	15.786	0.179	15.738	
	∞	1.553	1.759	1.553	1.202	0.758	0.259	
7	0.900	0.299	0.711	1.404	1.489	2.125	1.727	1.296
	0.800	0.322	0.606	1.517	1.278	2.334	1.546	1.652
	0.700	0.357	0.515	1.688	1.091	2.618	1.350	2.028
	0.600	0.408	0.432	1.928	0.917	3.005	1.150	2.477
	0.500	0.480	0.354	2.273	0.751	3.553	0.951	3.064
	0.400	0.590	0.278	2.795	0.592	4.380	0.754	3.904
	0.300	0.775	0.206	3.671	0.437	5.761	0.560	5.258
	0.200	1.145	0.135	5.427	0.287	8.526	0.369	7.908
	0.100	2.257	0.067	10.700	0.142	16.822	0.182	15.748
	∞	1.558	1.799	1.659	1.397	1.055	0.656	0.223
n	R_L/R_S	L_1	C_2	L_3	C_4	L_5	C_6	L_7

The attenuation of a Chebyshev filter can be found by making a few simple but tiresome calculations, and can be expressed as:

$$A_{dB} = 10 \log \left[1 + \epsilon^2 C_n{}^2 \left(\frac{\omega}{\omega_c}\right)'\right] \qquad \text{(Eq. 3-7)}$$

where,

$C_n{}^2 \left(\frac{\omega}{\omega_c}\right)'$ is the Chebyshev polynomial to the order n evaluated at $\left(\frac{\omega}{\omega_c}\right)'$.

The Chebyshev polynomials for the first seven orders are given in Table 3-3. The parameter ϵ is given by:

$$\epsilon = \sqrt{10^{R_{dB}/10} - 1} \qquad \text{(Eq. 3-8)}$$

where,

R_{dB} is the passband ripple in decibels.

Note that $\left(\frac{\omega}{\omega_c}\right)'$ is not the same as $\left(\frac{\omega}{\omega_c}\right)$. The quantity $\left(\frac{\omega}{\omega_c}\right)'$ can be found by defining another parameter:

$$B = \frac{1}{n} \cosh^{-1}\left(\frac{1}{\epsilon}\right) \qquad \text{(Eq. 3-9)}$$

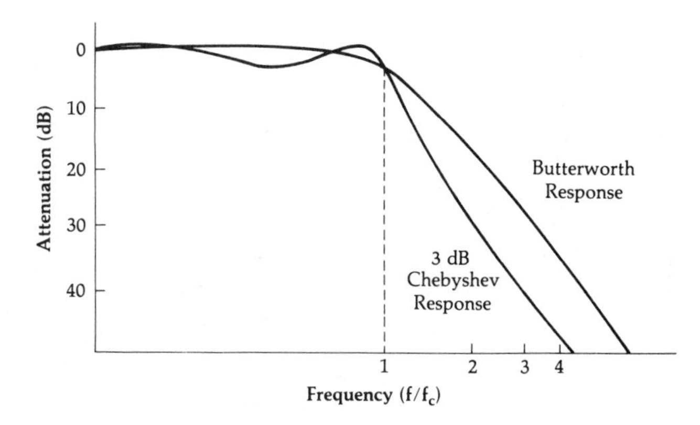

Fig. 3-14. Comparison of three-element Chebyshev and Butterworth responses.

Table 3-3. Chebyshev Polynomials to the Order n

n	*Chebyshev Polynomial*
1	$\frac{\omega}{\omega_c}$
2	$2\left(\frac{\omega}{\omega_c}\right)^2 - 1$
3	$4\left(\frac{\omega}{\omega_c}\right)^3 - 3\left(\frac{\omega}{\omega_c}\right)$
4	$8\left(\frac{\omega}{\omega_c}\right)^4 - 8\left(\frac{\omega}{\omega_c}\right)^2 + 1$
5	$16\left(\frac{\omega}{\omega_c}\right)^5 - 20\left(\frac{\omega}{\omega_c}\right)^3 + 5\left(\frac{\omega}{\omega_c}\right)$
6	$32\left(\frac{\omega}{\omega_c}\right)^6 - 48\left(\frac{\omega}{\omega_c}\right)^4 + 18\left(\frac{\omega}{\omega_c}\right)^2 - 1$
7	$64\left(\frac{\omega}{\omega_c}\right)^7 - 112\left(\frac{\omega}{\omega_c}\right)^5 + 56\left(\frac{\omega}{\omega_c}\right)^3 - 7\left(\frac{\omega}{\omega_c}\right)$

where,

n = the order of the filter,
ϵ = the parameter defined in Equation 3-8,
$\cosh^{-1}$ = the inverse hyperbolic cosine of the quantity in parentheses.

Finally, we have:

$$\left(\frac{\omega}{\omega_c}\right)' = \left(\frac{\omega}{\omega_c}\right) \cosh B \qquad \text{(Eq. 3-10)}$$

where,

$\left(\frac{\omega}{\omega_c}\right)$ = the ratio of the frequency of interest to the cutoff frequency,
cosh = the hyperbolic cosine.

If your calculator does not have hyperbolic and inverse hyperbolic functions, they can be manually determined from the following relations:

$$\cosh x = 0.5(e^x + e^{-x})$$

and

$$\cosh^{-1} x = \ln(x \pm \sqrt{x^2 - 1})$$

The preceding equations yield families of attenuation curves, each classified according to the amount of ripple allowed in the passband. Several of these families of curves are shown in Figs. 3-15 through 3-18, and include 0.01-dB, 0.1-dB, 0.5-dB, and 1.0-dB ripple. Each curve begins at $\omega/\omega_c = 1$, which is the normalized cutoff, or 3-dB frequency. The passband ripple is, therefore, not shown.

If other families of attenuation curves are needed with different values of passband ripple, the preceding Chebyshev equations can be used to derive them. The problem in Example 3-3 illustrates this.

Obviously, performing the calculations of Example 3-3 for various values of ω/ω_c, ripple, and filter order is a very time-consuming chore unless a programmable calculator or computer is available to do most of the work for you.

The low-pass prototype element values corresponding to the Chebyshev responses of Figs. 3-15 through 3-18 are given in Tables 3-4 through 3-7. Note that the Chebyshev prototype values could not be separated into two distinct sets of tables covering the equal and

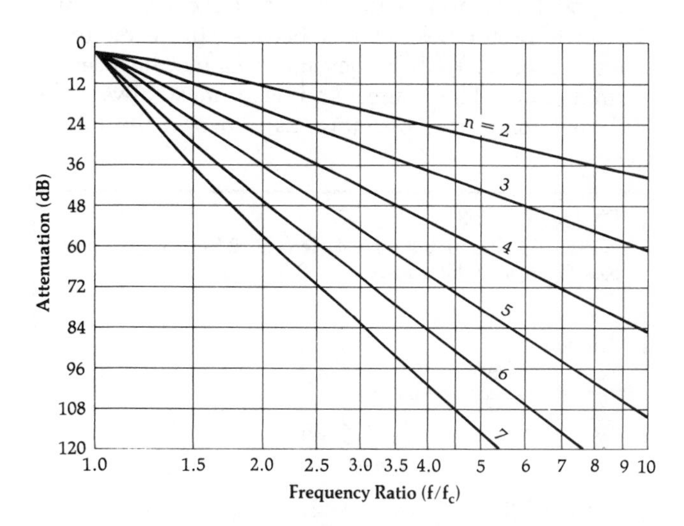

Fig. 3-15. Attenuation characteristics for a Chebyshev filter with 0.01-dB ripple.

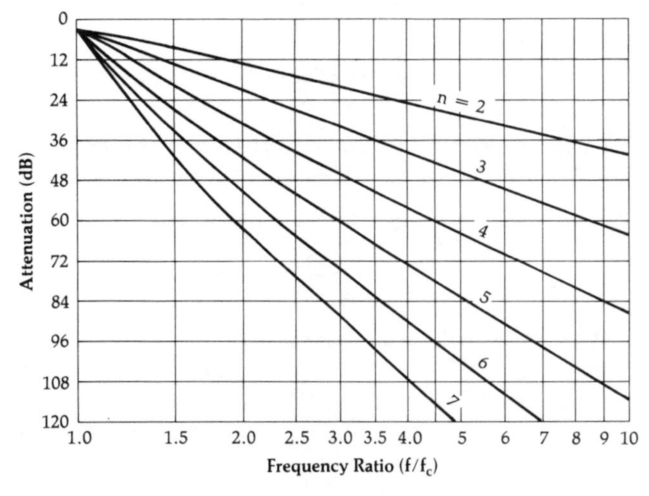

Fig. 3-16. Attenuation characteristics for a Chebyshev filter with 0.1-dB ripple.

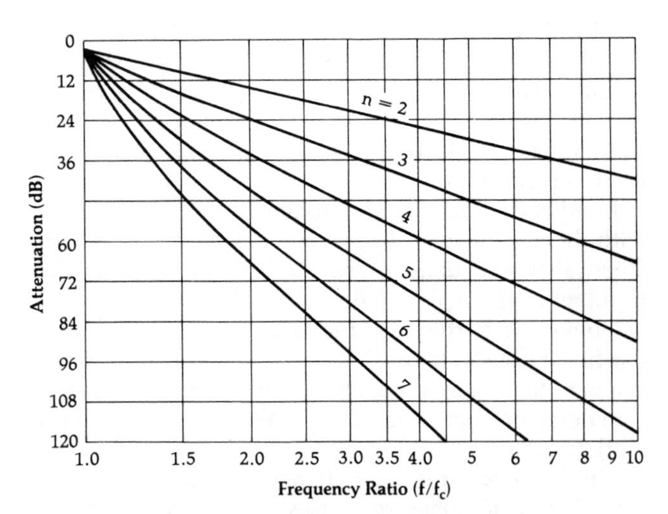

Fig. 3-17. Attenuation characteristics for a Chebyshev filter with 0.5-dB ripple.

unequal termination cases, as was done for the Butterworth prototypes. This is because the even order ($n = 2, 4, 6, \ldots$) Chebyshev filters cannot have equal terminations. The source and load must always be different for proper operation as shown in the tables.

EXAMPLE 3-3

Find the attenuation of a 4-element, 2.5-dB ripple, low-pass Chebyshev filter at $\omega/\omega_c = 2.5$.

Solution

First evaluate the parameter:

$$\epsilon = \sqrt{10^{2.5/10} - 1}$$
$$= 0.882$$

Next, find B.

$$B = \tfrac{1}{4}\left[\cosh^{-1}\left(\frac{1}{0.882}\right)\right]$$
$$= 0.1279$$

Then, $(\omega/\omega_c)'$ is:

$$(\omega/\omega_c)' = 2.5 \cosh .1279$$
$$= 2.5204$$

Finally, we evaluate the fourth order ($n = 4$) Chebyshev polynomial at $(\omega/\omega_c)' = 2.52$.

$$C_n\left(\frac{\omega}{\omega_c}\right) = 8\left(\frac{\omega}{\omega_c}\right)^4 - 8\left(\frac{\omega}{\omega_c}\right)^2 + 1$$
$$= 8(2.5204)^4 - 8(2.5204)^2 + 1$$
$$= 273.05$$

We can now evaluate the final equation.

$$A_{dB} = 10 \log_{10}\left[1 + \epsilon^2 C_n^2\left(\frac{\omega}{\omega_c}\right)'\right]$$
$$= 10 \log_{10}[1 + (0.882)^2(273.05)^2]$$
$$= 47.63 \text{ dB}$$

Thus, at an ω/ω_c of 2.5, you can expect 47.63 dB of attenuation for this filter.

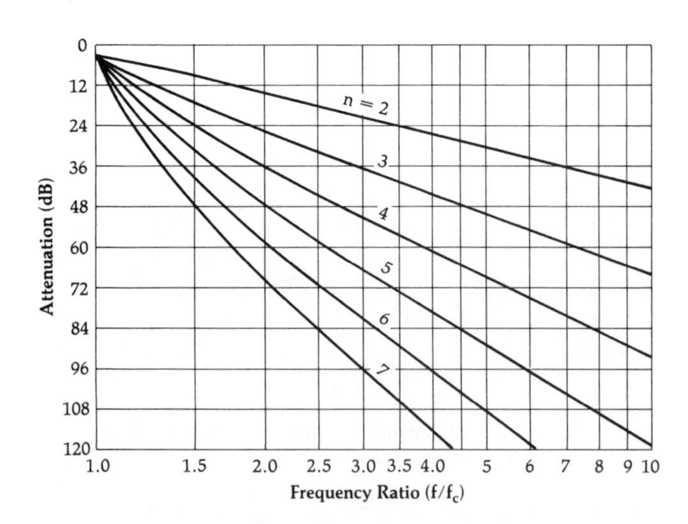

Fig. 3-18. Attenuation characteristics for a Chebyshev filter with 1-dB ripple.

The rules used for interpreting the Butterworth tables apply here also. The schematic shown above the table is used, and the element designators are read down from the top, when the ratio R_S/R_L is calculated as a design criteria. Alternately, with R_L/R_S calculations, use the schematic given below the table and read the element designators upwards from the bottom of the table. Example 3-4 is a practice problem for use in understanding the procedure.

EXAMPLE 3-4

Find the low-pass prototype values for an $n = 5$, 0.1-dB ripple, Chebyshev filter if the source resistance you are designing for is 50 ohms and the load resistance is 250 ohms.

Solution

Normalization of the source and load resistors yields an $R_S/R_L = 0.2$. A look at Table 3-5, for a 0.1-dB ripple filter with an $n = 5$ and an $R_S/R_L = 0.2$, yields the circuit values shown in Fig. 3-19.

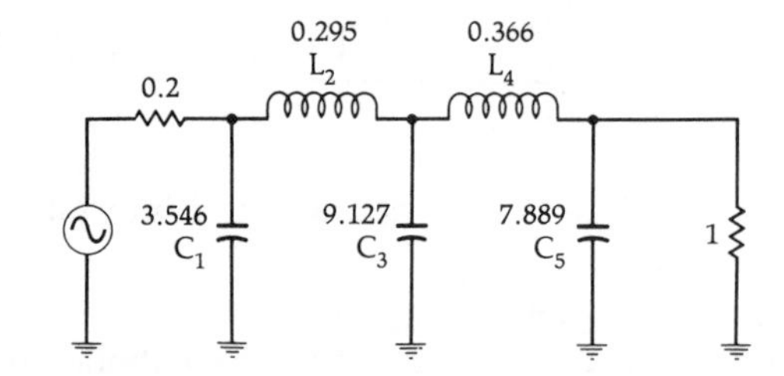

Fig. 3-19. Low-pass prototype circuit for Example 3-4.

It should be mentioned here that equations could have been presented in this section for deriving the element values for the Chebyshev low-pass prototypes. The equations are extremely long and tedious, however, and there would be little to be gained from their presentation.

Table 3-4A. Chebyshev Low-Pass Element Values for 0.01-dB Ripple

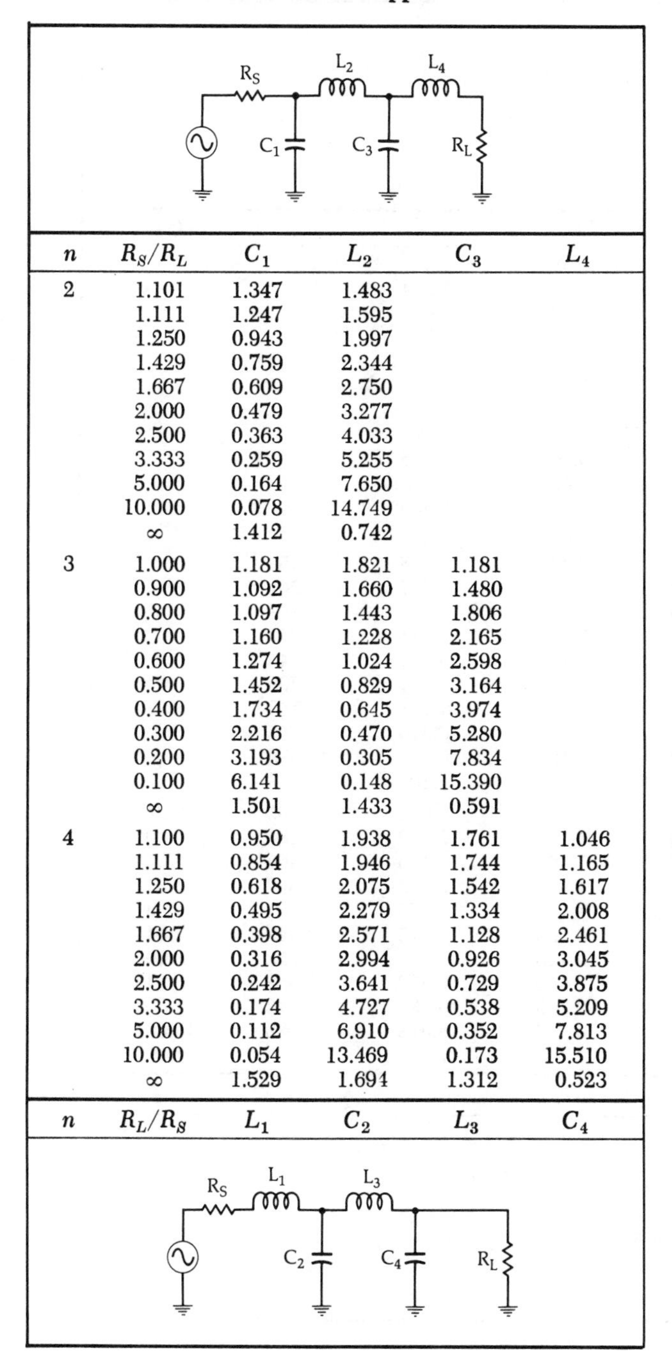

n	R_S/R_L	C_1	L_2	C_3	L_4
2	1.101	1.347	1.483		
	1.111	1.247	1.595		
	1.250	0.943	1.997		
	1.429	0.759	2.344		
	1.667	0.609	2.750		
	2.000	0.479	3.277		
	2.500	0.363	4.033		
	3.333	0.259	5.255		
	5.000	0.164	7.650		
	10.000	0.078	14.749		
	∞	1.412	0.742		
3	1.000	1.181	1.821	1.181	
	0.900	1.092	1.660	1.480	
	0.800	1.097	1.443	1.806	
	0.700	1.160	1.228	2.165	
	0.600	1.274	1.024	2.598	
	0.500	1.452	0.829	3.164	
	0.400	1.734	0.645	3.974	
	0.300	2.216	0.470	5.280	
	0.200	3.193	0.305	7.834	
	0.100	6.141	0.148	15.390	
	∞	1.501	1.433	0.591	
4	1.100	0.950	1.938	1.761	1.046
	1.111	0.854	1.946	1.744	1.165
	1.250	0.618	2.075	1.542	1.617
	1.429	0.495	2.279	1.334	2.008
	1.667	0.398	2.571	1.128	2.461
	2.000	0.316	2.994	0.926	3.045
	2.500	0.242	3.641	0.729	3.875
	3.333	0.174	4.727	0.538	5.209
	5.000	0.112	6.910	0.352	7.813
	10.000	0.054	13.469	0.173	15.510
	∞	1.529	1.694	1.312	0.523
n	R_L/R_S	L_1	C_2	L_3	C_4

The Bessel Filter

The initial stopband attenuation of the Bessel filter is very poor and can be approximated by:

$$A_{dB} = 3\left(\frac{\omega}{\omega_c}\right)^2 \qquad \text{(Eq. 3-11)}$$

This expression, however, is not very accurate above an ω/ω_c that is equal to about 2. For values of ω/ω_c greater than 2, a straight-line approximation of 6 dB per octave per element can be made. This yields the family of curves shown in Fig. 3-20.

But why would anyone deliberately design a filter with very poor initial stopband attenuation characteristics? The Bessel filter was originally optimized to obtain a *maximally flat group delay* or *linear phase* characteristic in the filter's passband. Thus, selectivity or stopband attenuation is not a primary concern when dealing with the Bessel filter. In high- and medium-Q filters, such as the Chebyshev and Butterworth filters, the phase response is extremely nonlinear over the filter's passband. This phase nonlinearity results in distortion of wideband signals due to the widely varying time delays associated with the different spectral components of the signal. Bessel filters, on the other hand, with their maximally flat (constant) group delay are able to pass wideband signals with a minimum of distortion, while still providing *some* selectivity.

The low-pass prototype element values for the Bessel filter are given in Table 3-8. Table 3-8 tabulates the prototype element values for various ratios of source to load resistance.

FREQUENCY AND IMPEDANCE SCALING

Once you specify the filter, choose the appropriate attenuation response, and write down the low-pass prototype values, the next step is to transform the prototype circuit into a usable filter. Remember, the cutoff frequency of the prototype circuit is 0.159 Hz ($\omega = 1$ rad/sec), and it operates between a source and load resistance that are normalized so that $R_L = 1$ ohm.

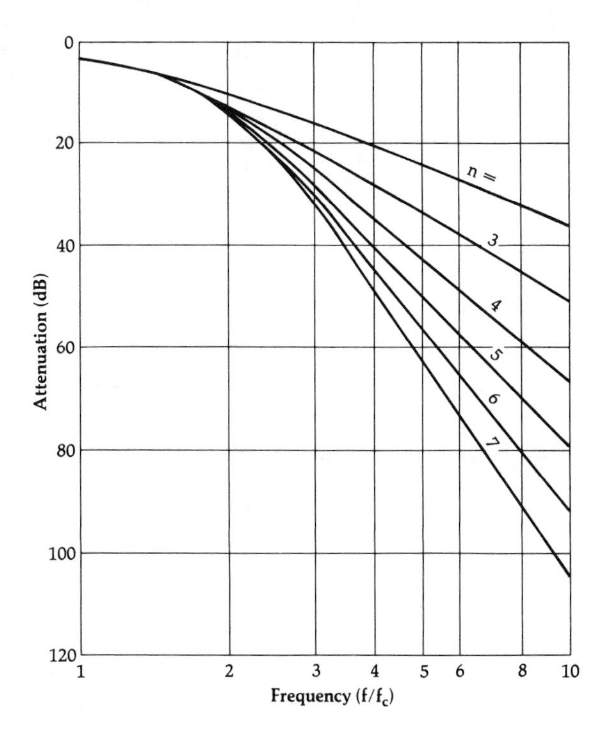

Fig. 3-20. Attenuation characteristics of Bessel filters.

Table 3-4B. Chebyshev Low-Pass Element Values for 0.01-dB Ripple

n	R_S/R_L	C_1	L_2	C_3	L_4	C_5	L_6	C_7
5	1.000	0.977	1.685	2.037	1.685	0.977		
	0.900	0.880	1.456	2.174	1.641	1.274		
	0.800	0.877	1.235	2.379	1.499	1.607		
	0.700	0.926	1.040	2.658	1.323	1.977		
	0.600	1.019	0.863	3.041	1.135	2.424		
	0.500	1.166	0.699	3.584	0.942	3.009		
	0.400	1.398	0.544	4.403	0.749	3.845		
	0.300	1.797	0.398	5.772	0.557	5.193		
	0.200	2.604	0.259	8.514	0.368	7.826		
	0.100	5.041	0.127	16.741	0.182	15.613		
	∞	1.547	1.795	1.645	1.237	0.488		
6	1.101	0.851	1.796	1.841	2.027	1.631	0.937	
	1.111	0.760	1.782	1.775	2.094	1.638	1.053	
	1.250	0.545	1.864	1.489	2.403	1.507	1.504	
	1.429	0.436	2.038	1.266	2.735	1.332	1.899	
	1.667	0.351	2.298	1.061	3.167	1.145	2.357	
	2.000	0.279	2.678	0.867	3.768	0.954	2.948	
	2.500	0.214	3.261	0.682	4.667	0.761	3.790	
	3.333	0.155	4.245	0.503	6.163	0.568	5.143	
	5.000	0.100	6.223	0.330	9.151	0.376	7.785	
	10.000	0.048	12.171	0.162	18.105	0.187	15.595	
	∞	1.551	1.847	1.790	1.598	1.190	0.469	
7	1.000	0.913	1.595	2.002	1.870	2.002	1.595	0.913
	0.900	0.816	1.362	2.089	1.722	2.202	1.581	1.206
	0.800	0.811	1.150	2.262	1.525	2.465	1.464	1.538
	0.700	0.857	0.967	2.516	1.323	2.802	1.307	1.910
	0.600	0.943	0.803	2.872	1.124	3.250	1.131	2.359
	0.500	1.080	0.650	3.382	0.928	3.875	0.947	2.948
	0.400	1.297	0.507	4.156	0.735	4.812	0.758	3.790
	0.300	1.669	0.372	5.454	0.546	6.370	0.568	5.148
	0.200	2.242	0.242	8.057	0.360	9.484	0.378	7.802
	0.100	4.701	0.119	15.872	0.178	18.818	0.188	15.652
	∞	1.559	1.867	1.866	1.765	1.563	1.161	0.456
n	R_L/R_S	L_1	C_2	L_3	C_4	L_5	C_6	L_7

The transformation is affected through the following formulas:

$$C = \frac{C_n}{2\pi f_c R} \quad \text{(Eq. 3-12)}$$

and

$$L = \frac{RL_n}{2\pi f_c} \quad \text{(Eq. 3-13)}$$

where,

C = the final capacitor value,
L = the final inductor value,
C_n = a low-pass prototype element value,
L_n = a low-pass prototype element value,
R = the final load resistor value,
f_c = the final cutoff frequency.

The normalized low-pass prototype source resistor must also be transformed to its final value by multiplying it by the final value of the load resistor (Example 3-5). Thus, the ratio of the two always remains the same.

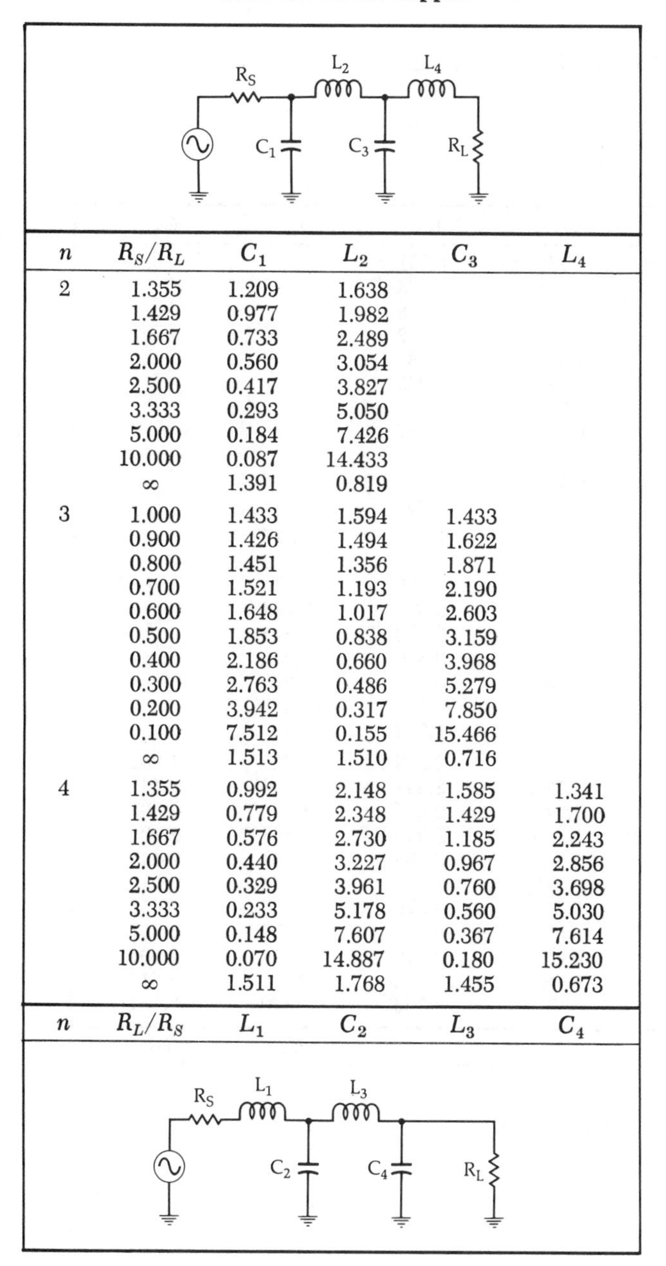

Table 3-5A. Chebyshev Low-Pass Prototype Element Values for 0.1-dB Ripple

n	R_S/R_L	C_1	L_2	C_3	L_4
2	1.355	1.209	1.638		
	1.429	0.977	1.982		
	1.667	0.733	2.489		
	2.000	0.560	3.054		
	2.500	0.417	3.827		
	3.333	0.293	5.050		
	5.000	0.184	7.426		
	10.000	0.087	14.433		
	∞	1.391	0.819		
3	1.000	1.433	1.594	1.433	
	0.900	1.426	1.494	1.622	
	0.800	1.451	1.356	1.871	
	0.700	1.521	1.193	2.190	
	0.600	1.648	1.017	2.603	
	0.500	1.853	0.838	3.159	
	0.400	2.186	0.660	3.968	
	0.300	2.763	0.486	5.279	
	0.200	3.942	0.317	7.850	
	0.100	7.512	0.155	15.466	
	∞	1.513	1.510	0.716	
4	1.355	0.992	2.148	1.585	1.341
	1.429	0.779	2.348	1.429	1.700
	1.667	0.576	2.730	1.185	2.243
	2.000	0.440	3.227	0.967	2.856
	2.500	0.329	3.961	0.760	3.698
	3.333	0.233	5.178	0.560	5.030
	5.000	0.148	7.607	0.367	7.614
	10.000	0.070	14.887	0.180	15.230
	∞	1.511	1.768	1.455	0.673
n	R_L/R_S	L_1	C_2	L_3	C_4

The process for designing a low-pass filter is a very simple one which involves the following procedure:

1. Define the response you need by specifying the required attenuation characteristics at selected frequencies.
2. Normalize the frequencies of interest by dividing them by the cutoff frequency of the filter. This step forces your data to be in the same form as that of the attenuation curves of this chapter, where the 3-dB point on the curve is:

$$\frac{f}{f_c} = 1$$

EXAMPLE 3-5

Scale the low-pass prototype values of Fig. 3-19 (Example 3-4) to a cutoff frequency of 50 MHz and a load resistance of 250 ohms.

Solution

Use Equations 3-12 and 3-13 to scale each component as follows:

$$C_1 = \frac{3.546}{2\pi(50 \times 10^6)(250)} = 45 \text{ pF}$$

$$C_3 = \frac{9.127}{2\pi(50 \times 10^6)(250)} = 116 \text{ pF}$$

$$C_5 = \frac{7.889}{2\pi(50 \times 10^6)(250)} = 100 \text{ pF}$$

$$L_2 = \frac{(250)(0.295)}{2\pi(50 \times 10^6)} = 235 \text{ nH}$$

$$L_4 = \frac{(250)(0.366)}{2\pi(50 \times 10^6)} = 291 \text{ nH}$$

The source resistance is scaled by multiplying its normalized value by the final value of the load resistor.

$$R_{s(final)} = 0.2(250) = 50 \text{ ohms}$$

The final circuit appears in Fig. 3-21.

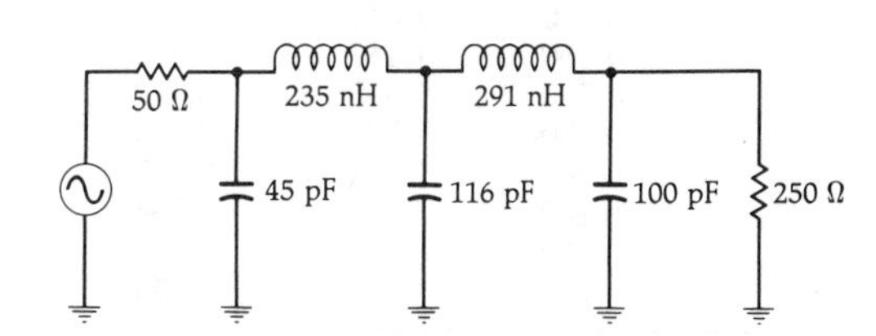

Fig. 3-21. Low-pass filter circuit for Example 3-5.

3. Determine the maximum amount of ripple that you can allow in the passband. Remember, the greater the amount of ripple allowed, the more selective the filter is. Higher values of ripple may allow you to eliminate a few components.
4. Match the normalized attenuation characteristics (Steps 1 and 2) with the attenuation curves provided in this chapter. Allow yourself a small "fudge-factor" for good measure. This step reveals the minimum number of circuit elements that you can get away with—given a certain filter type.
5. Find the low-pass prototype values in the tables.
6. Scale all elements to the frequency and impedance of the final design.

Example 3-6 diagrams the process of designing a low-pass filter using the preceding steps.

Table 3-5B. Chebyshev Low-Pass Prototype Element Values for 0.1-dB Ripple

n	R_S/R_L	C_1	L_2	C_3	L_4	C_5	L_6	C_7
5	1.000	1.301	1.556	2.241	1.556	1.301		
	0.900	1.285	1.433	2.380	1.488	1.488		
	0.800	1.300	1.282	2.582	1.382	1.738		
	0.700	1.358	1.117	2.868	1.244	2.062		
	0.600	1.470	0.947	3.269	1.085	2.484		
	0.500	1.654	0.778	3.845	0.913	3.055		
	0.400	1.954	0.612	4.720	0.733	3.886		
	0.300	2.477	0.451	6.196	0.550	5.237		
	0.200	3.546	0.295	9.127	0.366	7.889		
	0.100	6.787	0.115	17.957	0.182	15.745		
	∞	1.561	1.807	1.766	1.417	0.651		
6	1.355	0.942	2.080	1.659	2.247	1.534	1.277	
	1.429	0.735	2.249	1.454	2.544	1.405	1.629	
	1.667	0.542	2.600	1.183	3.064	1.185	2.174	
	2.000	0.414	3.068	0.958	3.712	0.979	2.794	
	2.500	0.310	3.765	0.749	4.651	0.778	3.645	
	3.333	0.220	4.927	0.551	6.195	0.580	4.996	
	5.000	0.139	7.250	0.361	9.261	0.384	7.618	
	10.000	0.067	14.220	0.178	18.427	0.190	15.350	
	∞	1.534	1.884	1.831	1.749	1.394	0.638	
7	1.000	1.262	1.520	2.239	1.680	2.239	1.520	1.262
	0.900	1.242	1.395	2.361	1.578	2.397	1.459	1.447
	0.800	1.255	1.245	2.548	1.443	2.624	1.362	1.697
	0.700	1.310	1.083	2.819	1.283	2.942	1.233	2.021
	0.600	1.417	0.917	3.205	1.209	3.384	1.081	2.444
	0.500	1.595	0.753	3.764	0.928	4.015	0.914	3.018
	0.400	1.885	0.593	4.618	0.742	4.970	0.738	3.855
	0.300	2.392	0.437	6.054	0.556	6.569	0.557	5.217
	0.200	3.428	0.286	8.937	0.369	9.770	0.372	7.890
	0.100	6.570	0.141	17.603	0.184	19.376	0.186	15.813
	∞	1.575	1.858	1.921	1.827	1.734	1.379	0.631
n	R_L/R_S	L_1	C_2	L_3	C_4	L_5	C_6	L_7

HIGH-PASS FILTER DESIGN

Once you have learned the mechanics of low-pass filter design, high-pass design becomes a snap. You can use all of the attenuation response curves presented, thus far, for the low-pass filters by simply inverting the f/f_c axis. For instance, a 5-element, 0.1–dB-ripple Chebyshev *low-pass filter* will produce an attenuation of about 60 dB at an f/f_c of 3 (Fig. 3-16). If you were working instead with a *high-pass filter* of the same size and type, you could still use Fig. 3-16 to tell you that at an f/f_c of 1/3 (or, $f_c/f = 3$) a 5-element, 0.1–dB-ripple Chebyshev *high-pass filter* will also produce an attenuation of 60 dB. This is obviously more convenient than having to refer to more than one set of curves.

After finding the response which satisfies all of the requirements, the next step is to simply refer to the tables of low-pass prototype values and copy down the prototype values that are called for. High-pass values for the elements are then obtained directly from the low-pass prototype values as follows (refer to Fig. 3-24):

EXAMPLE 3-6

Design a low-pass filter to meet the following specifications:

$f_c = 35$ MHz,
Response greater than 60 dB down at 105 MHz,
Maximally flat passband—no ripple,
$R_s = 50$ ohms,
$R_L = 500$ ohms.

Solution

The need for a maximally flat passband automatically indicates that the design must be a Butterworth response. The first step in the design process is to normalize everything. Thus,

$$\frac{R_s}{R_L} = \frac{50}{500} = 0.1$$

Next, normalize the frequencies of interest so that they may be found in the graph of Fig. 3-9. Thus, we have:

$$\frac{f_{60\,dB}}{f_{3\,dB}} = \frac{105\text{ MHz}}{35\text{ MHz}} = 3$$

We next look at Fig. 3-9 and find a response that is down at least 60 dB at a frequency ratio of $f/f_c = 3$. Fig. 3-9 indicates that it will take a minimum of 7 elements to provide the attenuation specified. Referring to the catalog of Butterworth low-pass prototype values given in Table 3-2 yields the prototype circuit of Fig. 3-22.

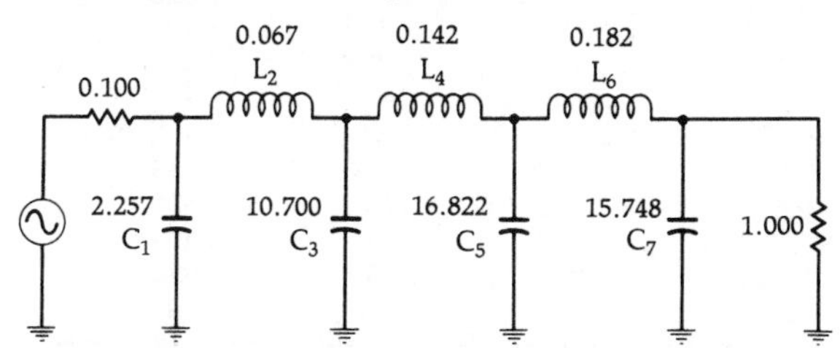

Fig. 3-22. Low-pass prototype circuit for Example 3-6.

We then scale these values using Equations 3-12 and 3-13. The first two values are worked out for you.

$$C_1 = \frac{2.257}{2\pi(35 \times 10^6)500} = 21\text{ pF}$$

$$L_2 = \frac{(500)(0.067)}{2\pi(35 \times 10^6)} = 152\text{ nH}$$

Similarly,

$C_3 = 97$ pF,
$C_5 = 153$ pF,
$C_7 = 143$ pF,
$L_4 = 323$ nH,
$L_6 = 414$ nH,
$R_s = 50$ ohms,
$R_L = 500$ ohms.

The final circuit is shown in Fig. 3-23.

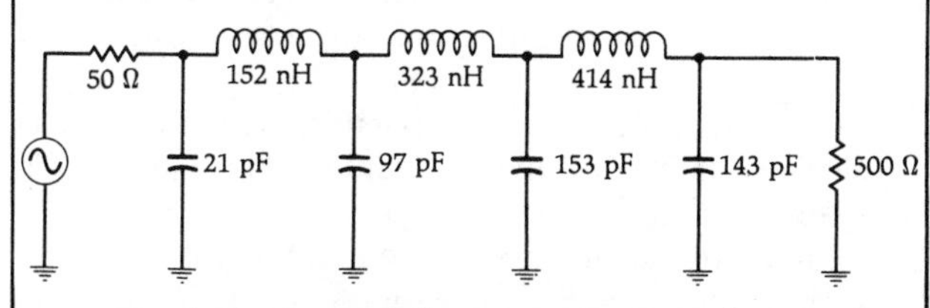

Fig. 3-23. Low-pass filter circuit for Example 3-6.

Table 3-6A. Chebyshev Low-Pass Prototype Element Values for 0.5-dB Ripple

n	R_S/R_L	C_1	L_2	C_3	L_4
2	1.984	0.983	1.950		
	2.000	0.909	2.103		
	2.500	0.564	3.165		
	3.333	0.375	4.411		
	5.000	0.228	6.700		
	10.000	0.105	13.322		
	∞	1.307	0.975		
3	1.000	1.864	1.280	1.834	
	0.900	1.918	1.209	2.026	
	0.800	1.997	1.120	2.237	
	0.700	2.114	1.015	2.517	
	0.500	2.557	0.759	3.436	
	0.400	2.985	0.615	4.242	
	0.300	3.729	0.463	5.576	
	0.200	5.254	0.309	8.225	
	0.100	9.890	0.153	16.118	
	∞	1.572	1.518	0.932	
4	1.984	0.920	2.586	1.304	1.826
	2.000	0.845	2.720	1.238	1.985
	2.500	0.516	3.766	0.869	3.121
	3.333	0.344	5.120	0.621	4.480
	5.000	0.210	7.708	0.400	6.987
	10.000	0.098	15.352	0.194	14.262
	∞	1.436	1.889	1.521	0.913
n	R_L/R_S	L_1	C_2	L_3	C_4

Simply replace each filter element with an element of the opposite type and with a reciprocal value. Thus, L_1 of Fig. 3-24B is equal to $1/C_1$ of Fig. 3-24A. Likewise, $C_2 = 1/L_2$ and $L_3 = 1/C_3$.

Stated another way, if the low-pass prototype indicates a capacitor of 1.181 farads, then, use an inductor with a value of $1/1.181 = 0.847$ henry, instead, for a high-pass design. However, the source and load resistors should not be altered.

The transformation process results in an attenuation characteristic for the high-pass filter that is an exact mirror image of the low-pass attenuation characteristic. The ripple, if there is any, remains the same and the magnitude of the slope of the stopband (or pass-

Table 3-6B. Chebyshev Low-Pass Prototype Element Values for 0.5-dB Ripple

n	R_S/R_L	C_1	L_2	C_3	L_4	C_5	L_6	C_7
5	1.000	1.807	1.303	2.691	1.303	1.807		
	0.900	1.854	1.222	2.849	1.238	1.970		
	0.800	1.926	1.126	3.060	1.157	2.185		
	0.700	2.035	1.015	3.353	1.058	2.470		
	0.600	2.200	0.890	3.765	0.942	2.861		
	0.500	2.457	0.754	4.367	0.810	3.414		
	0.400	2.870	0.609	5.296	0.664	4.245		
	0.300	3.588	0.459	6.871	0.508	5.625		
	0.200	5.064	0.306	10.054	0.343	8.367		
	0.100	9.556	0.153	19.647	0.173	16.574		
	∞	1.630	1.740	1.922	1.514	0.903		
6	1.984	0.905	2.577	1.368	2.713	1.299	1.796	
	2.000	0.830	2.704	1.291	2.872	1.237	1.956	
	2.500	0.506	3.722	0.890	4.109	0.881	3.103	
	3.333	0.337	5.055	0.632	5.699	0.635	4.481	
	5.000	0.206	7.615	0.406	8.732	0.412	7.031	
	10.000	0.096	15.186	0.197	17.681	0.202	14.433	
7	1.000	1.790	1.296	2.718	1.385	2.718	1.296	1.790
	0.900	1.835	1.215	2.869	1.308	2.883	1.234	1.953
	0.800	1.905	1.118	3.076	1.215	3.107	1.155	2.168
	0.700	2.011	1.007	3.364	1.105	3.416	1.058	2.455
	0.600	2.174	0.882	3.772	0.979	3.852	0.944	2.848
	0.500	2.428	0.747	4.370	0.838	2.289	0.814	3.405
	0.400	2.835	0.604	5.295	0.685	5.470	0.669	4.243
	0.300	3.546	0.455	6.867	0.522	7.134	0.513	5.635
	0.200	5.007	0.303	10.049	0.352	10.496	0.348	8.404
	0.100	9.456	0.151	19.649	0.178	20.631	0.176	16.665
	∞	1.646	1.777	2.031	1.789	1.924	1.503	0.895
n	R_L/R_S	L_1	C_2	L_3	C_4	L_5	C_6	L_7

band) skirts remains the same. Example 3-7 illustrates the design of high-pass filters.

A closer look at the filter designed in Example 3-7 reveals that it is symmetric. Indeed, all filters given for the equal termination class are symmetric. The equal termination class of filter thus yields a circuit that is easier to design (fewer calculations) and, in most cases, cheaper to build for a high-volume product, due to the number of equal valued components.

THE DUAL NETWORK

Thus far, we have been referring to the group of low-pass prototype element value tables presented and, then, we choose the schematic that is located either above or below the tables for the form of the filter that we are designing, depending on the value of R_L/R_S. Either form of the filter will produce exactly the same attenuation, phase, and group-delay characteristics, and each form is called the *dual* of the other.

Any filter network in a ladder arrangement, such as the ones presented in this chapter, can be changed into its dual form by application of the following rules:

1. Change all inductors to capacitors, and vice-versa, without changing element values. Thus, 3 henries becomes 3 farads.
2. Change all resistances into conductances, and vice-versa, with the value unchanged. Thus, 3 ohms becomes 3 mhos, or ⅓ ohm.

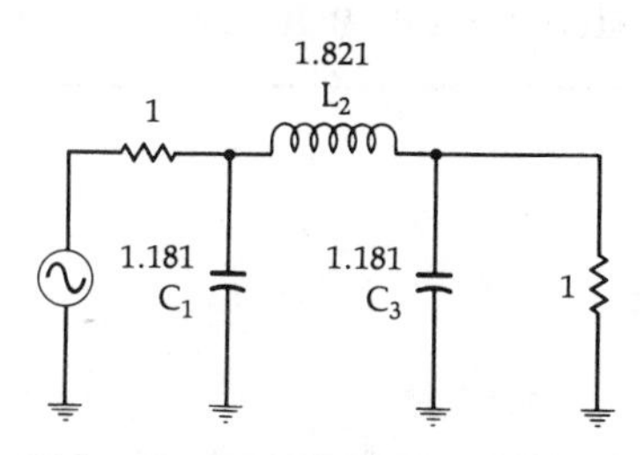

(A) *Low-pass prototype circuit.*

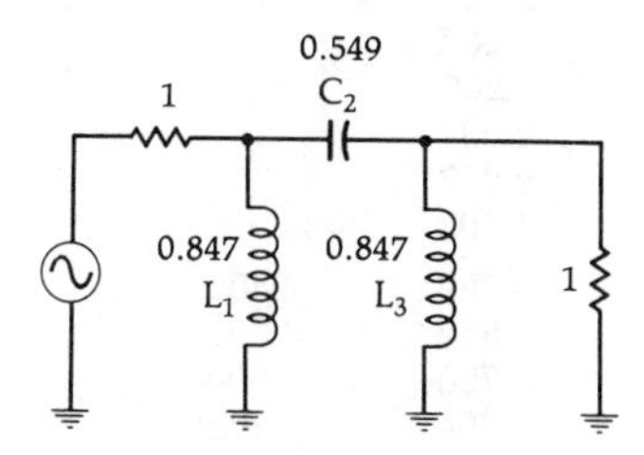

(B) *Equivalent high-pass prototype circuit.*

Fig. 3-24. Low-pass to high-pass filter transformation.

3. Change all shunt branches to series branches, and vice-versa.
4. Change all elements in series with each other into elements that are in parallel with each other.
5. Change all voltage sources into current sources, and vice-versa.

Fig. 3-26 shows a ladder network and its dual representation.

Dual networks are convenient, in the case of equal terminations, if you desire to change the topology of the filter without changing the response. It is most often used, as shown in Example 3-7, to eliminate an unnecessary inductor which might have crept into the design through some other transformation process. Inductors are typically more lower-Q devices than capacitors and, therefore, exhibit higher losses. These losses tend to cause insertion loss, in addition to generally degrading the overall performance of the filter. The number of inductors in any network should, therefore, be reduced whenever possible.

A little experimentation with dual networks having unequal terminations will reveal that you can quickly get yourself into trouble if you are not careful. This is especially true if the load and source resistance are a design criteria and cannot be changed to suit the needs of your filter. Remember, when the dual of a network with unequal terminations is taken, then, the terminations *must*, by definition, change value as shown in Fig. 3-26.

BANDPASS FILTER DESIGN

The low-pass prototype circuits and response curves given in this chapter can also be used in the design of bandpass filters. This is done through a simple

Table 3-7A. Chebyshev Low-Pass Prototype Element Values for 1.0-dB Ripple

n	R_S/R_L	C_1	L_2	C_3	L_4
2	3.000	0.572	3.132		
	4.000	0.365	4.600		
	8.000	0.157	9.658		
	∞	1.213	1.109		
3	1.000	2.216	1.088	2.216	
	0.500	4.431	0.817	2.216	
	0.333	6.647	0.726	2.216	
	0.250	8.862	0.680	2.216	
	0.125	17.725	0.612	2.216	
	∞	1.652	1.460	1.108	
4	3.000	0.653	4.411	0.814	2.535
	4.000	0.452	7.083	0.612	2.848
	8.000	0.209	17.164	0.428	3.281
	∞	1.350	2.010	1.488	1.106
n	R_L/R_S	L_1	C_2	L_3	C_4

transformation process similar to what was done in the high-pass case.

The most difficult task awaiting the designer of a bandpass filter, if the design is to be derived from the low-pass prototype, is in specifying the bandpass attenuation characteristics in terms of the low-pass response curves. A method for doing this is shown by the curves in Fig. 3-27. As you can see, when a low-pass design is transformed into a bandpass design, the attenuation bandwidth ratios remain the same. This means that a low-pass filter with a 3-dB cutoff frequency, or a bandwidth of 2 kHz, would transform into a bandpass filter with a 3-dB bandwidth of 2 kHz. If the response of the low-pass network were down 30 dB at a frequency or bandwidth of 4 kHz ($f/f_c = 2$), then the response of the bandpass network would be down 30 dB at a bandwidth of 4 kHz. Thus, the normalized f/f_c axis of the low-pass attenuation curves becomes a ratio of bandwidths rather than frequencies, such that:

$$\frac{BW}{BW_c} = \frac{f}{f_c} \qquad \text{(Eq. 3-14)}$$

where,

BW = the bandwidth at the required value of attenuation,

BW_c = the 3-dB bandwidth of the bandpass filter.

Table 3-7B. Chebyshev Low-Pass Prototype Element Values for 1.0-dB Ripple

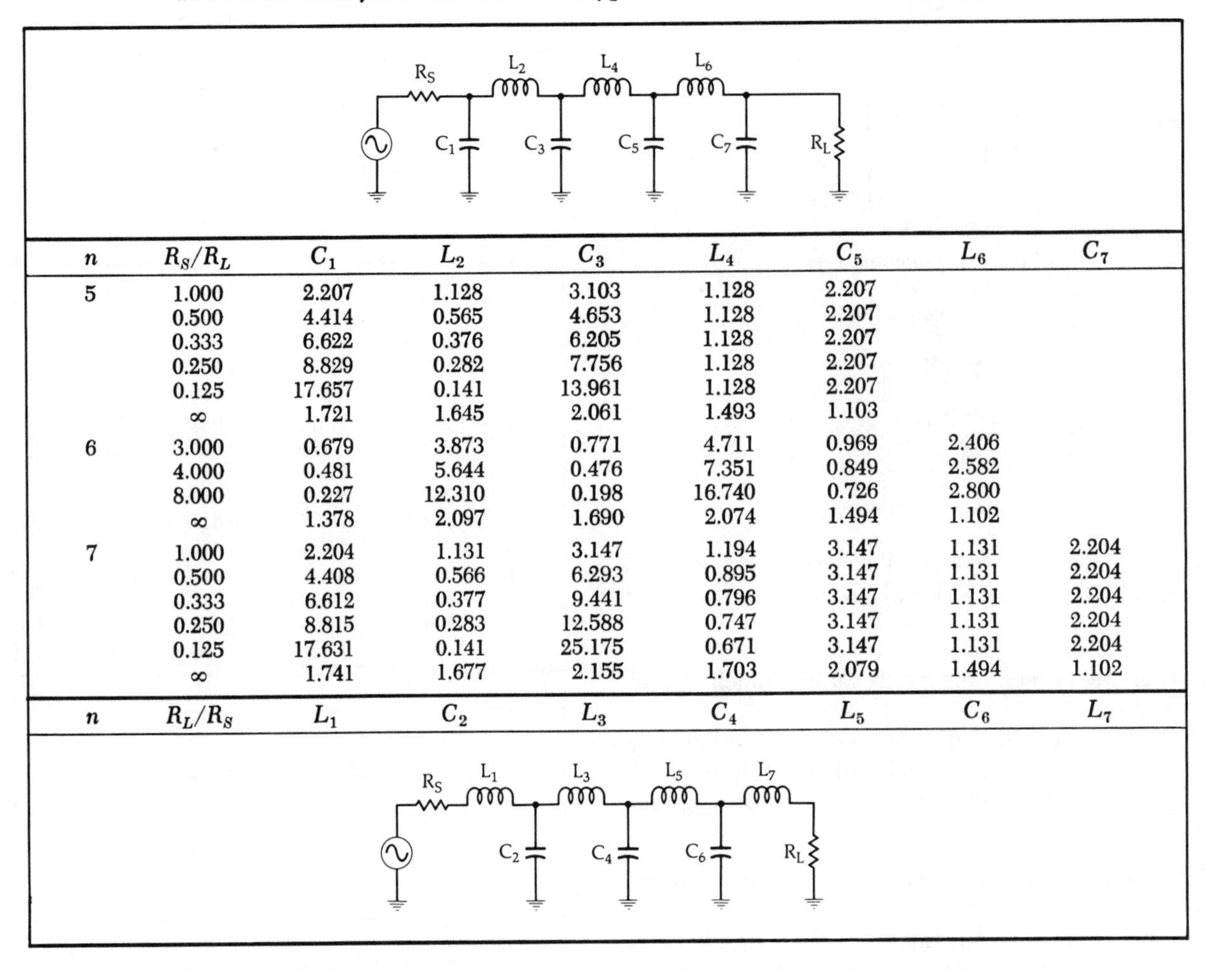

n	R_S/R_L	C_1	L_2	C_3	L_4	C_5	L_6	C_7
5	1.000	2.207	1.128	3.103	1.128	2.207		
	0.500	4.414	0.565	4.653	1.128	2.207		
	0.333	6.622	0.376	6.205	1.128	2.207		
	0.250	8.829	0.282	7.756	1.128	2.207		
	0.125	17.657	0.141	13.961	1.128	2.207		
	∞	1.721	1.645	2.061	1.493	1.103		
6	3.000	0.679	3.873	0.771	4.711	0.969	2.406	
	4.000	0.481	5.644	0.476	7.351	0.849	2.582	
	8.000	0.227	12.310	0.198	16.740	0.726	2.800	
	∞	1.378	2.097	1.690	2.074	1.494	1.102	
7	1.000	2.204	1.131	3.147	1.194	3.147	1.131	2.204
	0.500	4.408	0.566	6.293	0.895	3.147	1.131	2.204
	0.333	6.612	0.377	9.441	0.796	3.147	1.131	2.204
	0.250	8.815	0.283	12.588	0.747	3.147	1.131	2.204
	0.125	17.631	0.141	25.175	0.671	3.147	1.131	2.204
	∞	1.741	1.677	2.155	1.703	2.079	1.494	1.102
n	R_L/R_S	L_1	C_2	L_3	C_4	L_5	C_6	L_7

Often a bandpass response is not specified, as in Example 3-8. Instead, the requirements are often given as attenuation values at specified frequencies as shown by the curve in Fig. 3-28. In this case, you must transform the stated requirements into information that takes the form of Equation 3-14. As an example, consider Fig. 3-28. How do we convert the data that is given into the bandwidth ratios we need? Before we can answer that, we have to find f_3. Use the following method.

The frequency response of a bandpass filter exhibits geometric symmetry. That is, it is only symmetric when plotted on a logarithmic scale. The center frequency of a geometrically symmetric filter is given by the formula:

$$f_o = \sqrt{f_a f_b} \qquad \text{(Eq. 3-15)}$$

where f_a and f_b are any two frequencies (one above and one below the passband) having equal attenuation. Therefore, the center frequency of the response curve shown in Fig. 3-28 must be

$$f_o = \sqrt{(45)(75)}\ \text{MHz}$$
$$= 58.1\ \text{MHz}$$

We can use Equation 3-15 again to find f_3.

$$58.1 = \sqrt{f_3(125)}$$

or,

$$f_3 = 27\ \text{MHz}$$

Now that f_3 is known, the data of Fig. 3-28 can be put into the form of Equation 3-14.

$$\frac{BW_{40\,dB}}{BW_{3\,dB}} = \frac{125\ \text{MHz} - 27\ \text{MHz}}{75\ \text{MHz} - 45\ \text{MHz}}$$
$$= 3.27$$

To find a low-pass prototype curve that will satisfy these requirements, simply refer to any of the pertinent graphs presented in this chapter and find a response which will provide 40 dB of attenuation at an f/f_c of 3.27. (A fourth-order or better Butterworth filter will do quite nicely.)

The actual transformation from the low-pass to the bandpass configuration is accomplished by resonating each low-pass element with an element of the opposite type and of the same value. All shunt elements of the low-pass prototype circuit become parallel-resonant

EXAMPLE 3-7

Design an LC high-pass filter with an f_c of 60 MHz and a minimum attenuation of 40 dB at 30 MHz. The source and load resistance are equal at 300 ohms. Assume that a 0.5-dB passband ripple is tolerable.

Solution

First, normalize the attenuation requirements so that the low-pass attenuation curves may be used.

$$\frac{f}{f_c} = \frac{30 \text{ MHz}}{60 \text{ MHz}}$$
$$= 0.5$$

Inverting, we get:

$$\frac{f_c}{f} = 2$$

Now, select a normalized low-pass filter that offers at least 40-dB attenuation at a ratio of $f_c/f = 2$. Reference to Fig. 3.17 (attenuation response of 0.5–dB-ripple Chebyshev filters) indicates that a normalized $n = 5$ Chebyshev will provide the needed attenuation. Table 3-6 contains the element values for the corresponding network. The normalized low-pass prototype circuit is shown in Fig. 3-25A. Note that the schematic below Table 3-6B was chosen as the low-pass prototype circuit rather than the schematic above the table. The reason for doing this will become obvious after the next step. Keep in mind, however, that the ratio of R_S/R_L is the same as the ratio of R_L/R_S, and is unity. Therefore, it does not matter which form is used for the prototype circuit.

Next, transform the low-pass circuit to a high-pass network by replacing each inductor with a capacitor, and vice-versa, using reciprocal element values as shown in Fig. 3-25B. Note here that had we begun with the low-pass prototype circuit shown above Table 3-6B, this transformation would have yielded a filter containing three inductors rather than the two shown in Fig. 3-25B. The object in any of these filter designs is to reduce the number of inductors in the final design. More on this later.

The final step in the design process is to scale the network in both impedance and frequency using Equations 3-12 and 3-13. The first two calculations are done for you.

$$C_1 = \frac{\frac{1}{1.807}}{2\pi(60 \times 10^6)(300)}$$
$$= 4.9 \text{ pF}$$

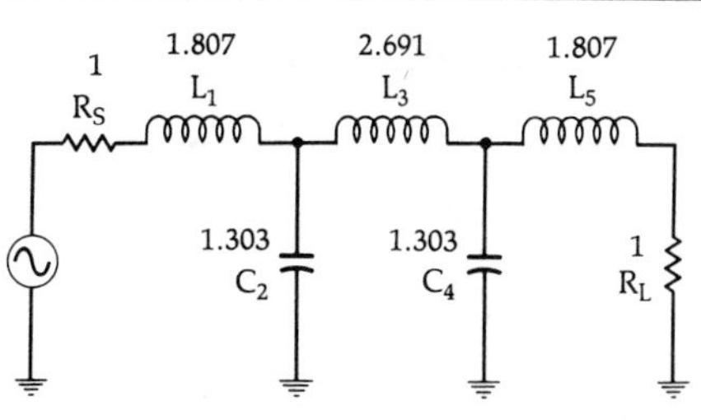

(A) *Normalized low-pass filter circuit.*

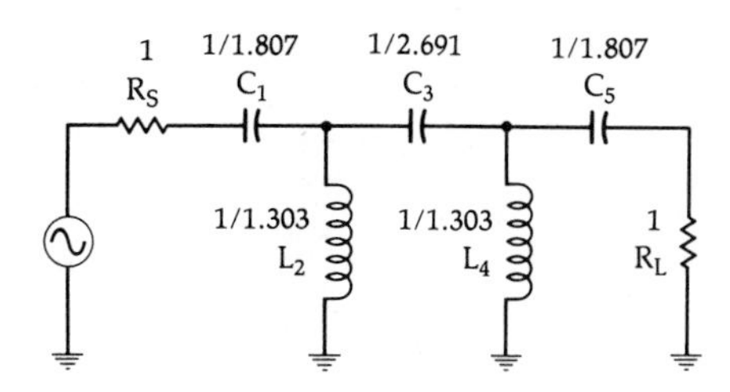

(B) *High-pass transformation.*

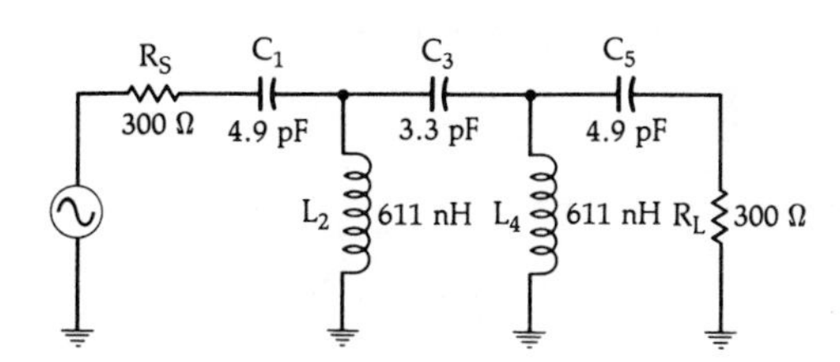

(C) *Frequency and impedance-scaled filter circuit.*

Fig. 3-25. High-pass filter design for Example 3-7.

$$L_2 = \frac{300\left(\frac{1}{1.303}\right)}{2\pi(60 \times 10^6)}$$
$$= 611 \text{ nH}$$

The remaining values are:

$C_3 = 3.3$ pF
$C_5 = 4.9$ pF
$L_4 = 611$ nH

The final filter circuit is given in Fig. 3-25C.

EXAMPLE 3-8

Find the Butterworth low-pass prototype circuit which, when transformed, would satisfy the following bandpass filter requirements:

$$BW_{3\,dB} = 2 \text{ MHz}$$
$$BW_{40\,dB} = 6 \text{ MHz}$$

Solution

Note that we are not concerned with the center frequency of the bandpass response just yet. We are only concerned with the relationship between the above requirements and the low-pass response curves. Using Equation 3-14, we have:

$$\frac{BW}{BW_c} = \frac{f}{f_c} = \frac{BW_{40\,dB}}{BW_{3\,dB}}$$
$$= \frac{6 \text{ MHz}}{2 \text{ MHz}}$$
$$= 3$$

Therefore, turn to the Butterworth response curves shown in Fig. 3-9 and find a prototype value that will provide 40 dB of attenuation at an $f/f_c = 3$. The curves indicate a 5-element Butterworth filter will provide the needed attenuation.

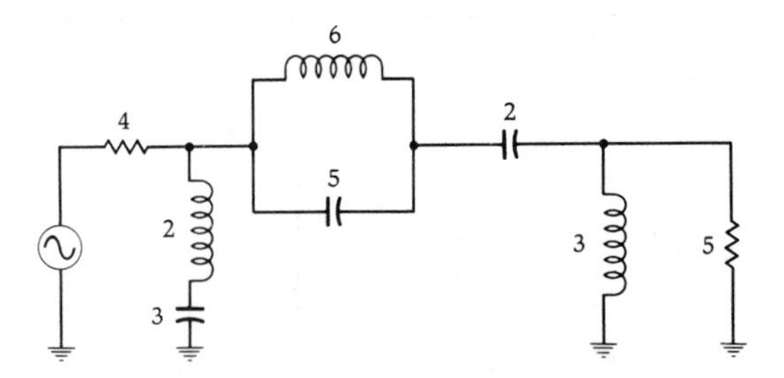

(*A*) *A representative ladder network.*

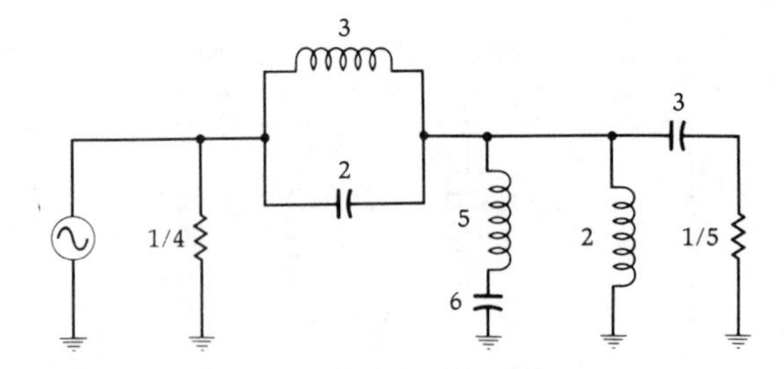

(*B*) *Its dual form.*

Fig. 3-26. Duality.

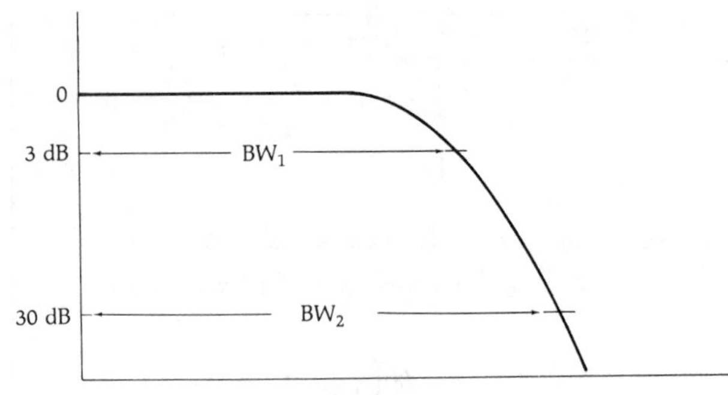

(*A*) *Low-pass prototype response.*

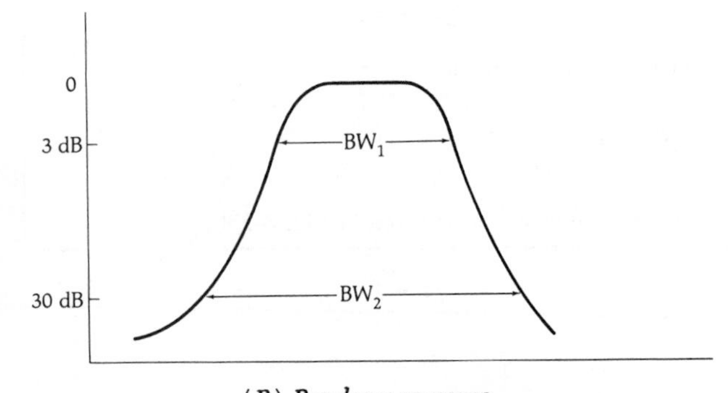

(*B*) *Bandpass response.*

Fig. 3-27. Low-pass to bandpass transformation bandwidths.

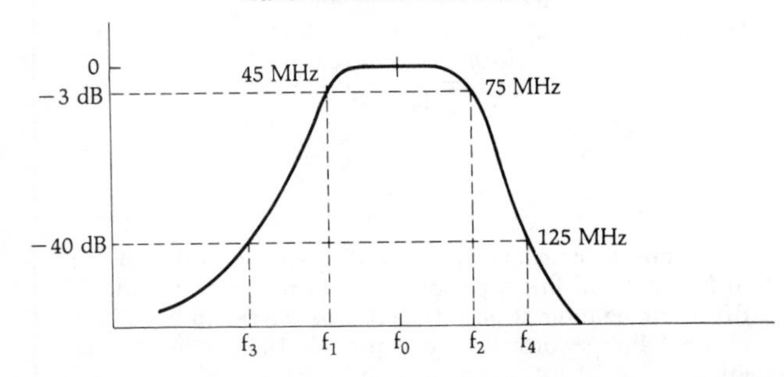

Fig. 3-28. Typical bandpass specifications.

Table 3-8A. Bessel Low-Pass Prototype Element Values

n	R_S/R_L	C_1	L_2	C_3	L_4
2	1.000	0.576	2.148		
	1.111	0.508	2.310		
	1.250	0.443	2.510		
	1.429	0.380	2.764		
	1.667	0.319	3.099		
	2.000	0.260	3.565		
	2.500	0.203	4.258		
	3.333	0.149	5.405		
	5.000	0.097	7.688		
	10.000	0.047	14.510		
	∞	1.362	0.454		
3	1.000	0.337	0.971	2.203	
	0.900	0.371	0.865	2.375	
	0.800	0.412	0.761	2.587	
	0.700	0.466	0.658	2.858	
	0.600	0.537	0.558	3.216	
	0.500	0.635	0.459	3.714	
	0.400	0.783	0.362	4.457	
	0.300	1.028	0.267	5.689	
	0.200	1.518	0.175	8.140	
	0.100	2.983	0.086	15.470	
	∞	1.463	0.843	0.293	
4	1.000	0.233	0.673	1.082	2.240
	1.111	0.209	0.742	0.967	2.414
	1.250	0.184	0.829	0.853	2.630
	1.429	0.160	0.941	0.741	2.907
	1.667	0.136	1.089	0.630	3.273
	2.000	0.112	1.295	0.520	3.782
	2.500	0.089	1.604	0.412	4.543
	3.333	0.066	2.117	0.306	5.805
	5.000	0.043	3.142	0.201	8.319
	10.000	0.021	6.209	0.099	15.837
	∞	1.501	0.978	0.613	0.211
n	R_L/R_S	L_1	C_2	L_3	C_4

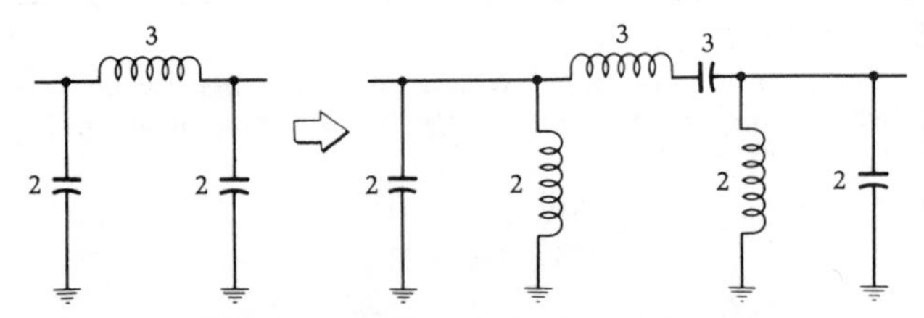

Fig. 3-29. Low-pass to bandpass circuit transformation.

circuits, and all series elements become series-resonant circuits. This process is illustrated in Fig. 3-30.

To complete the design, the transformed filter is

Table 3-8B. Bessel Low-Pass Prototype Element Values

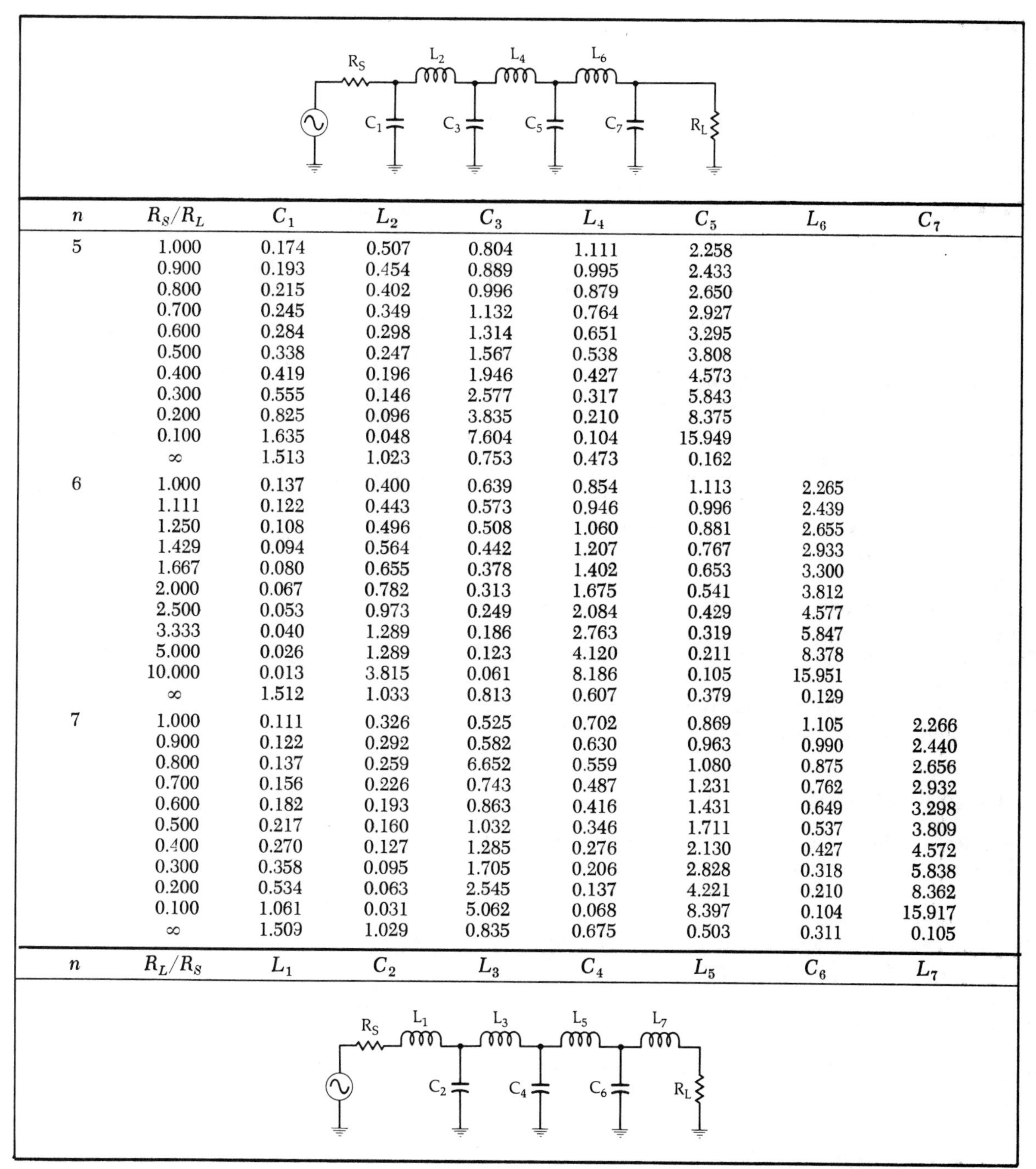

n	R_S/R_L	C_1	L_2	C_3	L_4	C_5	L_6	C_7
5	1.000	0.174	0.507	0.804	1.111	2.258		
	0.900	0.193	0.454	0.889	0.995	2.433		
	0.800	0.215	0.402	0.996	0.879	2.650		
	0.700	0.245	0.349	1.132	0.764	2.927		
	0.600	0.284	0.298	1.314	0.651	3.295		
	0.500	0.338	0.247	1.567	0.538	3.808		
	0.400	0.419	0.196	1.946	0.427	4.573		
	0.300	0.555	0.146	2.577	0.317	5.843		
	0.200	0.825	0.096	3.835	0.210	8.375		
	0.100	1.635	0.048	7.604	0.104	15.949		
	∞	1.513	1.023	0.753	0.473	0.162		
6	1.000	0.137	0.400	0.639	0.854	1.113	2.265	
	1.111	0.122	0.443	0.573	0.946	0.996	2.439	
	1.250	0.108	0.496	0.508	1.060	0.881	2.655	
	1.429	0.094	0.564	0.442	1.207	0.767	2.933	
	1.667	0.080	0.655	0.378	1.402	0.653	3.300	
	2.000	0.067	0.782	0.313	1.675	0.541	3.812	
	2.500	0.053	0.973	0.249	2.084	0.429	4.577	
	3.333	0.040	1.289	0.186	2.763	0.319	5.847	
	5.000	0.026	1.289	0.123	4.120	0.211	8.378	
	10.000	0.013	3.815	0.061	8.186	0.105	15.951	
	∞	1.512	1.033	0.813	0.607	0.379	0.129	
7	1.000	0.111	0.326	0.525	0.702	0.869	1.105	2.266
	0.900	0.122	0.292	0.582	0.630	0.963	0.990	2.440
	0.800	0.137	0.259	6.652	0.559	1.080	0.875	2.656
	0.700	0.156	0.226	0.743	0.487	1.231	0.762	2.932
	0.600	0.182	0.193	0.863	0.416	1.431	0.649	3.298
	0.500	0.217	0.160	1.032	0.346	1.711	0.537	3.809
	0.400	0.270	0.127	1.285	0.276	2.130	0.427	4.572
	0.300	0.358	0.095	1.705	0.206	2.828	0.318	5.838
	0.200	0.534	0.063	2.545	0.137	4.221	0.210	8.362
	0.100	1.061	0.031	5.062	0.068	8.397	0.104	15.917
	∞	1.509	1.029	0.835	0.675	0.503	0.311	0.105
n	R_L/R_S	L_1	C_2	L_3	C_4	L_5	C_6	L_7

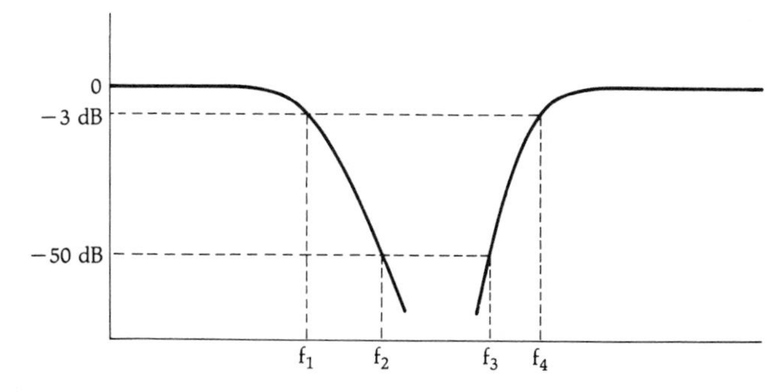

Fig. 3-30. Typical band-rejection filter curves.

then frequency- and impedance-scaled using the following formulas. For the parallel-resonant branches,

$$C = \frac{C_n}{2\pi RB} \qquad \text{(Eq. 3-16)}$$

$$L = \frac{RB}{2\pi f_o^2 L_n} \qquad \text{(Eq. 3-17)}$$

and, for the series-resonant branches,

$$C = \frac{B}{2\pi f_o^2 C_n R} \qquad \text{(Eq. 3-18)}$$

$$L = \frac{RL_n}{2\pi B} \qquad \text{(Eq. 3-19)}$$

where, in all cases,
R = the final load impedance,
B = the 3-dB bandwidth of the final design,
f_o = the geometric center frequency of the final design,
L_n = the normalized inductor *bandpass* element values,
C_n = the normalized capacitor *bandpass* element values.

Example 3-9 furnishes one final example of the procedure for designing a bandpass filter.

SUMMARY OF THE BANDPASS FILTER DESIGN PROCEDURE

1. Transform the bandpass requirements into an equivalent low-pass requirement using Equation 3-14.
2. Refer to the low-pass attenuation curves provided in order to find a response that meets the requirements of Step 1.
3. Find the corresponding low-pass prototype and write it down.
4. Transform the low-pass network into a bandpass configuration.
5. Scale the bandpass configuration in both impedance and frequency using Equations 3-16 through 3-19.

BAND-REJECTION FILTER DESIGN

Band-rejection filters are very similar in design approach to the bandpass filter of the last section. Only, in this case, we want to *reject* a certain group of frequencies as shown by the curves in Fig. 3-30.

The band-reject filter lends itself well to the low-pass prototype design approach using the same procedures as were used for the bandpass design. First, define the bandstop requirements in terms of the low-pass attenuation curves. This is done by using the inverse of Equation 3-14. Thus, referring to Fig. 3-30, we have:

$$\frac{BW_c}{BW} = \frac{f_4 - f_1}{f_3 - f_2}$$

This sets the attenuation characteristic that is needed and allows you to read directly off the low-pass attenuation curves by substituting BW_c/BW for f_c/f on the normalized frequency axis. Once the number of elements that are required in the low-pass prototype circuit is determined, the low-pass network is transformed into a band-reject configuration as follows:

> Each shunt element in the low-pass prototype circuit is replaced by a shunt *series-resonant circuit,* and each series-element is replaced by a series *parallel-resonant circuit.*

EXAMPLE 3-9

Design a bandpass filter with the following requirements:

$f_o = 75$ MHz	Passband Ripple = 1 dB
$BW_{3dB} = 7$ MHz	$R_s = 50$ ohms
$BW_{45dB} = 35$ MHz	$R_L = 100$ ohms

Solution

Using Equation 3-14:

$$\frac{BW_{45dB}}{BW_{3dB}} = \frac{35}{7} = 5$$

Substitute this value for f/f_c in the low-pass attenuation curves for the 1-dB-ripple Chebyshev response shown in Fig. 3-18. This reveals that a 3-element filter will provide about 50 dB of attenuation at an $f/f_c = 5$, which is more than adequate. The corresponding element values for this filter can be found in Table 3-7 for an $R_S/R_L = 0.5$ and an $n = 3$. This yields the low-pass prototype circuit of Fig. 3-32A which is transformed into the bandpass prototype circuit of Fig. 3-32B. Finally, using Equations 3-16 through 3-19, we obtain the final circuit that is shown in Fig. 3-32C. The calculations follow. Using Equations 3-16 and 3-17:

$$C_1 = \frac{4.431}{2\pi(100)(7 \times 10^6)} = 1007 \text{ pF}$$

$$L_1 = \frac{(100)(7 \times 10^6)}{2\pi(75 \times 10^6)^2(4.431)} = 4.47 \text{ nH}$$

Using Equations 3-18 and 3-19:

$$C_2 = \frac{7 \times 10^6}{2\pi(75 \times 10^6)^2(0.817)100} = 2.4 \text{ pF}$$

$$L_2 = \frac{(100)(0.817)}{2\pi(7 \times 10^6)} = 1.86\ \mu\text{H}$$

Similarly,
$C_3 = 504$ pF
$L_3 = 8.93$ nH

This is shown in Fig. 3-31. Note that both elements in each of the resonant circuits have the same normalized value.

Once the prototype circuit has been transformed into its band-reject configuration, it is then scaled in impedance and frequency using the following formulas. For all series-resonant circuits:

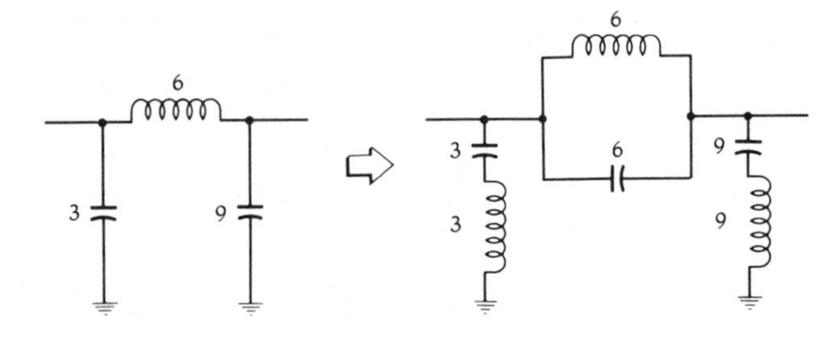

Fig. 3-31. Low-pass to band-reject transformation.

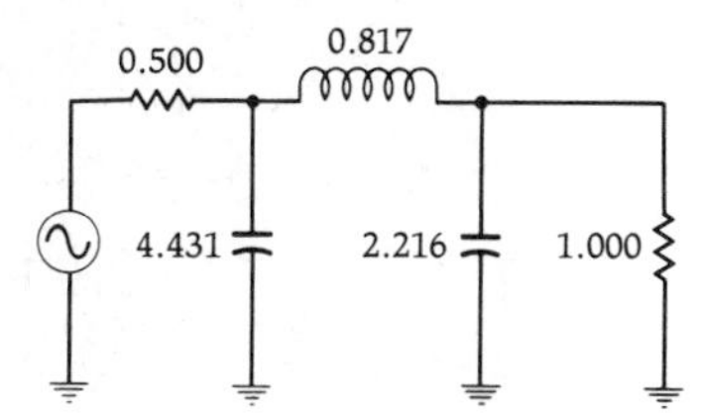

(A) *Low-pass prototype circuit.*

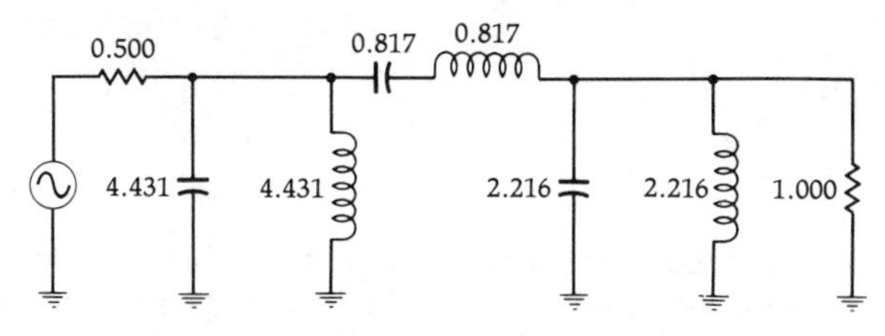

(B) *Bandpass transformation.*

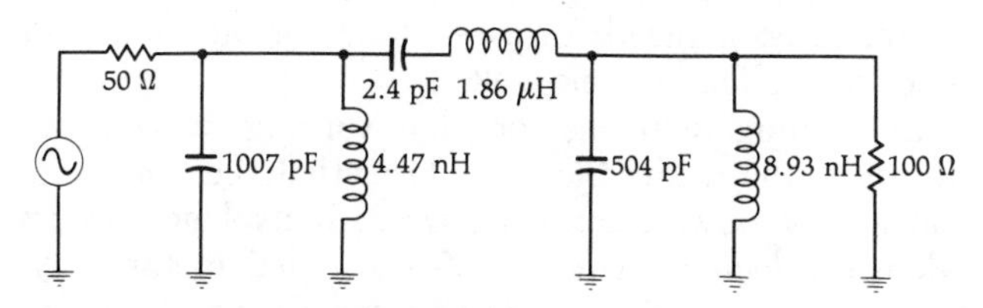

(C) *Final circuit with frequency and impedance scaled.*

Fig. 3-32. Bandpass filter design for Example 3-9.

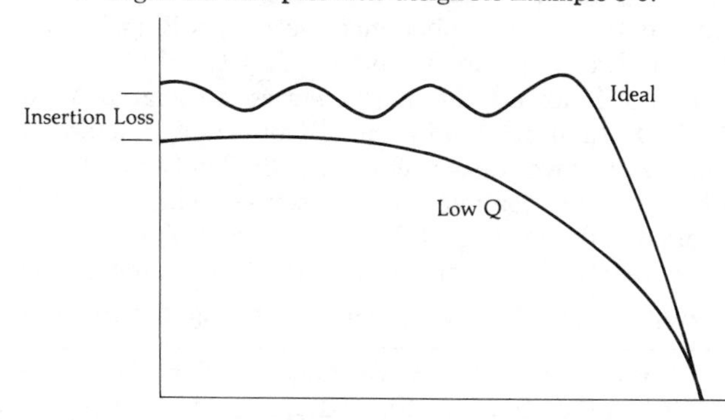

Fig. 3-33. The effect of finite-Q elements on filter response.

$$C = \frac{C_n}{2\pi RB} \quad \text{(Eq. 3-20)}$$

$$L = \frac{RB}{2\pi f_o^2 L_n} \quad \text{(Eq. 3-21)}$$

For all parallel-resonant circuits:

$$C = \frac{B}{2\pi f_o^2 RC_n} \quad \text{(Eq. 3-22)}$$

$$L = \frac{RL_n}{2\pi B} \quad \text{(Eq. 3-23)}$$

where, in all cases,

B = the 3-dB bandwidth,
R = the final load resistance,
f_o = the geometric center frequency,
C_n = the normalized capacitor band-reject element value,
L_n = the normalized inductor band-reject element value.

THE EFFECTS OF FINITE Q

Thus far in this chapter, we have assumed the inductors and capacitors used in the designs to be lossless. Indeed, all of the response curves presented in this chapter are based on that assumption. But we know from our previous study of Chapters 1 and 2 that even though capacitors can be approximated as having infinite Q, inductors cannot, and the effects of the finite-Q inductor must be taken into account in any filter design.

The use of finite element Q in a design intended for lossless elements causes the following unwanted effects (refer to Fig. 3-33):

1. Insertion loss of the filter is increased whereas the final stopband attenuation does not change. The relative attenuation between the two is decreased.
2. At frequencies in the vicinity of cutoff (f_c), the response becomes more rounded and usually results in an attenuation greater than the 3 dB that was originally intended.
3. Ripple that was designed into the passband will be reduced. If the element Q is sufficiently low, ripple will be totally eliminated.
4. For band-reject filters, the attenuation in the stopband becomes finite. This, coupled with an increase in *passband* insertion loss, decreases the relative attenuation significantly.

Regardless of the gloomy predictions outlined above, however, it is possible to design filters, using the approach outlined in this chapter, that very closely resemble the ideal response of each network. The key is to use the highest-Q inductors available for the given task. Table 3-9 outlines the recommended minimum element-Q requirements for the filters presented in this chapter. Keep in mind, however, that anytime a low-Q component is used, the actual attenuation response of the network strays from the ideal response to a degree depending upon the element Q. It is, therefore, highly recommended that you make it a habit to use only the highest-Q components available.

Table 3-9. Filter Element-Q Requirements

Filter Type	*Minimum Element Q Required*
Bessel	3
Butterworth	15
0.01-dB Chebyshev	24
0.1-dB Chebyshev	39
0.5-dB Chebyshev	57
1-dB Chebyshev	75

The insertion loss of the filters presented in this chapter can be calculated in the same manner as was used in Chapter 2. Simply replace each reactive element with resistor values corresponding to the Q of the element and, then, exercise the voltage division rule from source to load.

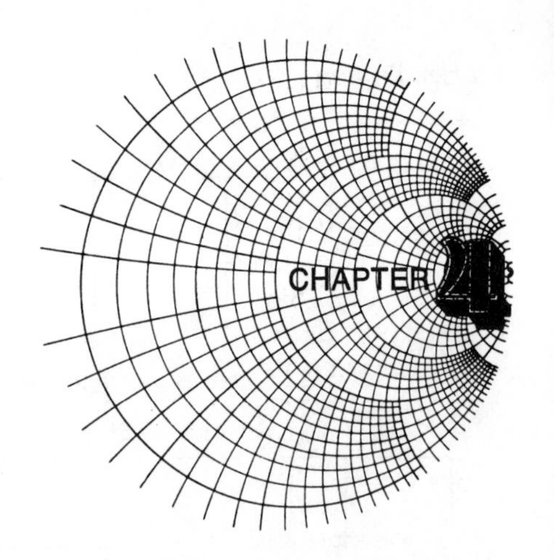

IMPEDANCE MATCHING

Impedance matching is often necessary in the design of rf circuitry to provide the maximum possible transfer of power between a source and its load. Probably the most vivid example of the need of such a transfer of power occurs in the front end of any sensitive receiver. Obviously, any *unnecessary* loss in a circuit that is already carrying extremely small signal levels simply cannot be tolerated. Therefore, in most instances, extreme care is taken during the initial design of such a front end to make sure that each device in the chain is matched to its load.

In this chapter, then, we will study several methods of matching a given source to a given load. This will be done both numerically and with the aid of the Smith Chart and, in both cases, exact step-by-step procedures will be presented making any calculations as painless as possible.

BACKGROUND

There is a well-known theorem which states that, for *dc circuits,* maximum power will be transferred from a source to its load if the *load resistance* equals the *source resistance.* A simple proof of this theorem is given by the calculations and the sketches shown in Fig. 4-1. In the calculation, for convenience, the source is normalized for a resistance of one ohm and a source voltage of one volt.

In dealing with ac or time-varying waveforms, however, that same theorem states that the maximum transfer of power, from a source to its load, occurs when the *load impedance* (Z_L) is equal to the *complex conjugate* of the *source impedance.* Complex conjugate simply refers to a complex impedance having the same *real part* with an opposite reactance. Thus, if the source impedance were $Z_s = R + jX$, then its complex conjugate would be $Z_s = R - jX$.

If you followed the mathematics associated with Fig. 4-1, then it should be obvious why maximum transfer of power does occur when the load impedance is the complex conjugate of the source. This is shown schematically in Fig. 4-2. The source (Z_s), with a series reactive component of $+jX$ (an inductor), is driving its complex conjugate load impedance con-

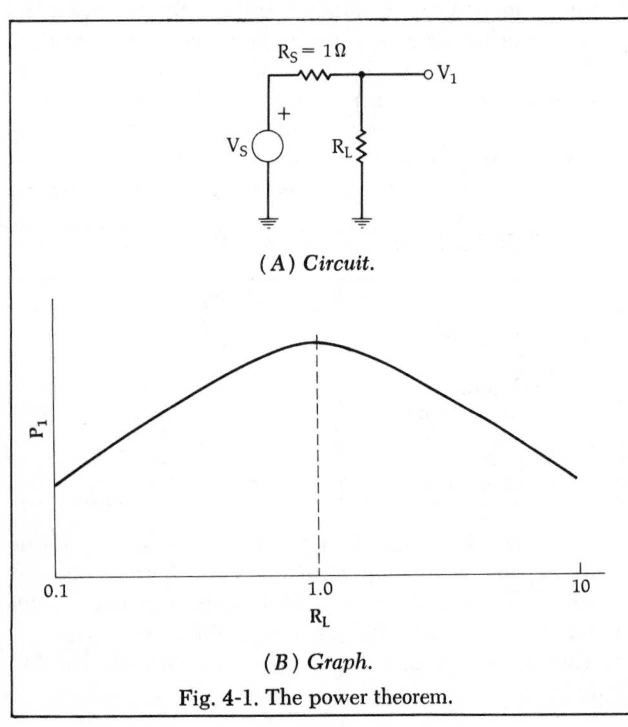

(*A*) *Circuit.*

(*B*) *Graph.*

Fig. 4-1. The power theorem.

Proof that P_{out} MAX occurs when $R_L = R_S$, in the circuit of Fig. 4-1A, is given by the formula:

$$V_1 = \frac{R_L}{R_S + R_L}(V_s)$$

Set $V_s = 1$ and $R_S = 1$, for convenience. Therefore,

$$V_1 = \frac{R_L}{1 + R_L}$$

Then, the power into R_L is:

$$P_1 = \frac{V_1^2}{R_L}$$

$$= \frac{\left(\frac{R_L}{1 + R_L}\right)^2}{R_L}$$

$$= \frac{R_L}{(1 + R_L)^2}$$

If you plot P_1 versus R_L, as in the preceding equation, the result is shown by the curve of the graph in Fig. 4-1B.

sisting of a −jX reactance (capacitor) in series with R_L. The +jX component of the source and the −jX component of the load are in series and, thus, cancel each other, leaving only R_S and R_L, which are equal by definition. Since R_S and R_L are equal, maximum power transfer will occur. So when we speak of a source driving its complex conjugate, we are simply referring to a condition in which any *source* reactance is resonated with an equal and opposite *load* reactance; thus, leaving only equal resistor values for the source and the load terminations.

The primary objective in any impedance *matching* scheme, then, is to force a load impedance to "look like" the complex conjugate of the source impedance so that maximum power may be transferred to the load. This is shown in Fig. 4-3 where a load impedance of 2 −j6 ohms is transformed by the impedance matching network to a value of 5 +j10 ohms. Therefore, the source "sees" a load impedance of 5 +j10 ohms, which just happens to be its complex conjugate. It should be noted here that because we are dealing with *reactances*, which are frequency dependent, the *perfect* impedance match can occur only at one frequency. That is the frequency at which the +jX component exactly equals the −jX component and, thus, cancellation or resonance occurs. At all other frequencies removed from the matching center frequency, the impedance match becomes progressively worse and eventually nonexistent. This can be a problem in broadband circuits where we would ideally like to provide a perfect match everywhere within the broad passband. There are methods, however, of increasing the bandwidth of the match and a few of these methods will be presented later in this chapter.

There are an infinite number of possible networks which could be used to perform the impedance matching function of Fig. 4-3. Something as simple as a 2-element LC network or as elaborate as a 7-element filter, depending on the application, would work equally well. The remainder of this chapter is devoted to providing you with an insight into a few of those infinite possibilities. After studying this chapter, you should be able to match almost any two complex loads with a minimum of effort.

THE L NETWORK

Probably the simplest and most widely used matching circuit is the L network shown in Fig. 4-4. This circuit receives its name because of the component orientation which resembles the shape of an L. As shown in the sketches, there are four possible arrangements of the two L and C components. Two of the arrangements (Figs. 4-4A and 4-4B) are in a low-pass configuration while the other two (Figs. 4-4C and 4-4D) are in a high-pass configuration. Both of these circuits should be recognized from Chapter 3.

Before we introduce equations which can be used to design the matching networks of Fig. 4-4, let's first analyze an existing matching network so that we can understand exactly how the impedance match occurs. Once this analysis is made, a little of the "black magic" surrounding impedance matching should subside.

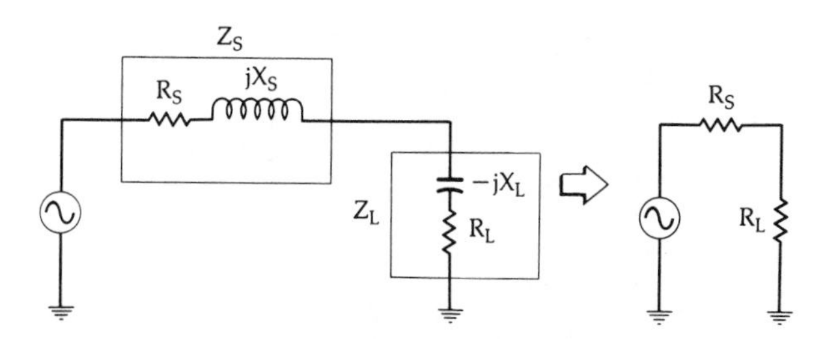

Fig. 4-2. Source impedance driving its complex conjugate and the resulting equivalent circuit.

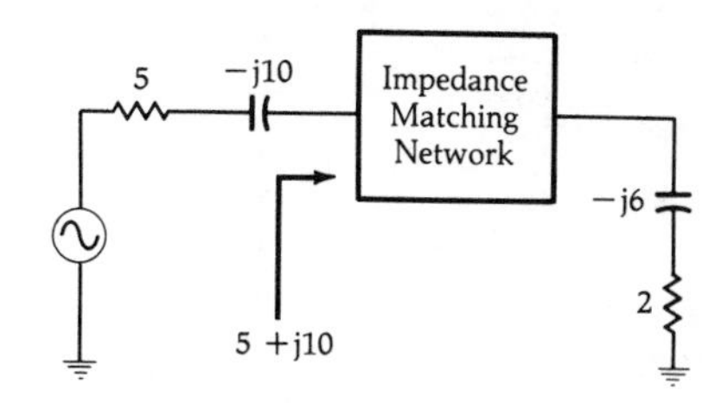

Fig. 4-3. Impedance transformation.

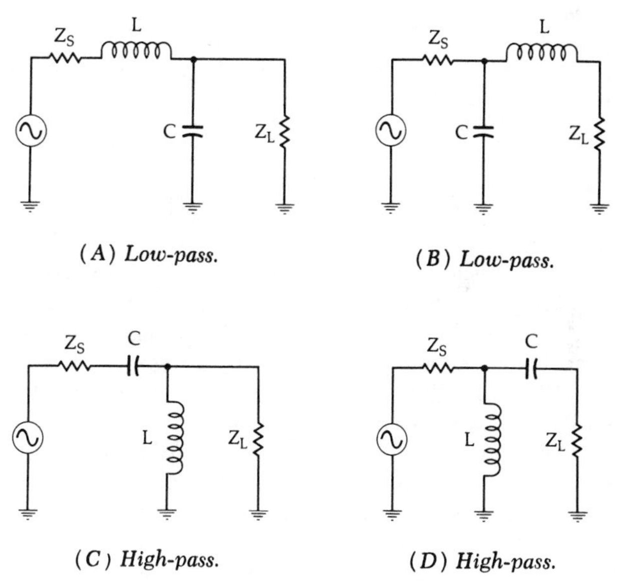

Fig. 4-4. The L network.

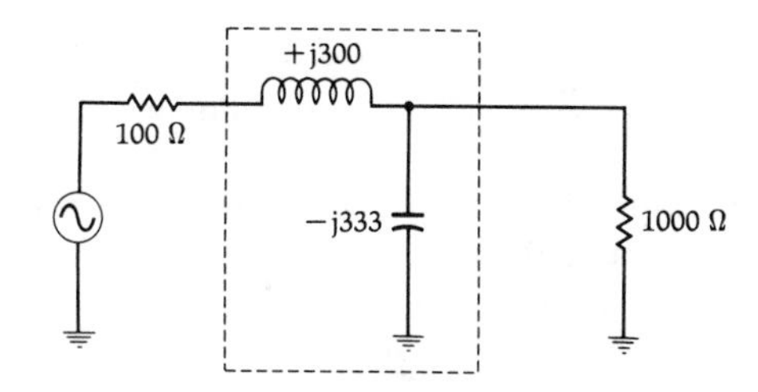

Fig. 4-5. Simple impedance-match network between a 100-ohm source and a 1000-ohm load.

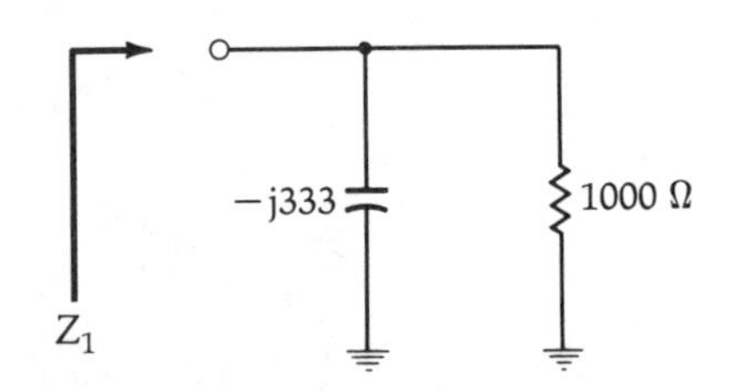

Fig. 4-6. Impedance looking into the parallel combination of R_L and X_c.

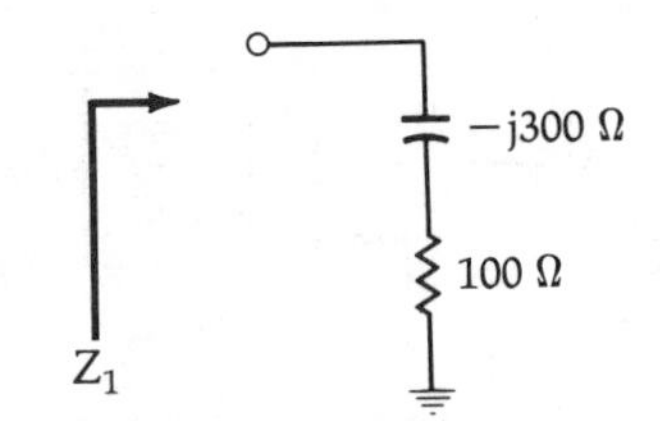

Fig. 4-7. Equivalent circuit of Fig. 4-6.

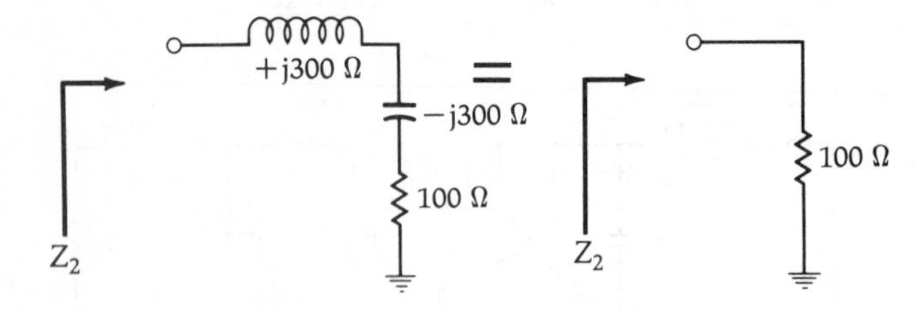

Fig. 4-8. Completing the match.

Fig. 4-5 shows a simple L network impedance-matching circuit between a 100-ohm source and a 1000-ohm load. Without the impedance-matching network installed, and with the 100-ohm source driving the 1000-ohm load directly, about 4.8 dB of the available power from the source would be lost. Thus, roughly one-third of the signal *available* from the source is gone before we even get started. The impedance-matching network eliminates this loss and allows for maximum power transfer to the load. This is done by forcing the 100-ohm source to see 100 ohms when it looks into the impedance-matching network. But how?

If you analyze Fig. 4-5, the simplicity of how the match occurs will amaze you. Take a look at Fig. 4-6. The first step in the analysis is to determine what the load impedance actually looks like when the −j333-ohm capacitor is placed across the 1000-ohm load resistor. This is easily calculated by:

$$\begin{aligned} Z &= \frac{X_c R_L}{X_c + R_L} \\ &= \frac{-j333(1000)}{-j333 + 1000} \\ &= 315 \angle -71.58^\circ \\ &= 100 - j300 \text{ ohms} \end{aligned}$$

Thus, the parallel combination of the −j333-ohm capacitor and the 1000-ohm resistor *looks like* an impedance of 100 −j300 ohms. This is a *series* combination of a 100-ohm resistor and a −j300-ohm capacitor as shown in Fig. 4-7. Indeed, if you hooked a signal generator up to circuits that are similar to Figs. 4-6 and 4-7, you would not be able to tell the difference between the two as they would exhibit the same characteristics (except at dc, obviously).

Now that we have an *apparent* series 100 −j300-ohm impedance for a load, all we must do to complete the impedance match to the 100-ohm source is to add an equal and opposite (+j300 ohm) reactance in series with the network of Fig. 4-7. The addition of the +j300-ohm inductor causes cancellation of the −j300-ohm capacitor leaving only an apparent 100-ohm load resistor. This is shown in Fig. 4-8. Keep in mind here that the actual network topology of Fig. 4-5 has not changed. All we have done is to analyze small portions of the network so that we can understand the function of each component.

To summarize then, the function of the *shunt* component of the impedance-matching network is to transform a larger impedance down to a smaller value with a real part equal to the real part of the other terminating impedance (in our case, the 100-ohm source). The series impedance-matching element then resonates with or cancels any reactive component present, thus leaving the source driving an apparently equal load for optimum power transfer. So you see, the impedance match isn't "black magic" at all but can be completely explained every step of the way.

Now, back to the *design* of the impedance-matching networks of Fig. 4-4. These circuits can be very easily designed using the following equations:

$$Q_s = Q_p = \sqrt{\frac{R_p}{R_s} - 1} \qquad \text{(Eq. 4-1)}$$

$$Q_s = \frac{X_s}{R_s} \qquad \text{(Eq. 4-2)}$$

$$Q_p = \frac{R_p}{X_p} \qquad \text{(Eq. 4-3)}$$

where, as shown in Fig. 4-9:

Q_s = the Q of the series leg,
Q_p = the Q of the shunt leg,
R_p = the shunt resistance,
X_p = the shunt reactance,
R_s = the series resistance,
X_s = the series reactance.

The quantities X_p and X_s may be either capacitive or inductive reactance but each must be of the opposite type. Once X_p is chosen as a capacitor, for example, X_s must be an inductor, and vice-versa. Example 4-1 illustrates the procedure.

DEALING WITH COMPLEX LOADS

The design of Example 4-1 was used for the simple case of matching two *real* impedances (pure resis-

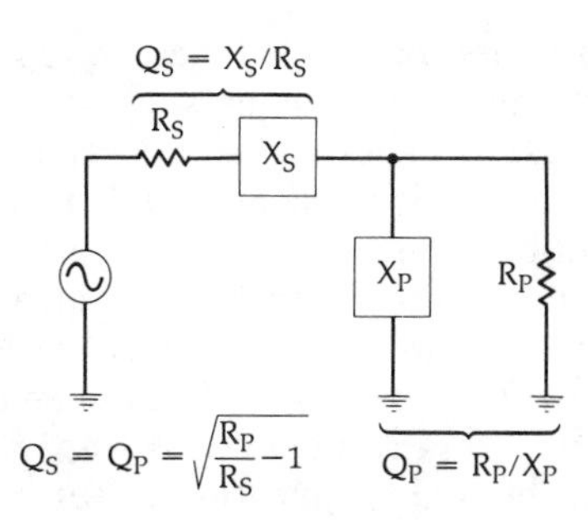

Fig. 4-9. Summary of the L-network design.

EXAMPLE 4-1

Design a circuit to match a 100-ohm source to a 1000-ohm load at 100 MHz. Assume that a dc voltage must also be transferred from the source to the load.

Solution

The need for a dc path between the source and load dictates the need for an inductor in the series leg, as shown in Fig. 4-4A. From Equation 4-1, we have:

$$Q_s = Q_p = \sqrt{\frac{1000}{100} - 1}$$
$$= \sqrt{9}$$
$$= 3$$

From Equation 4-2, we get:

$$X_s = Q_s R_s$$
$$= (3)(100)$$
$$= 300 \text{ ohms (inductive)}$$

Then, from Equation 4-3,

$$X_p = \frac{R_p}{Q_p}$$
$$= \frac{1000}{3}$$
$$= 333 \text{ ohms (capacitive)}$$

Thus, the component values at 100 MHz are:

$$L = \frac{X_s}{\omega}$$
$$= \frac{300}{2\pi(100 \times 10^6)}$$
$$= 477 \text{ nH}$$
$$C = \frac{1}{\omega X_p}$$
$$= \frac{1}{2\pi(100 \times 10^6)(333)}$$
$$= 4.8 \text{ pF}$$

This yields the circuit shown in Fig. 4-10. Notice that what you have done is to design the circuit that was previously given in Fig. 4-5 and, then, analyzed.

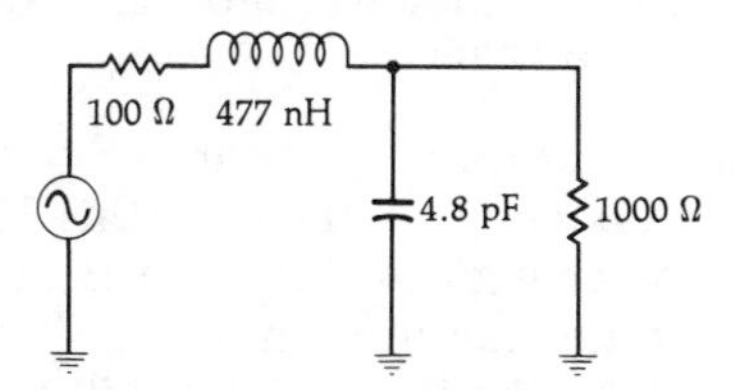

Fig. 4-10. Final circuit for Example 4-1.

tances). It is very rare when such an occurrence actually exists in the real world. Transistor input and output impedances are almost always *complex;* that is they contain both resistive and reactive components ($R \pm jX$). Transmission lines, mixers, antennas, and most other sources and loads are no different in that respect. Most will always have some reactive component which must be dealt with. It is, therefore, necessary to know how to handle these stray reactances and, in some instances, to actually put them to work for you.

There are two basic approaches in handling complex impedances:

1. Absorption—To actually absorb any stray reactances into the impedance-matching network itself. This can be done through prudent placement of each matching element such that element capacitors are placed in parallel with stray capacitances, and element inductors are placed in series with any stray inductances. The *stray* component values are then subtracted from the *calculated* element values leaving new element values (C′, L′), which are smaller than the calculated element values.
2. Resonance—To resonate any stray reactance with an equal and opposite reactance at the frequency of interest. Once this is done the matching network design can proceed as shown for two pure resistances in Example 4-1.

Of course, it is possible to use both of the approaches outlined above at the same time. In fact, the majority of impedance-matching designs probably do utilize a little of both. Let's take a look at two simple examples to help clarify matters.

Notice that nowhere in Example 4-2 was a *conjugate* match even mentioned. However, you can rest assured that if you perform the simple analysis outlined in the previous section of this chapter, the impedance looking into the matching network, as seen by the source, will be 100 −j126 ohms, which is indeed the complex conjugate of 100 +j126 ohms.

Obviously, if the *stray* element values are larger than the calculated element values, absorption cannot take place. If, for instance, the *stray* capacitance of Fig. 4-11 were 20 pF, we could not have added a *shunt* element capacitor to give us the total needed shunt capacitance of 4.8 pF. In a situation such as this, when absorption is not possible, the concept of resonance coupled with absorption will often do the trick.

Examples 4-2 and 4-3 detail some very important concepts in the design of impedance-matching networks. With a little planning and preparation, the design of simple impedance-matching networks between complex loads becomes a simple number-crunching task using elementary algebra. Any stray reactances present in the source and load can usually be absorbed in the matching network (Example 4-2), or they can

EXAMPLE 4-2

Use the absorption approach to match the source and load shown in Fig. 4-11 (at 100 MHz).

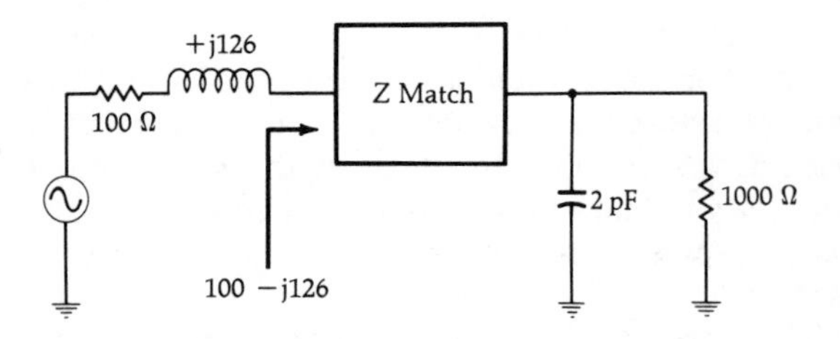

Fig. 4-11. Complex source and load circuit for Example 4-2.

Solution

The first step in the design process is to totally ignore the reactances and simply match the 100-ohm real part of the source to the 1000-ohm real part of the load (at 100 MHz). Keep in mind that you would like to use a matching network that will place element inductances in series with stray inductance and element capacitances in parallel with stray capacitances. Thus, conveniently, the network circuit shown in Fig. 4-4A is again chosen for the design and, again, Example 4-1 is used to provide the details of the procedure. Thus, the calculated values for the network, if we ignore stray reactances, are shown in the circuit of Fig. 4-10. But, since the stray reactances really do exist, the design is not yet finished as we must now somehow absorb the stray reactances into the matching network. This is done as follows. At the load end, we need 4.8 pF of capacitance for the matching network. We already have a stray 2 pF available at the load so why not use it. Thus, if we use a 2.8-pF *element* capacitor, the *total* shunt capacitance becomes 4.8 pF, the design value. Similarly, at the source, the matching network calls for a series 477-nH inductor. We already have a +j126-ohm, or 200-nH, inductor available in the source. Thus, if we use an actual element inductance of 477 nH − 200 nH = 277 nH, then the *total* series inductance will be 477 nH—which is the calculated design value. The final design circuit is shown in Fig. 4-12.

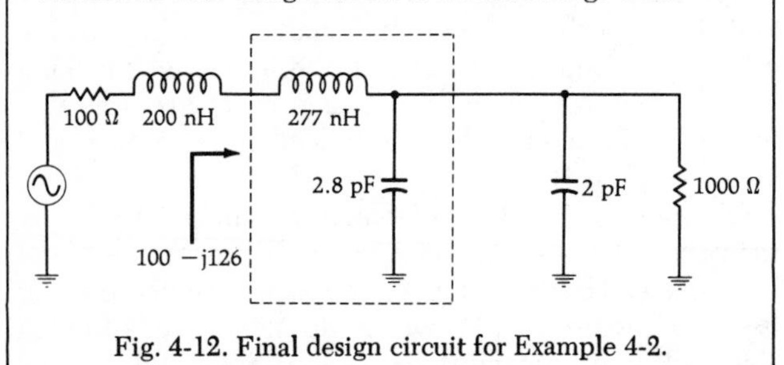

Fig. 4-12. Final design circuit for Example 4-2.

be resonated with an equal and opposite reactance, which is then absorbed into the network (Example 4-3).

THREE-ELEMENT MATCHING

Equation 4-1 reveals a potential disadvantage of the 2-element L networks described in the previous sections. It is a fact that once R_s and R_p, or the source and load impedance, are determined, the Q of the network is defined. In other words, with the L network, the designer does not have a choice of circuit Q and simply must take what he gets. This is, of course, usually the case because the source and load impedance are typically given in any design and, thus, R_p and R_s cannot be changed.

The lack of circuit-Q versatility in a matching network can be a hindrance, however, especially if a *narrow* bandwidth is required. The 3-element network overcomes this disadvantage and can be used for narrow-band high-Q applications. Furthermore, the designer can *select* any practical circuit Q that he wishes as long as it is *greater than* that Q which is possible with the L-matching network alone. In other words, the circuit Q established with an L-matching network is the *minimum* circuit Q available in the 3-element matching arrangement.

The 3-element network (shown in Fig. 4-17) is called a *Pi network* because it closely resembles the Greek letter π. Its companion network (shown in Fig. 4-18) is called a T network for equally obvious reasons.

The Pi Network

The Pi network can best be described as two "back-to-back" L networks that are both configured to match the load and the source to an invisible or "virtual" resistance located at the junction between the two networks. This is illustrated in Fig. 4-19. The significance of the negative signs for $-X_{s1}$ and $-X_{s2}$ is symbolic. They are used merely to indicate that the X_s values are the opposite type of reactance from X_{p1} and X_{p2}, respectively. Thus, if X_{p1} is a capacitor, X_{s1} must be an inductor, and vice-versa. Similarly, if X_{p2} is an inductor, X_{s2} must be a capacitor, and vice-versa. They do *not* indicate negative reactances (capacitors).

The design of each section of the Pi network proceeds exactly as was done for the L networks in the previous sections. The virtual resistance (R) must be smaller than either R_S or R_L because it is connected to the series arm of each L section but, otherwise, it can be any value you wish. Most of the time, however, R is defined by the desired loaded Q of the circuit that you specify at the beginning of the design process. For our purposes, the loaded Q of this network will be defined as:

$$Q = \sqrt{\frac{R_H}{R} - 1} \qquad \text{(Eq. 4-4)}$$

where,

R_H = the largest terminating impedance of R_s or R_L,
R = the virtual resistance.

Although this is not entirely accurate, it is a widely accepted Q-determining formula for this circuit, and is certainly close enough for most practical work. Example 4-4 illustrates the procedure.

Any of the networks in Fig. 4-21 will perform the impedance match between the 100-ohm source and the

EXAMPLE 4-3

Design an impedance matching network that will block the flow of dc from the source to the load in Fig. 4-13. The frequency of operation is 75 MHz. Try the resonant approach.

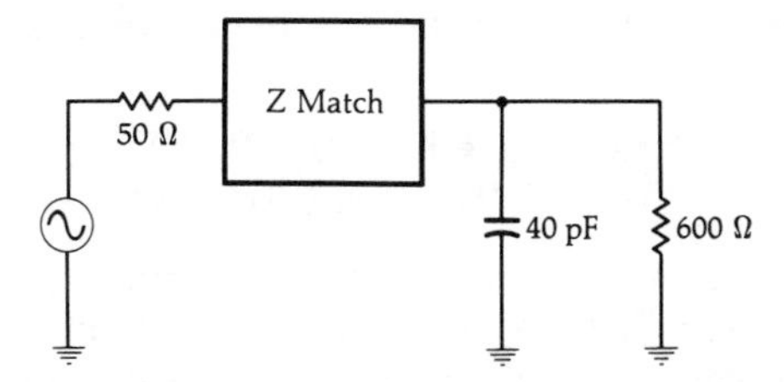

Fig. 4-13. Complex load circuit for Example 4-3.

Solution

The need to block the flow of dc from the source to the load dictates the use of the matching network of Fig. 4-4C. But, first, let's get rid of the stray 40-pF capacitor by resonating it with a shunt inductor at 75 MHz.

$$L = \frac{1}{\omega^2 C_{stray}}$$

$$= \frac{1}{[2\pi(75 \times 10^6)]^2(40 \times 10^{-12})}$$

$$= 112.6 \text{ nH}$$

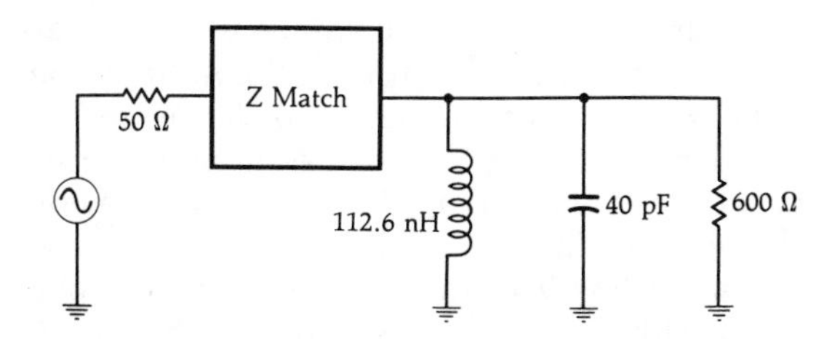

Fig. 4-14. Resonating the stray load capacitance.

This leaves us with the circuit shown in Fig. 4-14. Now that we have eliminated the stray capacitance, we can proceed with matching the network between the 50-ohm load and the apparent 600-ohm load. Thus,

$$Q_s = Q_p = \sqrt{\frac{R_p}{R_s} - 1}$$

$$= \sqrt{\frac{600}{50} - 1}$$

$$= 3.32$$

$$X_s = Q_s R_s$$

$$= (3.32)(50)$$

$$= 166 \text{ ohms}$$

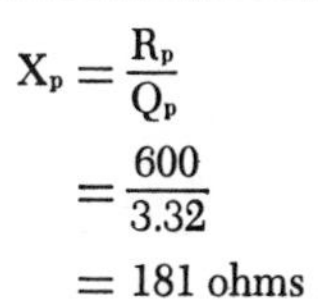

$$X_p = \frac{R_p}{Q_p}$$

$$= \frac{600}{3.32}$$

$$= 181 \text{ ohms}$$

Therefore, the element values are:

$$C = \frac{1}{\omega X_s}$$

$$= \frac{1}{2\pi(75 \times 10^6)(166)}$$

$$= 12.78 \text{ pF}$$

$$L = \frac{X_p}{\omega}$$

$$= \frac{181}{2\pi(75 \times 10^6)}$$

$$= 384 \text{ nH}$$

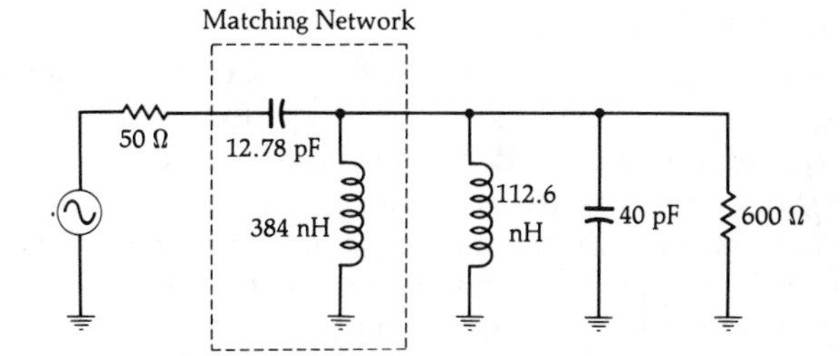

Fig. 4-15. The circuit of Fig. 4-14 after impedance matching.

These values, then, yield the circuit of Fig. 4-15. But notice that this circuit can be further simplified by simply replacing the two shunt inductors with a single inductor. Therefore,

$$L_{new} = \frac{L_1 L_2}{L_1 + L_2}$$

$$= \frac{(384)(112.6)}{384 + 112.6}$$

$$= 87 \text{ nH}$$

The final circuit design appears in Fig. 4-16.

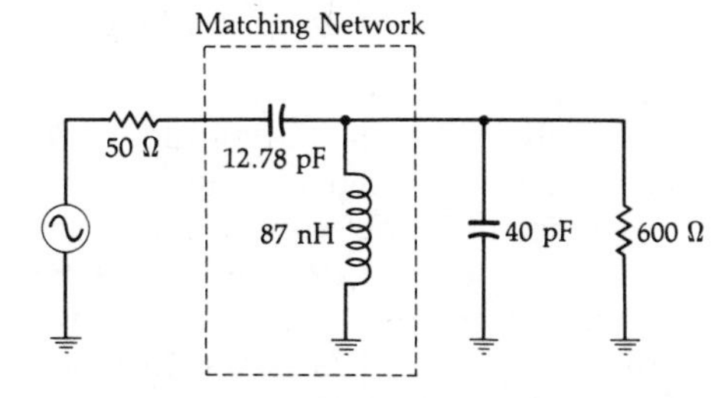

Fig. 4-16. Final design circuit for Example 4-3.

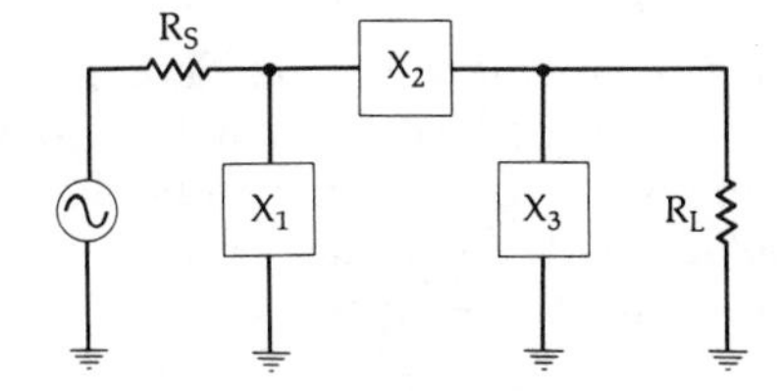

Fig. 4-17. The three-element Pi network.

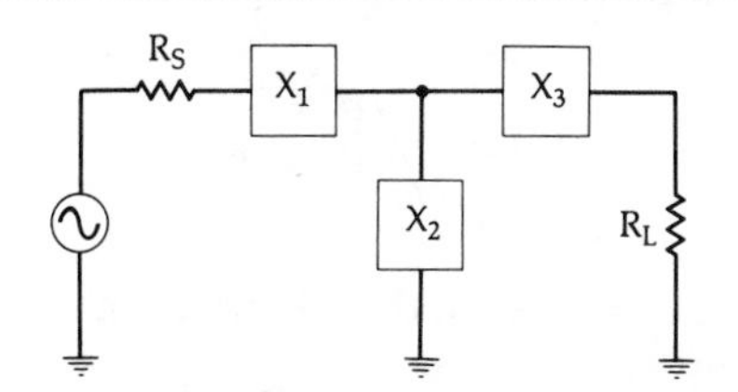

Fig. 4-18. The three-element T network.

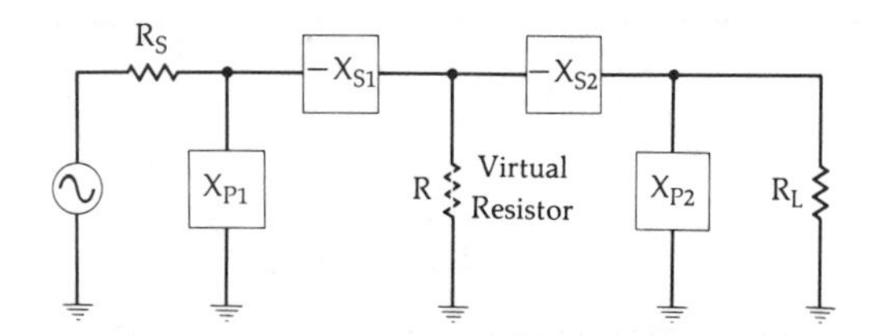

Fig. 4-19. The Pi network shown as two back-to-back L networks.

1000-ohm load. The one that you choose for each particular application will depend on any number of factors including:

1. The elimination of stray reactances.
2. The need for harmonic filtering.
3. The need to pass or block dc voltage.

The T network

The design of the 3-element T network is exactly the same as for the Pi network except that with the T, you match the load and the source, through two L-type networks, to a virtual resistance which is *larger* than either the load or source resistance. This means that the two L-type networks will then have their shunt legs connected together as shown in Fig. 4-22.

The T network is often used to match two low-valued impedances when a high-Q arrangement is needed. The loaded Q of the T network is determined by the L section that has the highest Q. By definition, the L section with the highest Q will occur on the end which has the *smallest* terminating resistor. Remember, too, that each terminating resistor is in the *series* leg of each network. Therefore, the formula for determining the loaded Q of the T network is:

$$Q = \sqrt{\frac{R}{R_{small}} - 1} \qquad \text{(Eq. 4-5)}$$

where,

R = the virtual resistance,
R_{small} = the smallest terminating resistance.

This formula is exactly the same as the Q formula that was previously given for the Pi-type networks. However, since we have reversed or "flip-flopped" the L sections to produce the T network, we must also make sure that we redefine the Q formula to account for the new resistor placement, in relation to those L networks. In other words, Equations 4-4 and 4-5 are only special applications of the general formula that is given in Equation 4-1 (and repeated here for convenience).

$$Q = \sqrt{\frac{R_p}{R_s} - 1} \qquad \text{(Eq. 4-1)}$$

where,

R_p = the resistance in the shunt branch of the L network,
R_s = the resistance in the series branch of the L network.

So, try not to get confused with the different definitions of circuit Q. They are all the same.

Each L network is calculated in exactly the same manner as was given in the previous examples and, as we shall soon see, we will also end up with four possible configurations for the T network (Example 4-5).

LOW-Q OR WIDEBAND MATCHING NETWORKS

Thus far in this chapter we have studied: (1) the L network, which has a circuit Q that is automatically defined when the source and load impedances are set, and (2) the Pi and T networks, which allow us to select a circuit Q independent of the source and load impedances *as long as the Q chosen is larger than that which is available with the L network.* This seems to indicate, and rightfully so, that the Pi and T networks are great for narrow-band matching networks. But what if an impedance match is required over a fairly broad range of frequencies. How do we handle that? The answer is to simply use two L sections in still another configuration as shown in Fig. 4-25. Notice here that the virtual resistor is in the shunt leg of one L section and in the series leg of the other L section. We, therefore, have two *series-connected* L sections rather than the back-to-back configuration of the Pi and T networks. In this new configuration, the value of the virtual resistor (R) must be larger than the smallest termination impedance and, also, smaller than the largest termination impedance. Of course, any virtual resistance that satisfies these criteria may be chosen. The net result is a range of loaded-Q values that is *less than* the range of Q values obtainable from either a single L section, or the Pi and T networks previously described.

The maximum bandwidth (minimum Q) available from this network is obtained when the virtual resistor (R) is made equal to the geometric mean of the two impedances being matched.

$$R = \sqrt{R_S R_L} \qquad \text{(Eq. 4-6)}$$

The loaded Q of the network, for our purposes, is defined as:

$$Q = \sqrt{\frac{R}{R_{smaller}} - 1} = \sqrt{\frac{R_{larger}}{R} - 1} \qquad \text{(Eq. 4-7)}$$

where,

R = the virtual resistance,
$R_{smaller}$ = the smallest terminating resistance,
R_{larger} = the largest terminating resistance.

If even wider bandwidths are needed, more L networks may be cascaded with virtual resistances between each network. Optimum bandwidths in these cases are obtained if the ratios of each of the two succeeding resistances are equal:

$$\frac{R_1}{R_{smaller}} = \frac{R_2}{R_1} = \frac{R_3}{R_2} \ldots = \frac{R_{larger}}{R_n} \qquad \text{(Eq. 4-8)}$$

EXAMPLE 4-4

Using Fig. 4-19 as a reference, design four different Pi networks to match a 100-ohm source to a 1000-ohm load. Each network must have a loaded Q of 15.

Solution

From Equation 4-4, we can find the virtual resistance we will be matching.

$$R = \frac{R_H}{Q^2 + 1} = \frac{1000}{226} = 4.42 \text{ ohms}$$

To find X_{p2} we have:

$$X_{p2} = \frac{R_p}{Q_p} = \frac{R_L}{Q} = \frac{1000}{15} = 66.7 \text{ ohms}$$

Similarly, to find X_{s2}:

$$X_{s2} = QR_{series} = 15(R) = (15)(4.42) = 66.3 \text{ ohms}$$

This completes the design of the L section on the *load* side of the network. Note that R_{series} in the above equation was substituted for the virtual resistor R, which by definition is in the series arm of the L section.

The Q for the other L network is now defined by the ratio of R_s to R, as per Equation 4-1, where:

$$Q_1 = \sqrt{\frac{R_s}{R} - 1} = \sqrt{\frac{100}{4.42} - 1} = 4.6$$

Notice here that the source resistor is now considered to be in the shunt leg of the L network. Therefore, R_s is defined as R_p, and

$$X_{p1} = \frac{R_p}{Q_1} = \frac{100}{4.6} = 21.7 \text{ ohms}$$

Similarly,

$$X_{s1} = Q_1 R_{series} = Q_1 R = (4.6)(4.46) = 20.51 \text{ ohms}$$

The actual network design is now complete and is shown in Fig. 4-20. Remember that the virtual resistor (R) is not really in the circuit and, therefore, is not shown. Reactances $-X_{s1}$ and $-X_{s2}$ are now in series and can simply be added together to form a single component.

So far in the design, we have dealt only with reactances and have not yet computed actual component values. This

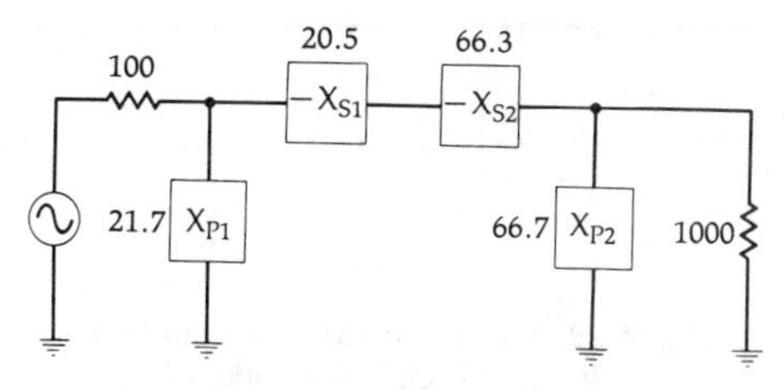

Fig. 4-20. Calculated reactances for Example 4-4.

is because of the need to maintain a general design approach so that four final networks can be generated quickly as per the problem statement.

Notice that X_{p1}, X_{s1}, X_{p2}, and X_{s2} can all be either capacitive or inductive reactances. The only constraint is that X_{p1} and X_{s1} are of opposite types, and X_{p2} and X_{s2} are of opposite types. This yields the four networks of Fig. 4-21 (the source and load have been omitted). Each component in Fig. 4-21 is shown as a reactance (in ohms). Therefore, to perform the transformation from the dual-L to the Pi network, the two series components are merely added if they are alike, and subtracted if the reactances are of opposite type. The final step, of course, is to change each reactance into a component value of capacitance and inductance at the frequency of operation.

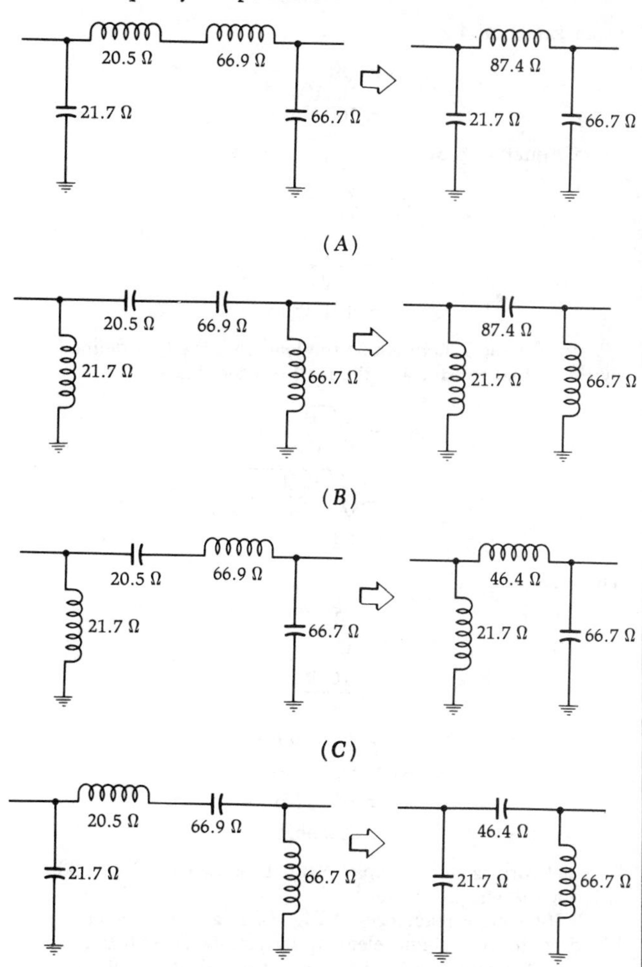

Fig. 4-21. The transformation from double-L to Pi networks.

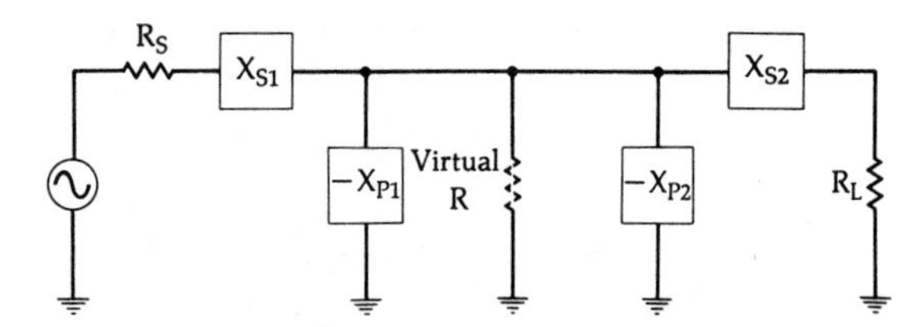

Fig. 4-22. The T network shown as two back-to-back L networks.

where,

$R_{smaller}$ = the smallest terminating resistance,
R_{larger} = the largest terminating resistance,
$R_1, R_2, \ldots R_n$ = virtual resistors.

This is shown in Fig. 4-26.

The design procedure for these wideband matching networks is precisely the same as was given for the previous examples. To design for a specific low Q, simply solve Equation 4-7 for R to find the virtual

EXAMPLE 4-5

Using Fig. 4-22 as a reference, design four different networks to match a 10-ohm source to a 50-ohm load. Each network is to have a loaded Q of 10.

Solution

Using Equation 4-5, we can find the virtual resistance we need for the match.

$$R = R_{small}(Q^2 + 1)$$
$$= 10(101)$$
$$= 1010 \text{ ohms}$$

From Equation 4-2:

$$X_{s1} = QR_s$$
$$= 10(10)$$
$$= 100 \text{ ohms}$$

From Equation 4-3:

$$X_{p1} = \frac{R}{Q}$$
$$= \frac{1010}{10}$$
$$= 101 \text{ ohms}$$

Now, for the L network on the load end, the Q is defined by the virtual resistor and the load resistor. Thus,

$$Q_2 = \sqrt{\frac{R}{R_L} - 1}$$
$$= \sqrt{\frac{1010}{50} - 1}$$
$$= 4.4$$

Therefore,

$$X_{p2} = \frac{R}{Q_2}$$
$$= \frac{1010}{4.4}$$
$$= 230 \text{ ohms}$$
$$X_{s2} = Q_2 R_L$$
$$= (4.4)(50)$$
$$= 220 \text{ ohms}$$

The network is now complete and is shown in Fig. 4-23 without the virtual resistor.

The two shunt reactances of Fig. 4-23 can again be combined to form a single element by simply substituting a value that is equal to the combined equivalent parallel reactance of the two.

The four possible T-type networks that can be used for matching the 10-ohm source to the 50-ohm load are shown in Fig. 4-24.

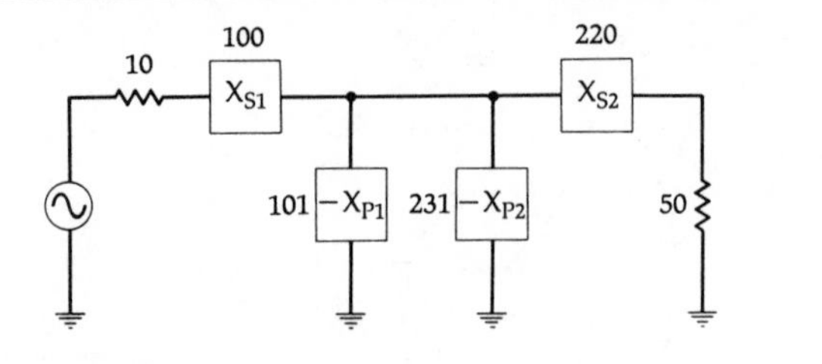

Fig. 4-23. The calculated reactances of Example 4-5.

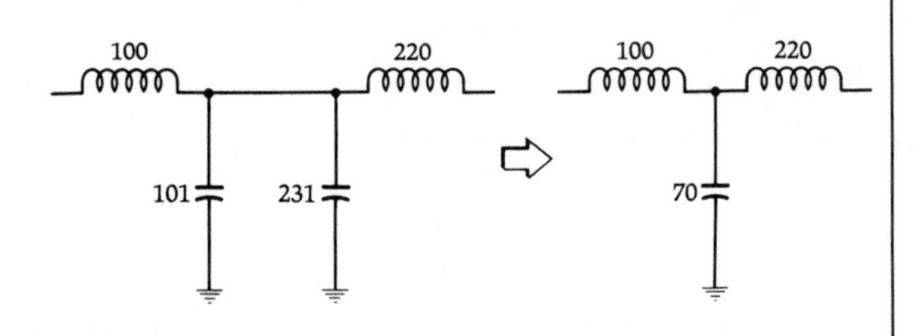

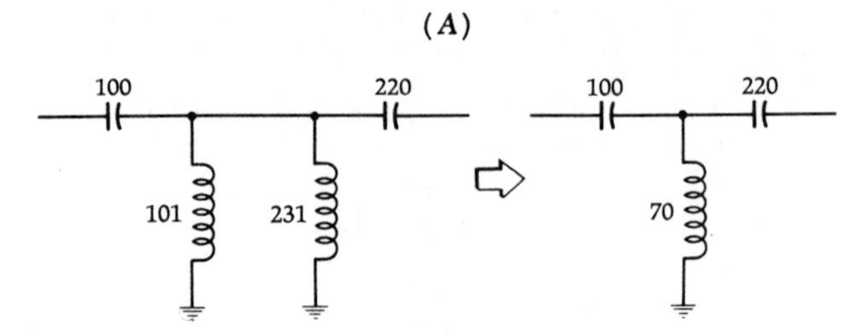

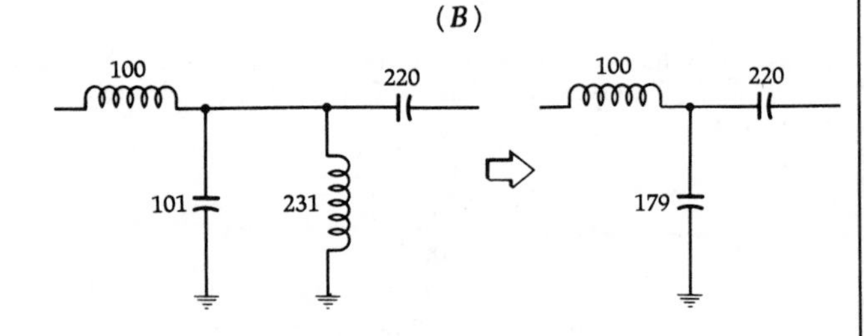

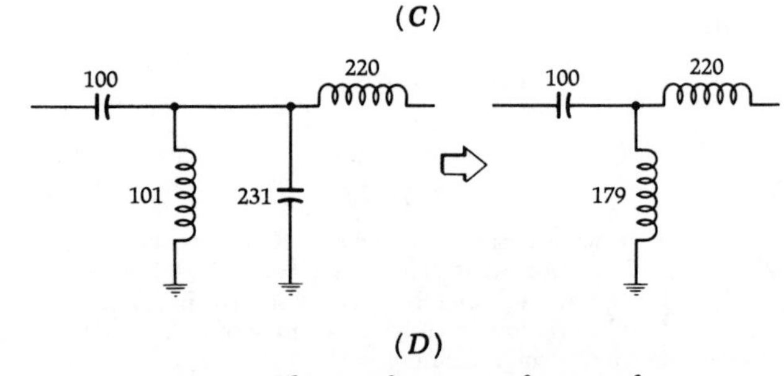

Fig. 4-24. The transformation of circuits from double-L to T-type networks.

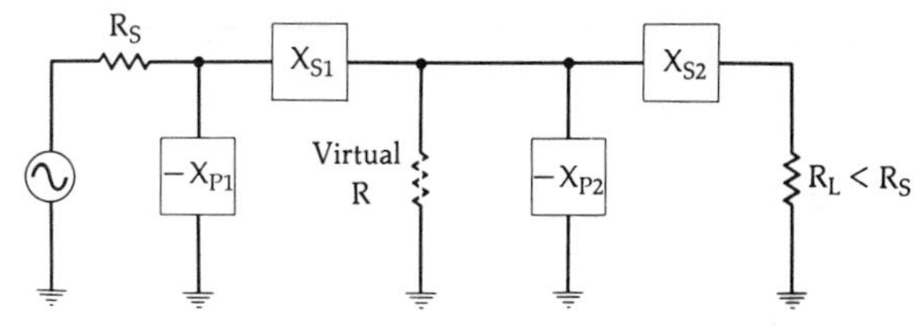

(A) *R in shunt leg.*

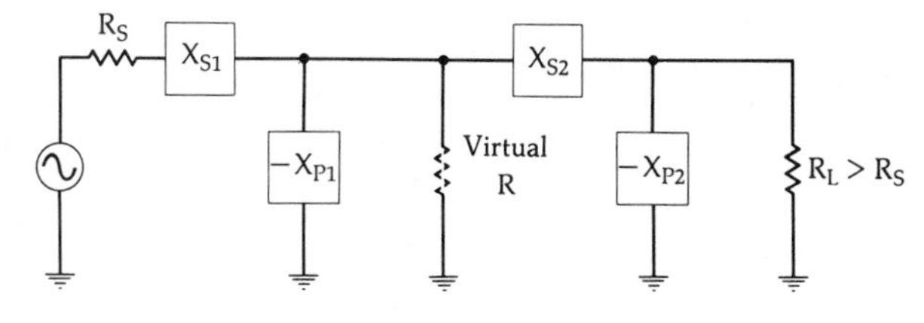

(B) *R in series leg.*

Fig. 4-25. Two series-connected L networks for lower Q applications.

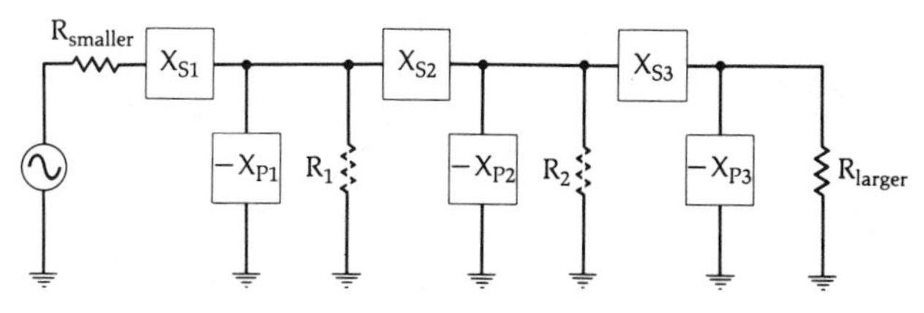

Fig. 4-26. Expanded version of Fig. 4-25 for even wider bandwidths.

resistance needed. Or, to design for an optimally wide bandwidth, solve Equation 4-6 for R. Once R is known, the design is straightforward.

THE SMITH CHART

Probably one of the most useful graphical tools available to the rf circuit designer today is the Smith Chart shown in Fig. 4-27. The chart was originally conceived back in the Thirties by a Bell Laboratories engineer named Phillip Smith, who wanted an easier method of solving the tedious repetitive equations that often appear in rf theory. His solution, appropriately named the Smith Chart, is still widely in use.

At first glance, a Smith Chart appears to be quite complex. Indeed, why would anyone of sound mind even care to look at such a chart? The answer is really quite simple; once the Smith Chart and its uses are understood, the rf circuit designer's job becomes much less tedious and time consuming. Very lengthy complex equations can be solved graphically on the chart in seconds, thus lessening the possibility of errors creeping into the calculations.

Smith Chart Construction

The mathematics behind the construction of a Smith Chart are given here for those that are interested. It is important to note, however, that you do not *need* to know or understand the mathematics surrounding the actual construction of a chart as long as you understand what the chart represents and how it can be used to your advantage. Indeed, there are so many uses for the chart that an entire volume has been written on the subject. In this chapter, we will concentrate mainly on the Smith Chart as an impedance matching tool and other uses will be covered in later chapters. The mathematics follow.

The reflection coefficient of a load impedance when given a source impedance can be found by the formula:

$$\rho = \frac{Z_s - Z_L}{Z_s + Z_L} \qquad \text{(Step 1)}$$

In normalized form, this equation becomes:

$$\rho = \frac{Z_o - 1}{Z_o + 1} \qquad \text{(Step 2)}$$

where Z_o is a complex impedance of the form $R + jX$.

The polar form of the reflection coefficient can also be represented in rectangular coordinates:

$$\rho = p + jq$$

Substituting into Step 2, we have:

$$p + jq = \frac{R + jX - 1}{R + jX + 1} \qquad \text{(Step 3)}$$

If we solve for the real and imaginary parts of $p + jq$, we get:

$$p = \frac{R^2 - 1 + X^2}{(R+1)^2 + X^2} \qquad \text{(Step 4)}$$

and,

$$q = \frac{2x}{(R+1)^2 + X^2} \qquad \text{(Step 5)}$$

Solve Step 5 for X:

$$X = \left(\frac{p(R+1)^2 - R^2 + 1}{1 - p}\right)^{1/2} \qquad \text{(Step 6)}$$

Then, substitute Step 6 into Step 5 to obtain:

$$\left(p - \frac{R}{R+1}\right)^2 + q^2 = \left(\frac{1}{R+1}\right)^2 \qquad \text{(Step 7)}$$

Step 7 is the equation for a family of circles whose centers are at:

$$p = \frac{R}{R+1}$$

$$q = 0$$

and whose radii are equal to:

$$\frac{1}{R+1}$$

These are the constant resistance circles, some of which are shown in Fig. 4-28A.

Similarly, we can eliminate R from Steps 4 and 5 to obtain:

$$(p-1)^2 + \left(q - \frac{1}{X}\right)^2 = \left(\frac{1}{X}\right)^2 \qquad \text{(Step 8)}$$

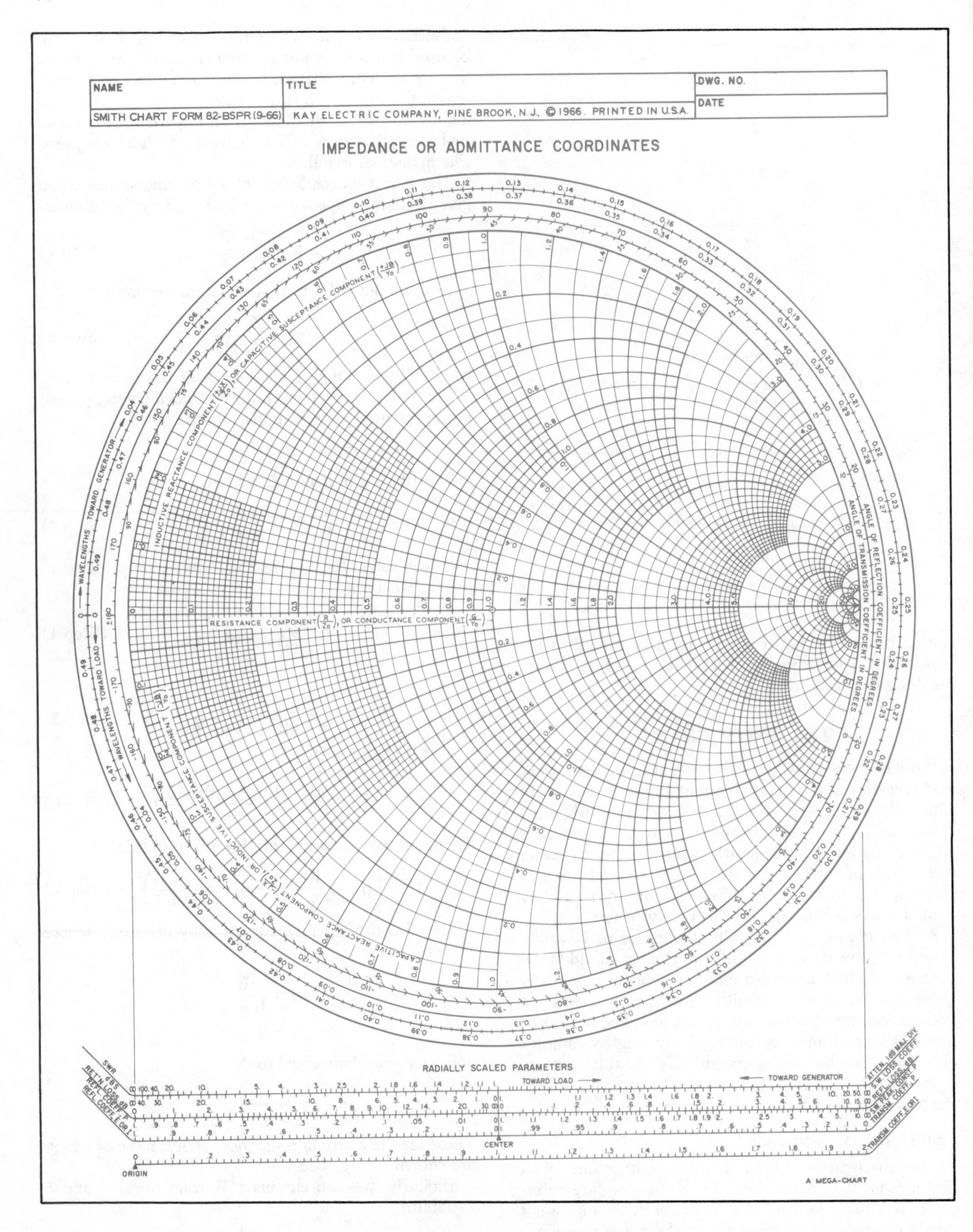

Fig. 4-27. The Smith Chart. (*Courtesy Analog Instruments Co.*)

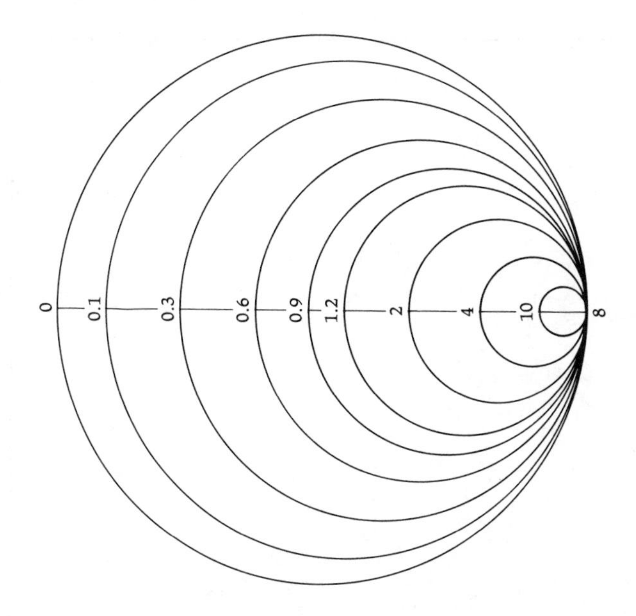

(A) *Constant resistance circles.*

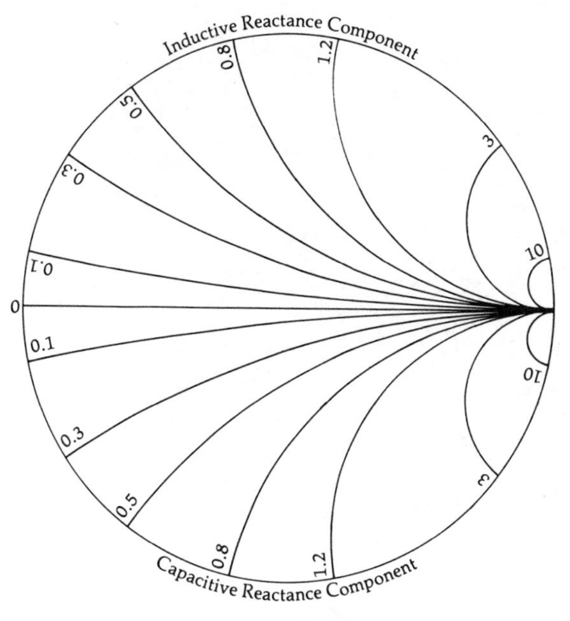

(B) *Constant reactance circles.*

Fig. 4-28. Smith Chart construction.

which represents a family of circles with centers at $p = 1, V = \frac{1}{X}$, and radii of $\frac{1}{X}$. These circles are shown plotted on the p, jq axis in Fig. 4-28B.

As the preceding mathematics indicate, the Smith Chart is basically a combination of a family of circles and a family of arc of circles—the centers and radii of which can be calculated using the equations given (Steps 1 through 8). Fig. 4-28 shows the chart broken down into these two families. The circles of Fig. 4-28A are known as *constant resistance circles*. Each point on a constant resistance circle has the same *resistance* as any other point on the circle. The arcs of circles shown in Fig. 4-28B are known as *constant reactance circles*, as each point on a circle has the same *reactance* as any other point on that circle. These circles are centered off of the chart and, therefore, only a small portion of each is contained within the boundary of the chart. All arcs above the centerline of the chart represent +jX, or inductive reactances, and all arcs below the centerline represent −jX, or capacitive reactances. The centerline must, therefore, represent an axis where X = 0 and is, therefore, called the *real axis*.

Notice in Fig. 4-28A that the "constant resistance = 0" circle defines the outer boundary of the chart. As the resistive component increases, the radius of each circle decreases and the center of each circle moves toward the right on the chart. Then, at infinite resistance, you end up with an infinitely small circle that is located at the extreme right-hand side of the chart. A similar thing happens for the constant reactance circles shown in Fig. 4-28B. As the magnitude of the reactive component increases (−jX or +jX), the radius of each circle decreases, and the center of each circle moves closer and closer to the extreme right side of the chart. Infinite resistance and infinite reactance are thus represented by the same point on the chart.

Since the outer boundary of the chart is defined as the "R = 0" circle, with higher values of R being contained within the chart, it follows then that any point outside of the chart must contain a negative resistance. The concept of negative resistance is useful in the study of oscillators and it is mentioned here only to state that the concept does exist, and if needed, the Smith Chart can be expanded to deal with it.

When the two charts of Fig. 4-28 are incorporated into a single version, the Smith Chart of Fig. 4-29 is born. If we add a few peripheral scales to aid us in other rf design tasks, such as determining *standing wave ratio* (*SWR*), *reflection coefficient*, and *transmission loss* along a transmission line, the basic chart of Fig. 4-27 is completed.

Plotting Impedance Values

Any point on the Smith Chart represents a *series* combination of resistance and reactance of the form Z = R + jX. Thus, to locate the impedance Z = 1 + j1, you would find the R = 1 constant resistance circle and follow it until it crossed the X = 1 constant reactance circle. The junction of these two circles would then represent the needed impedance value. This particular point, shown in Fig. 4-30, is located in the upper half of the chart because X is a positive reactance or an inductor. On the other hand, the point 1 − j1 is located in the *lower* half of the chart because, in this instance, X is a negative quantity and represents a capacitor. Thus, the junction of the R = 1 constant resistance circle and the X = −1 constant reactance circle defines that point.

In general, then, to find any *series* impedance of the form R ± jX on a Smith Chart, you simply find the junction of the R = constant and X = constant circles.

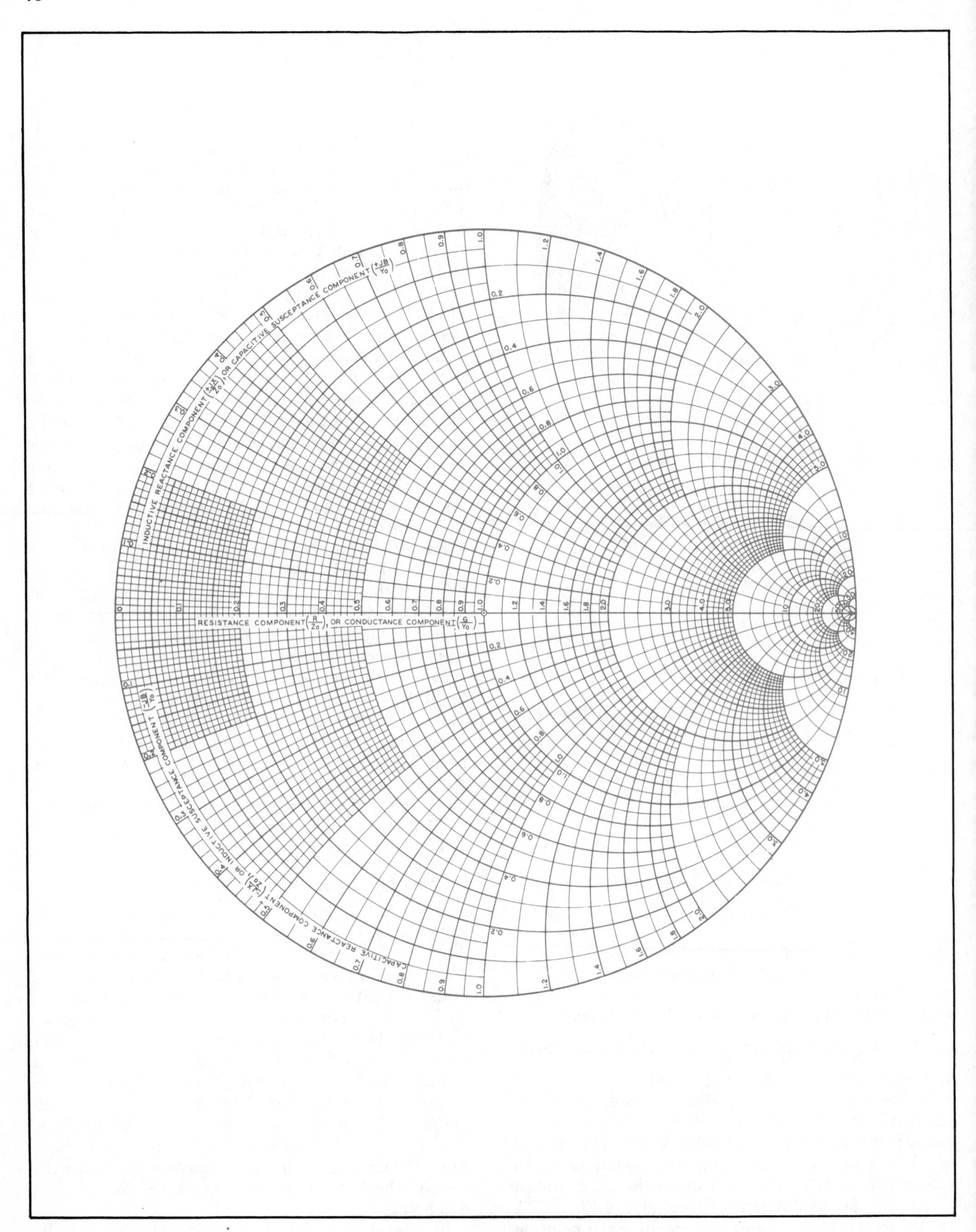

Fig. 4-29. The basic Smith Chart.

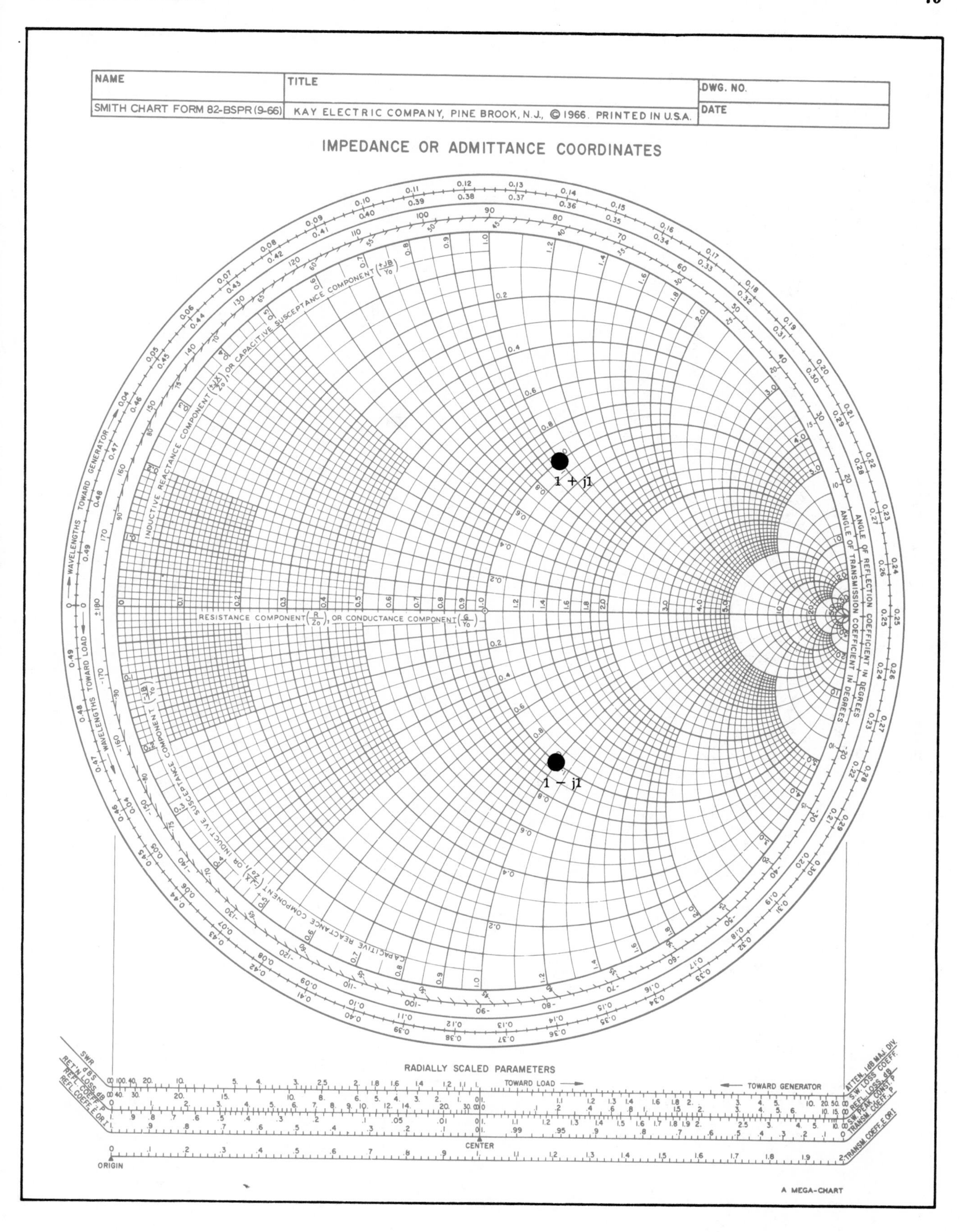

Fig. 4-30. Plotting impedances on the chart.

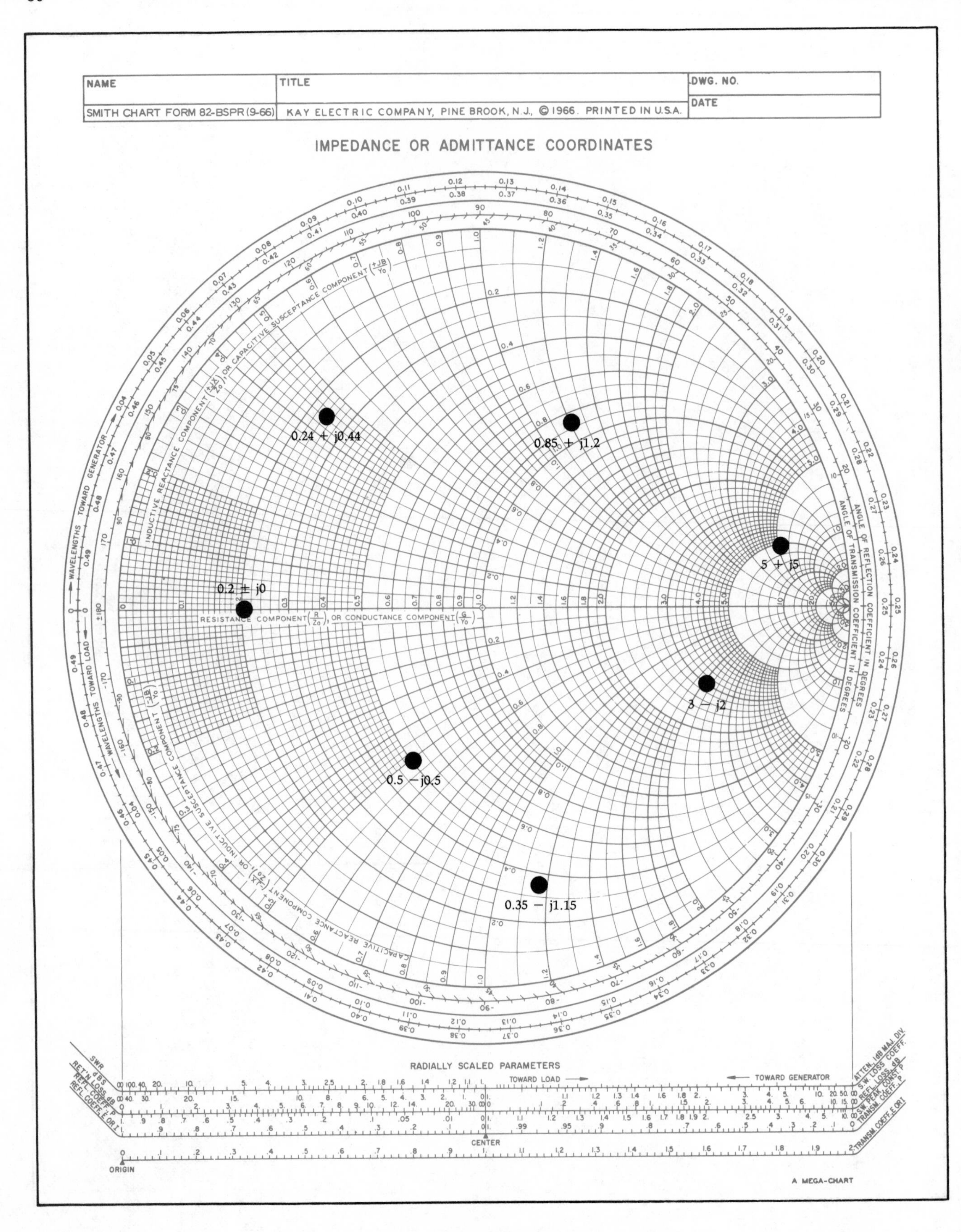

Fig. 4-31. More impedances are plotted on the chart.

In many cases, the actual circles will not be present on the chart and you will have to interpolate between two that are shown. Thus, plotting impedances and, therefore, any manipulation of those impedances must be considered an inexact procedure which is subject to "pilot error." Most of the time, however, the error introduced by subjective judgements on the part of the user, in plotting impedances on the chart, is so small as to be negligible for practical work. Fig. 4-31 shows a few more impedances plotted on the chart.

Notice that all of the impedance values plotted in Fig. 4-31 are very small numbers. Indeed, if you try to plot an impedance of $Z = 100 + j150$ ohms, you will not be able to do it accurately because the $R = 100$ and $X = 150$ ohm circles would be (if they were drawn) on the extreme right edge of the chart—very close to infinity. In order to facilitate the plotting of larger impedances, *normalization* must be used. That is, each impedance to be plotted is divided by a convenient number that will place the new *normalized* impedance near the center of the chart where increased accuracy in plotting is obtained. Thus, for the preceding example, where $Z = 100 + j150$ ohms, it would be convenient to divide Z by 100, which yields the value $Z = 1 + j1.5$. This is very easily found on the chart. Once a chart is normalized in this manner, all impedances plotted on that chart *must be* divided by the *same* number in the normalization process. Otherwise, you will be left with a bunch of impedances with which nothing can be done.

Impedance Manipulation on the Chart

Fig. 4-32 graphically indicates what happens when a series capacitive reactance of $-j1.0$ ohm is added to an impedance of $Z = 0.5 + j0.7$ ohm. Mathematically, the result is

$$\begin{aligned} Z &= 0.5 + j0.7 - j1.0 \\ &= 0.5 - j0.3 \text{ ohm} \end{aligned}$$

which represents a series RC quantity. Graphically, what we have done is move *downward* along the $R =$ 0.5-ohm constant resistance circle for a distance of $X = -j1.0$ ohm. This is the plotted impedance point of $Z = 0.5 - j0.3$ ohm, as shown. In a similar manner, as shown in Fig. 4-33, adding a series inductance to a plotted impedance value simply causes a move *upward* along a constant resistance circle to the new impedance value. This type of construction is very important in the design of impedance-matching networks using the Smith Chart and must be understood. In general then, the addition of a series capacitor to an impedance moves that impedance *downward* (counterclockwise) along a constant resistance circle for a distance that is equal to the reactance of the capacitor. The addition of any series inductor to a plotted impedance moves that impedance *upward* (clockwise) along a constant resistance circle for a distance that is equal to the reactance of the inductor.

Conversion of Impedance to Admittance

The Smith Chart, although described thus far as a family of impedance coordinates, can easily be used to convert any impedance (Z) to an admittance (Y), and vice-versa. In mathematical terms, an admittance is simply the inverse of an impedance, or

$$Y = \frac{1}{Z} \qquad \text{(Eq. 4-9)}$$

where, the admittance (Y) contains both a real and an imaginary part, similar to the impedance (Z). Thus,

$$Y = G \pm jB \qquad \text{(Eq. 4-10)}$$

where,
G = the conductance in mhos,
B = the susceptance in mhos.

The circuit representation is shown in Fig. 4-34. Notice that the *susceptance is positive for a capacitor and negative for an inductor,* whereas, for reactance, the opposite is true.

To find the inverse of a series impedance of the form $Z = R + jX$ mathematically, you would simply use Equation 4-9 and perform the resulting calculation. But, how can you use the Smith Chart to perform the calculation for you without the need for a calculator? The easiest way of describing the use of the chart in performing this function is to first work a problem out mathematically and, then, plot the results on the chart to see how the two functions are related. Take, for example, the series impedance $Z = 1 + j1$. The inverse of Z is:

$$\begin{aligned} Y &= \frac{1}{1 + j1} \\ &= \frac{1}{1.414 \angle 45^\circ} \\ &= 0.7071 \angle -45^\circ \\ &= 0.5 - j0.5 \text{ mho} \end{aligned}$$

If we plot the points $1 + j1$ and $0.5 - j0.5$ on the Smith Chart, we can easily see the graphical relationship between the two. This construction is shown in Fig. 4-35. Notice that the two points are located at exactly the same distance (d) from the center of the chart but in opposite directions (180°) from each other. Indeed, the same relationship holds true for *any* impedance and its inverse. Therefore, without the aid of a calculator, you can find the reciprocal of an impedance or an admittance by simply plotting the point on the chart, measuring the distance (d) from the center of the chart to that point, and, then, plotting the measured result the same distance from the center but in the opposite direction (180°) from the original point. This is a very simple construction technique that can be done in seconds.

Another approach that we could take to achieve the same result involves the manipulation of the actual chart rather than the performing of a construc-

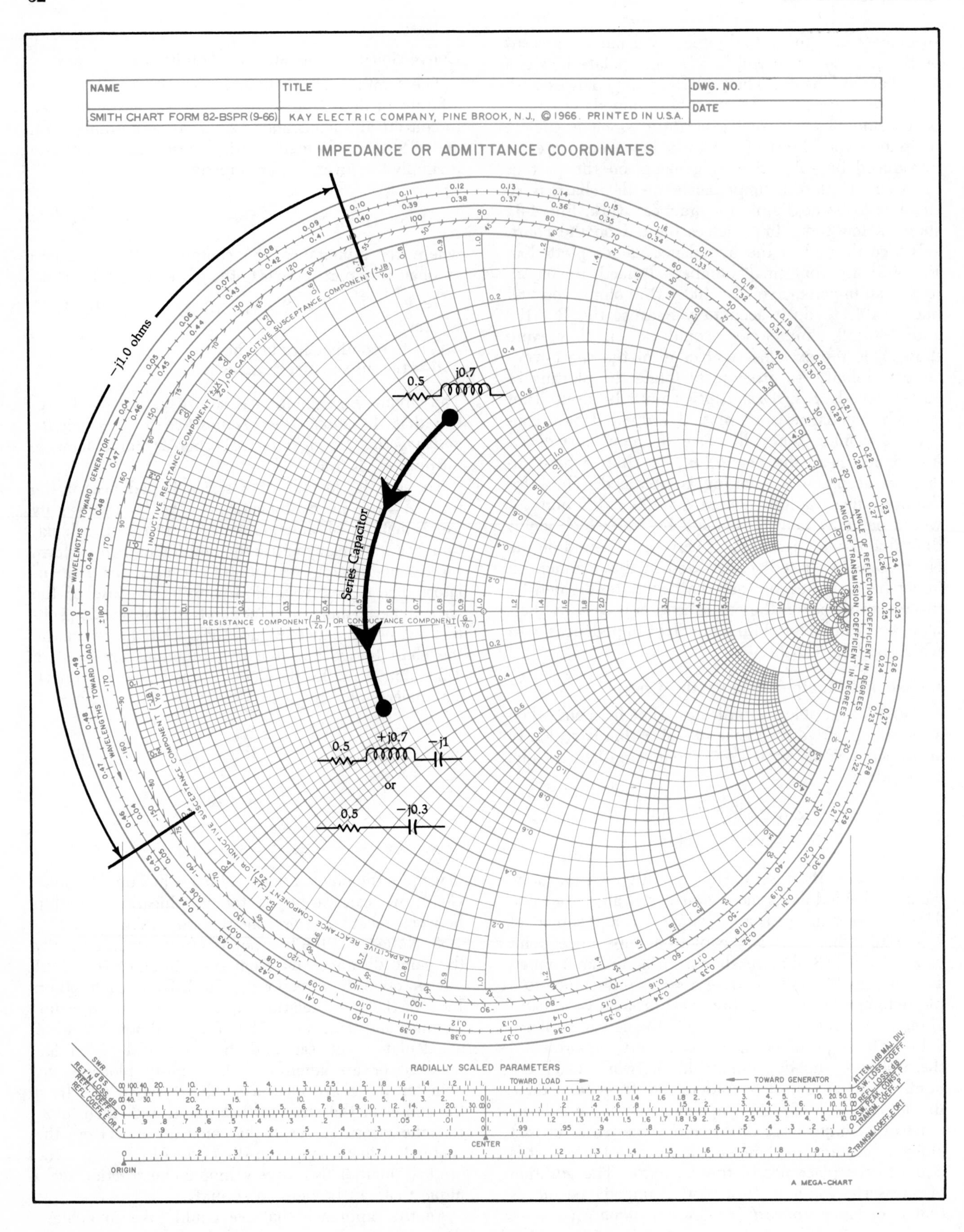

Fig. 4-32. Addition of a series capacitor.

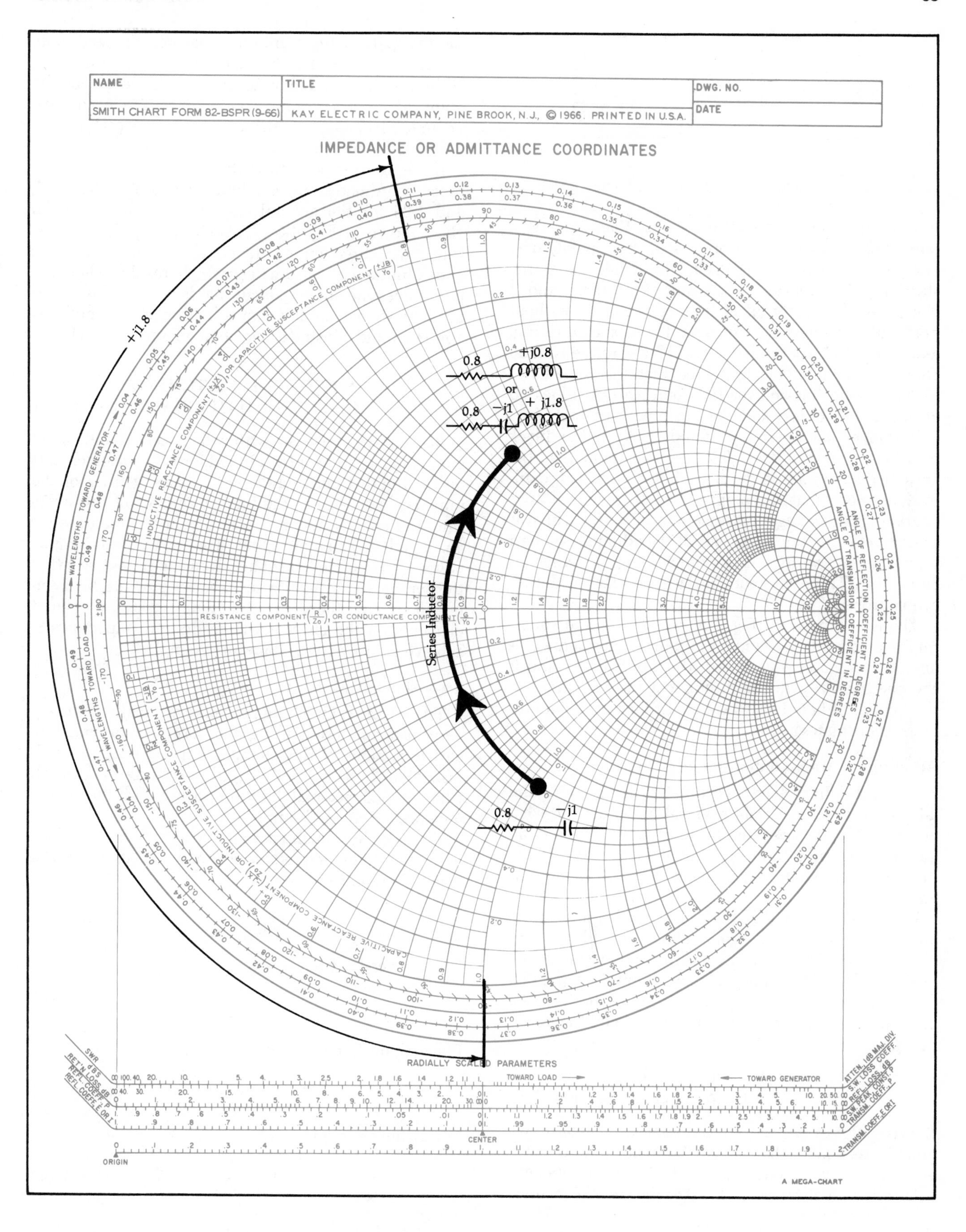

Fig. 4-33. Addition of a series inductor.

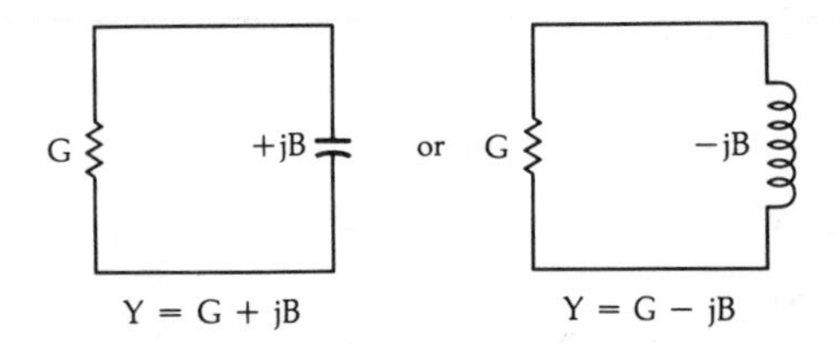

Fig. 4-34. Circuit representation for admittance.

tion on the chart. For instance, rather than locating a point 180° away from our original starting point, why not just rotate the chart itself 180° while fixing the starting point in space? The result is the same, and it can be read directly off of the rotated chart without performing a single construction. This is shown in Fig. 4-36 (Smith Chart Form ZY-01-N)* where the rotated chart is shown in black. Notice that the impedance plotted (solid lines on the red coordinates) is located at $Z = 1 + j1$ ohms, and the reciprocal of that (the admittance) is shown by dotted lines on the black coordinates as $Y = 0.5 - j0.5$. Keep in mind that because we have rotated the chart 180° to obtain the admittance coordinates, the upper half of the admittance chart represents *negative susceptance* (−jB) which is *inductive*, while the lower half of the admittance chart represents a *positive susceptance* (+jB) which is *capacitive*. Therefore, nothing has been lost in the rotation process.

The chart shown in Fig. 4-36, containing the superimposed impedance and admittance coordinates, is an extremely useful version of the Smith Chart and is the one that we will use throughout the remainder of the book. But first, let's take a closer look at the admittance coordinates alone.

Admittance Manipulation on the Chart

Just as the impedance coordinates of Figs. 4-32 and 4-33 were used to obtain a visual indication of what occurs when a *series* reactance is added to an *impedance*, the admittance coordinates provide a visual indication of what occurs when a *shunt* element is added to an *admittance*. The addition of a shunt capacitor is shown in Fig. 4-37. Here we begin with an admittance of $Y = 0.2 - j0.5$ mho and add a shunt capacitor with a susceptance (reciprocal of reactance) of +j0.8 mho. Mathematically, we know that parallel susceptances are simply added together to find the equivalent susceptance. When this is done, the result becomes:

$$\begin{aligned} Y &= 0.2 - j0.5 + j0.8 \\ &= 0.2 + j0.3 \text{ mho} \end{aligned}$$

If this point is plotted on the admittance chart, we quickly recognize that all we have done is to move along a constant conductance circle (G) *downward* (clockwise) a distance of jB = 0.8 mho. In other words, the real part of the admittance has not changed, only the imaginary part has. Similarly, as Fig. 4-38 indicates, adding a shunt inductor to an admittance moves the point along a constant conductance circle upward (counterclockwise) a distance (−jB) equal to the value of its susceptance.

If we again superimpose the impedance and admittance coordinates and combine Figs. 4-32, 4-33, 4-37, and 4-38 for the general case, we obtain the useful chart shown in Fig. 4-39. This chart graphically illustrates the direction of travel, along the impedance and admittance coordinates, which results when the particular type of component that is indicated is added to an existing impedance or admittance. A simple example should illustrate the point (Example 4-6).

IMPEDANCE MATCHING ON THE SMITH CHART

Because of the ease with which series and shunt components can be added in ladder-type arrangements on the Smith Chart, while easily keeping track of the impedance as seen at the input terminals of the structure, the chart seems to be an excellent candidate for an impedance-matching tool. The idea here is simple. Given a load impedance and given the impedance that the source would like to see, simply plot the load impedance and, then, begin adding series and shunt elements on the chart until the desired impedance is achieved—just as was done in Example 4-6.

Two-Element Matching

Two-element matching networks are mathematically very easy to design using the formulas provided in earlier sections of this chapter. For the purpose of illustration, however, let's begin our study of a Smith Chart impedance-matching procedure with the simple network given in Example 4-7.

To make life much easier for you as a Smith Chart user, the following equations may be used. For a series-C component:

$$C = \frac{1}{\omega XN} \qquad \text{(Eq. 4-11)}$$

For a series-L component:

$$L = \frac{XN}{\omega} \qquad \text{(Eq. 4-12)}$$

For a shunt-C component:

$$C = \frac{B}{\omega N} \qquad \text{(Eq. 4-13)}$$

For a shunt-L component:

$$L = \frac{N}{\omega B} \qquad \text{(Eq. 4-14)}$$

where,

$\omega = 2\pi f$,

X = the reactance as read from the chart,

* Smith Chart Form ZY-01-N is a copyright of Analog Instruments Company, P.O. Box 808, New Providence, NJ 07974. It and other Smith Chart accessories are available from the company.

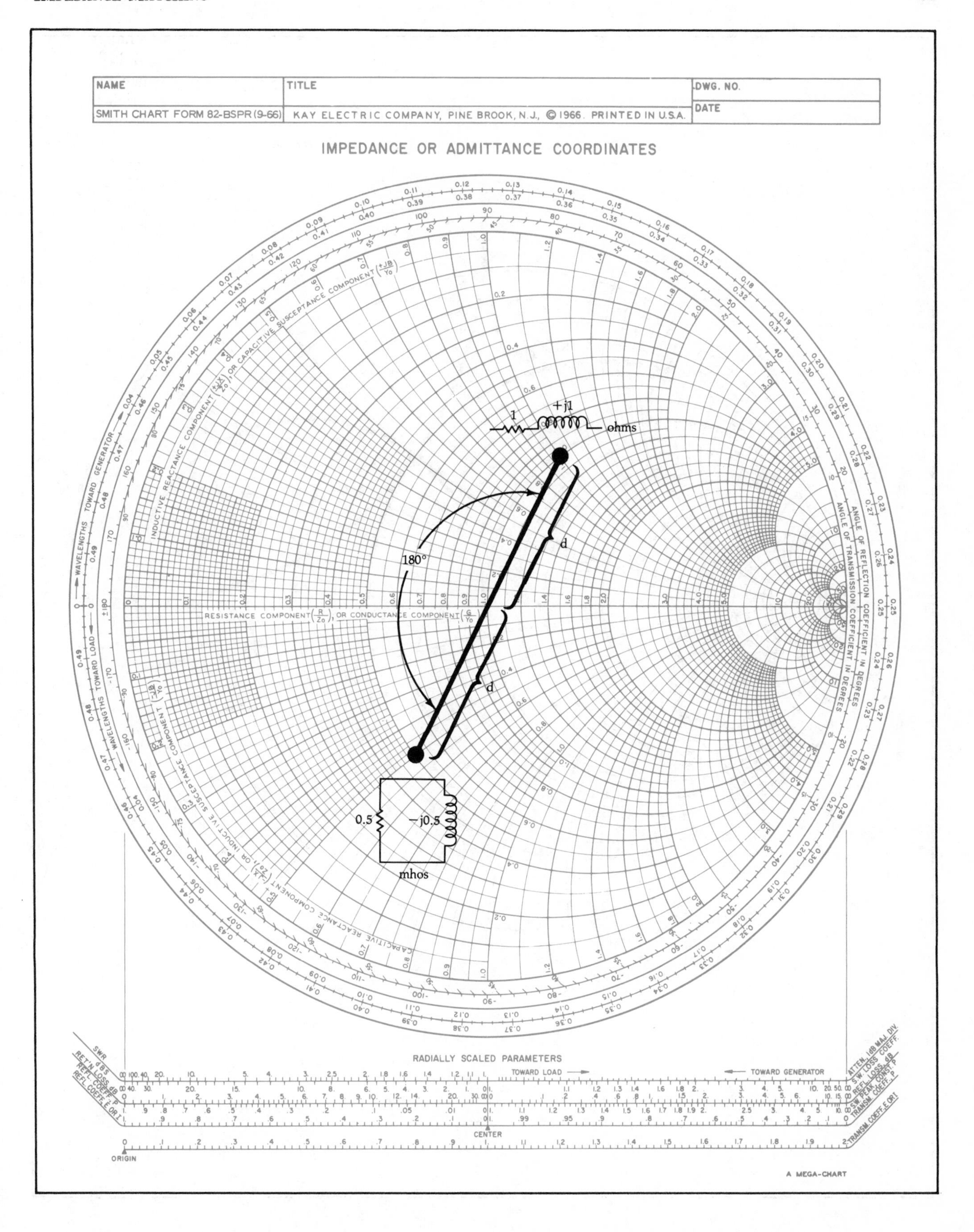

Fig. 4-35. Impedance-admittance conversion on the Smith Chart.

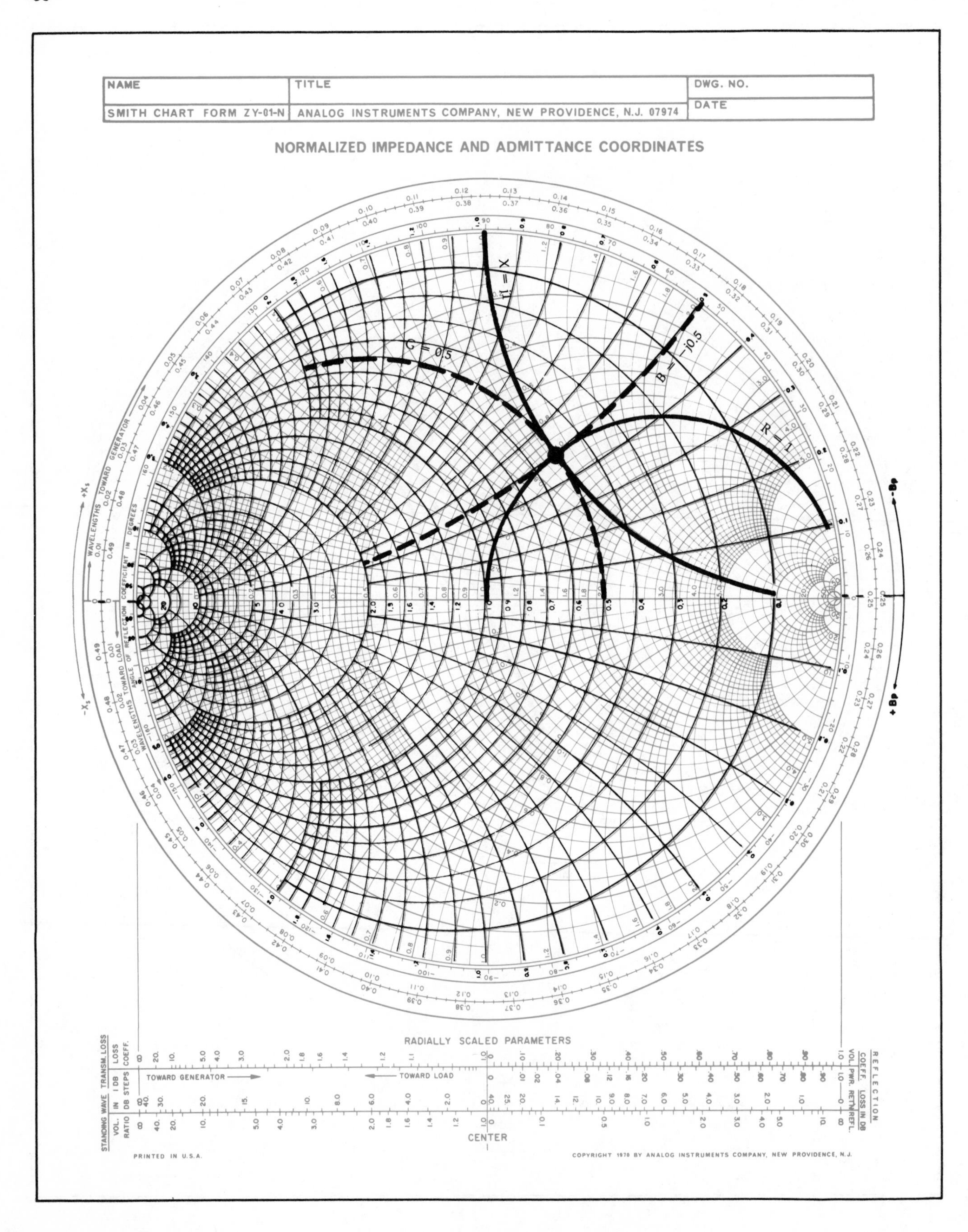

Fig. 4-36. Superimposed admittance coordinates.

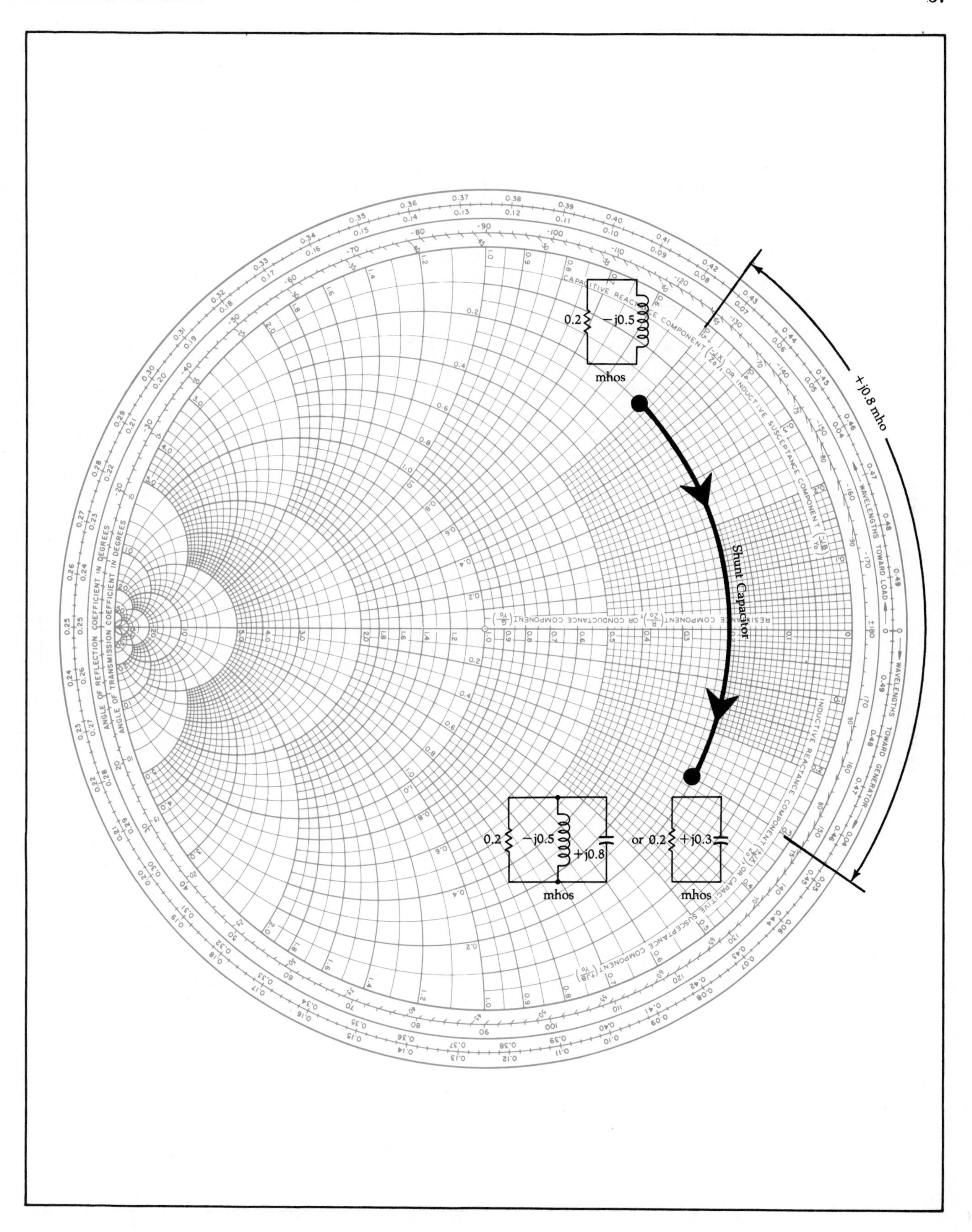

Fig. 4-37. Addition of a shunt capacitor.

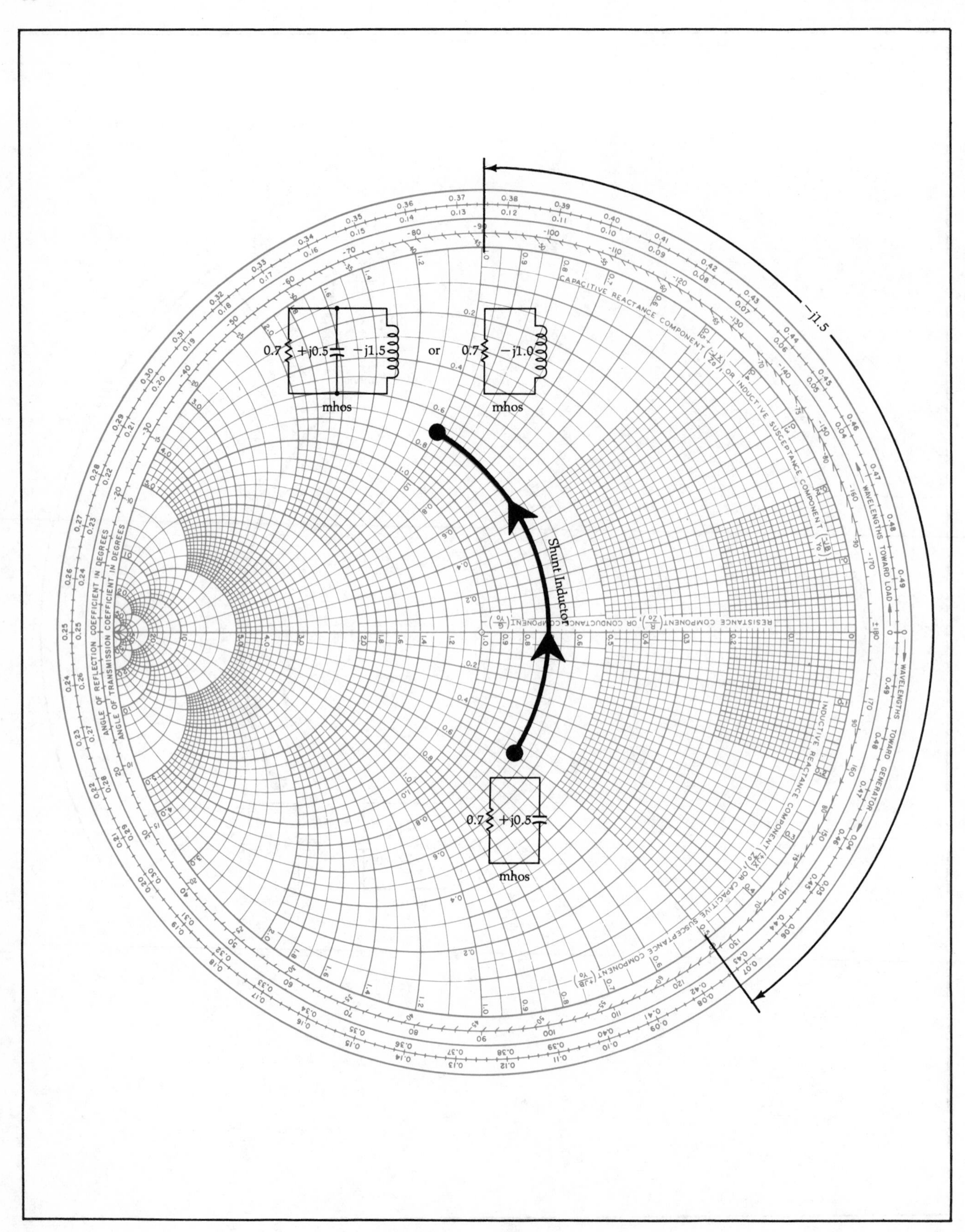

Fig. 4-38. Addition of a shunt inductor.

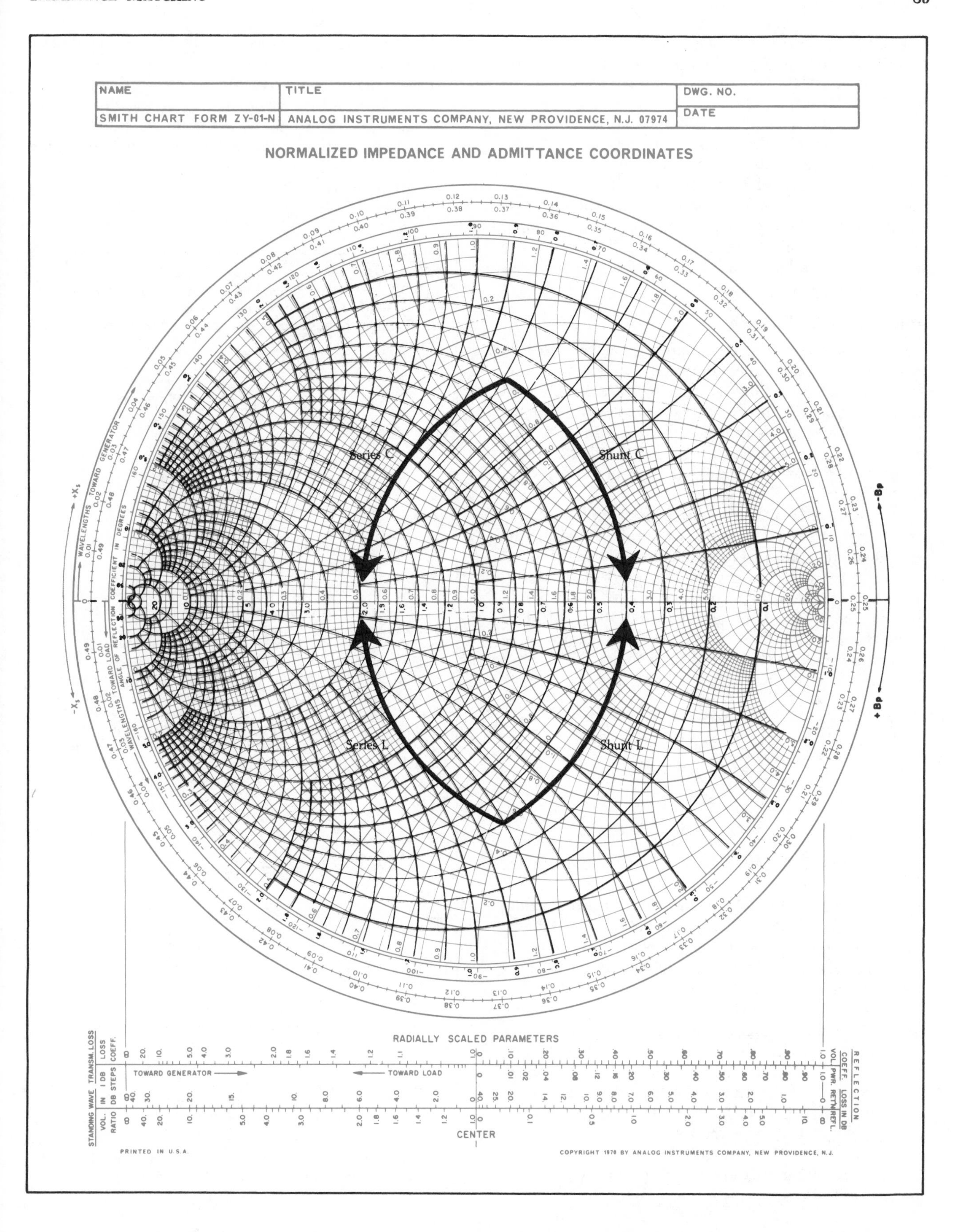

Fig. 4-39. Summary of component addition on a Smith Chart.

B = the susceptance as read from the chart,
N = the number used to normalize the original impedances that are to be matched.

If you use the preceding equations, you will never have to worry about changing susceptances into reactances before unnormalizing the impedances. The equations take care of both operations. The only thing you have to do is read the value of susceptance (for shunt components) or reactance (for series components) directly off of the chart, plug this value into the equation used, and wait for your actual component values to pop out.

Three-Element Matching

In earlier sections of this chapter, you learned that the only real difference between two-element and three-element matching is that with three-element matching, you are able to choose the loaded Q for the network. That was easy enough to do in a mathematical-design approach due to the virtual resistance concept. But how can circuit Q be represented on a Smith Chart?

As you have seen before, in earlier chapters, the Q of a series-impedance circuit is simply equal to the ratio of its reactance to its resistance. Thus, any point on a Smith Chart has a Q associated with it. Alternately, if you were to specify a certain Q, you could find an infinite number of points on the chart that could satisfy that Q requirement. For example, the following impedances located on a Smith Chart have a Q of 5:

$$
\begin{aligned}
R + jX &= 1 \pm j5 \\
&= 0.5 \pm j2.5 \\
&= 0.2 \pm j1 \\
&= 0.1 \pm j0.5 \\
&= 0.05 \pm j0.25
\end{aligned}
$$

These values are plotted in Fig. 4-45 and form the arcs shown. Thus, any impedance located on these arcs must have a Q of 5. Similar arcs for other values of Q can be drawn with the arc of infinite Q being located along the perimeter of the chart and the Q = 0 arc (actually a straight line) lying along the pure resistance line located at the center of the chart.

The design of high-Q three-element matching networks on a Smith Chart is approached in much the same manner as in the mathematical methods presented earlier in this chapter. Namely, one branch of the network will determine the loaded Q of the circuit, and it is this branch that will set the characteristics of the rest of the circuit.

The procedure for designing a three-element impedance-matching network for a specified Q is summarized as follows:

1. Plot the constant-Q arcs for the specified Q.
2. Plot the load impedance and the complex conjugate of the source impedance.
3. Determine the end of the network that will be used to establish the loaded Q of the design. For T networks, the end with the *smaller* terminating resistance determines the Q. For Pi networks, the end with the *larger* terminating resistor sets the Q.
4. For T networks:

$$R_s > R_L$$

EXAMPLE 4-6

What is the impedance looking into the network shown in Fig. 4-40? Note that the task has been simplified due to the fact that shunt susceptances are shown rather than shunt reactances.

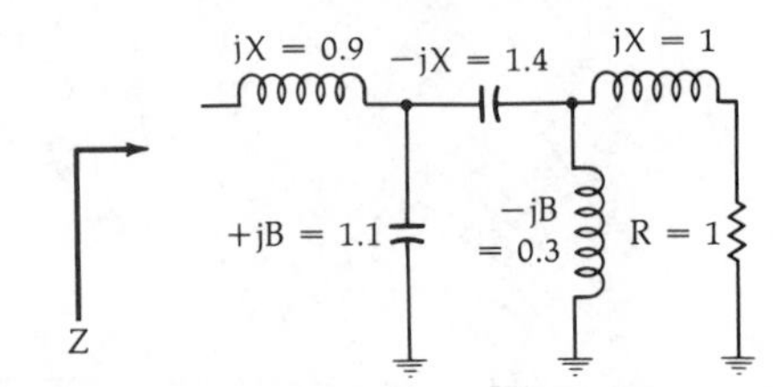

Fig. 4-40. Circuit for Example 4-6.

Solution

This problem is very easily handled on a Smith Chart and not a single calculation needs to be performed. The solution is shown in Fig. 4-42. It is accomplished as follows.

First, break the circuit down into individual branches as shown in Fig. 4-41. Plot the impedance of the series RL branch where Z = 1 + j1 ohm. This is point A in Fig. 4-42. Next, following the rules diagrammed in Fig. 4-39, begin adding each component back into the circuit—one at a time. Thus, the following constructions (Fig. 4-42) should be noted:

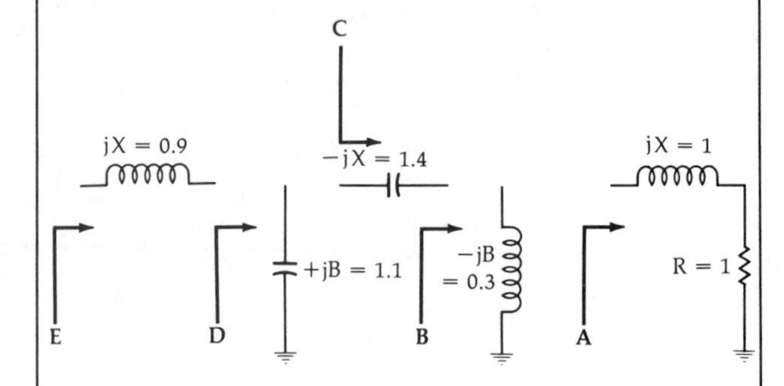

Fig. 4-41. Circuit is broken down into individual branch elements.

Arc AB = shunt L = −jB = 0.3 mho
Arc BC = series C = −jX = 1.4 ohms
Arc CD = shunt C = +jB = 1.1 mhos
Arc DE = series L = +jX = 0.9 ohm

The impedance at point E (Fig. 4-42) can then be read directly off of the chart as Z = 0.2 + j0.5 ohm.

Continued on next page

EXAMPLE 4-6—Cont.

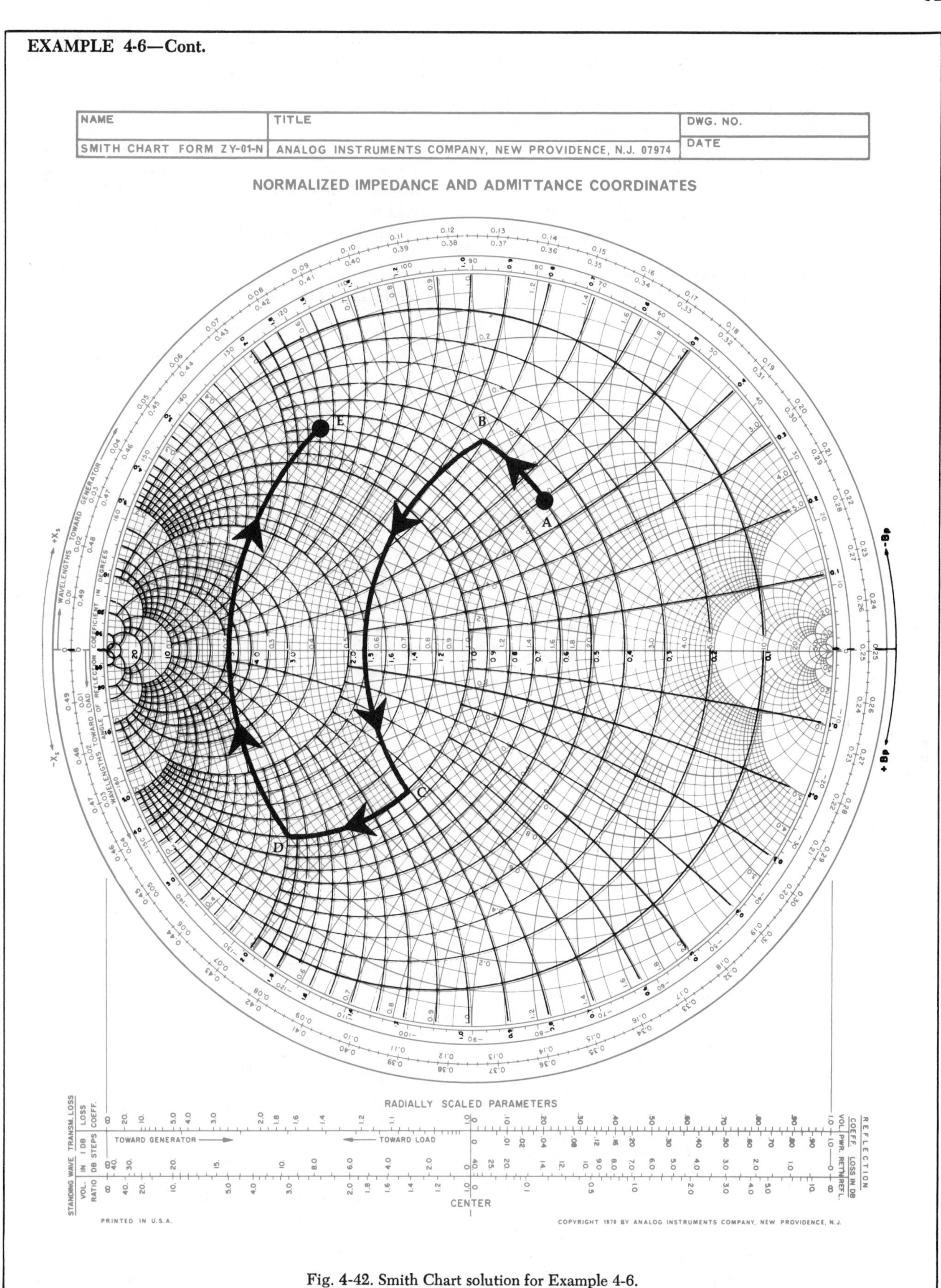

Fig. 4-42. Smith Chart solution for Example 4-6.

Move from the load along a constant-R circle (series element) and intersect the Q curve. The length of this move determines your first element. Then, proceed from this point to Z_s^* ($Z_s^* = Z_s$ conjugate) in two moves—first with a shunt and, then, with a series element.

$$R_s < R_L$$

Find the intersection (I) of the Q curve and the source impedance's R = constant circle, and plot that point. Move *from the load impedance* to point I with two elements—first, a series element and, then, a shunt element. Move from point I to Z_s^* along the R = constant circle with another series element.

5. For Pi networks:

$$R_S > R_L$$

Find the intersection (I) of the Q curve and the source impedance's G = constant circle, and plot that point. Move from the load impedance to point I with two elements—first, a shunt element and, then, a series element. Move from point I to Z_s^* along the G = constant circle with another shunt element.

$$R_S < R_L$$

Move from the load along a constant G circle (shunt element) and intersect the Q curve. The length of this move determines your first element. Then, proceed from this point to Z_s^* in two moves—first, with a series element and, then, with a shunt element.

The above procedures might seem complicated to the neophyte but remember that we are only forcing the constant-resistance or constant-conductance arc, located between the Q-determining termination and the specified-Q curve, to be one of our matching elements. An example may help to clarify matters (Example 4-8).

Multielement Matching

In multielement matching networks where there is no Q constraint, the Smith Chart becomes a veritable

---

EXAMPLE 4-7

Design a two-element impedance-matching network on a Smith Chart so as to match a 25 − j15-ohm source to a 100 − j25-ohm load at 60 MHz. The matching network must also act as a low-pass filter between the source and the load.

Solution

Since the source is a complex impedance, it wants to "see" a load impedance that is equal to its complex conjugate (as discussed in earlier sections of this chapter). Thus, the task before us is to force the 100 − j25-ohm load to look like an impedance of 25 + j15 ohms.

Obviously, the source and load impedances are both too large to plot on the chart, so normalization is necessary. Let's choose a convenient number (N = 50) and divide all impedances by this number. The results are 0.5 + j0.3 ohm for the impedance the source would like to see and 2 − j0.5 ohms for the actual load impedance. These two values are easily plotted on the Smith Chart, as shown in Fig. 4-44, where, at point A, Z_L is the *normalized* load impedance and, at point C, Z_s^* is the *normalized* complex conjugate of the source impedance.

The requirement that the matching network also be a low-pass filter forces us to use some form of series-L, shunt-C arrangement. The only way we can get from the impedance at point A to the impedance at point C and still fulfill this requirement is along the path shown in Fig. 4-44. Thus, following the rules of Fig. 4-39, the arc AB of Fig. 4-44 is a shunt capacitor with a value of +jB = 0.73 mho. The arc BC is a series inductor with a value of +jX = 1.2 ohms.

The shunt capacitor as read from the Smith Chart is a susceptance and can be changed into an equivalent reactance by simply taking the reciprocal.

$$X_C = \frac{1}{+jB} = \frac{1}{j0.73 \text{ mho}} = -j1.37 \text{ ohms}$$

To complete the network, we must now unnormalize all impedance values by *multiplying* them by the number N = 50—the value originally used in the normalization process. Therefore:

$$X_L = 60 \text{ ohms}$$
$$X_C = 68.5 \text{ ohms}$$

The component values are:

$$L = \frac{X_L}{\omega} = \frac{60}{2\pi(60 \times 10^6)} = 159 \text{ nH}$$

$$C = \frac{1}{\omega X_C} = \frac{1}{2\pi(60 \times 10^6)(68.5)} = 38.7 \text{ pF}$$

The final circuit is shown in Fig. 4-43.

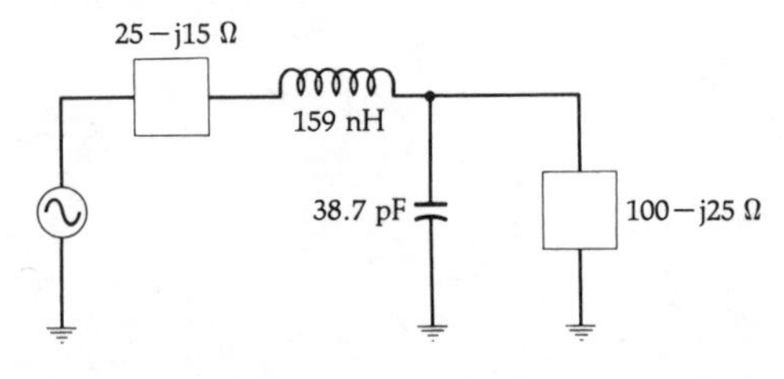

Fig. 4-43. Final circuit for Example 4-7.

Continued on next page

EXAMPLE 4-7—Cont.

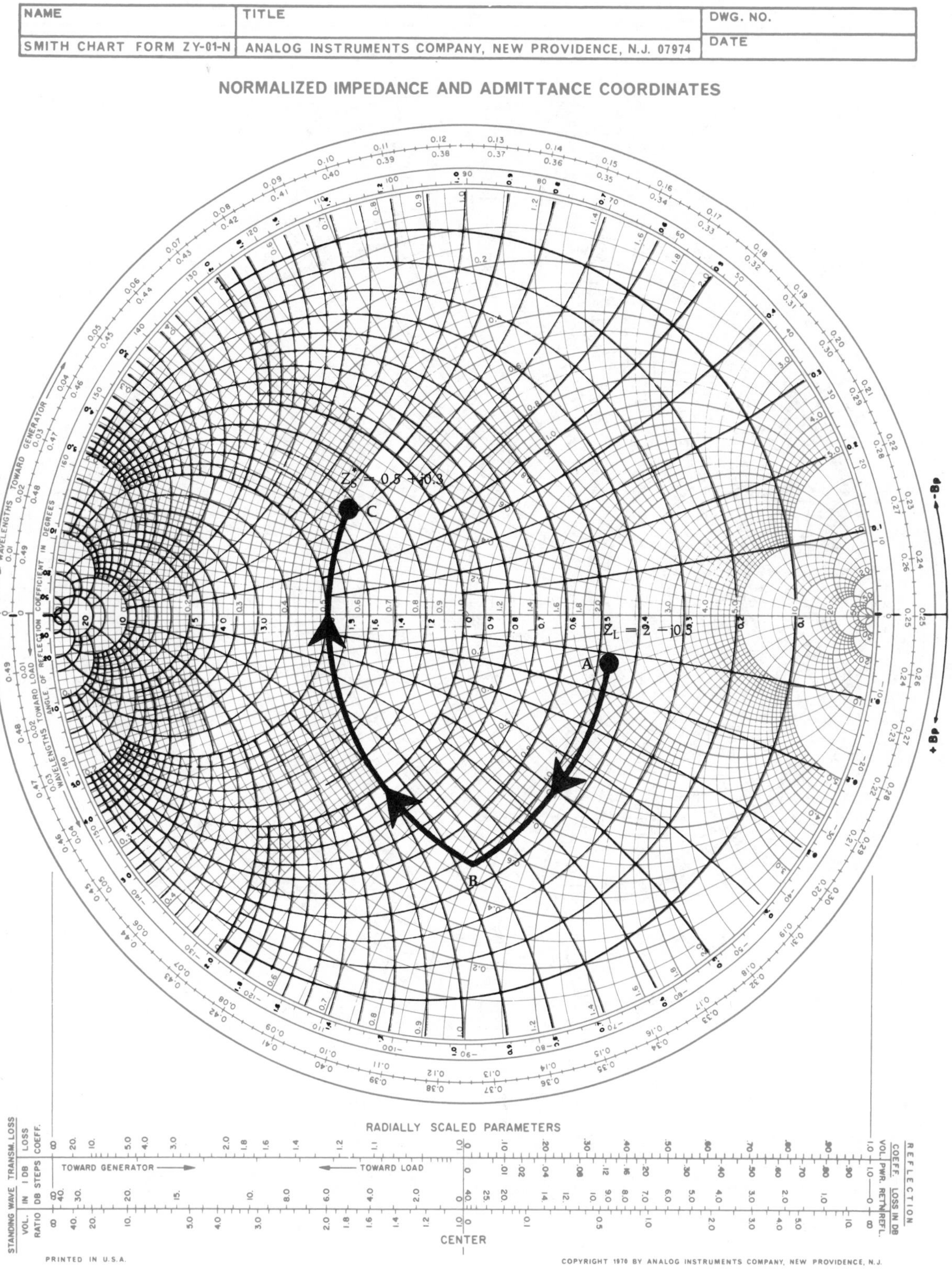

NAME
TITLE
DWG. NO.
SMITH CHART FORM ZY-01-N
ANALOG INSTRUMENTS COMPANY, NEW PROVIDENCE, N.J. 07974
DATE
NORMALIZED IMPEDANCE AND ADMITTANCE COORDINATES
Zs = 0.5 + j0.3
C
ZL = 2 − j0.5
A
B
WAVELENGTHS TOWARD GENERATOR
WAVELENGTHS TOWARD LOAD
ANGLE OF REFLECTION COEFFICIENT IN DEGREES
RADIALLY SCALED PARAMETERS
TOWARD GENERATOR
TOWARD LOAD
CENTER
STANDING WAVE VOL. RATIO
IN DB
TRANSM. LOSS 1 DB STEPS
LOSS COEFF.
REFLECTION COEFF. VOL. PWR.
LOSS IN DB RETN. REFL.
PRINTED IN U.S.A.
COPYRIGHT 1970 BY ANALOG INSTRUMENTS COMPANY, NEW PROVIDENCE, N.J.

Fig. 4-44. Solution to Example 4-7.

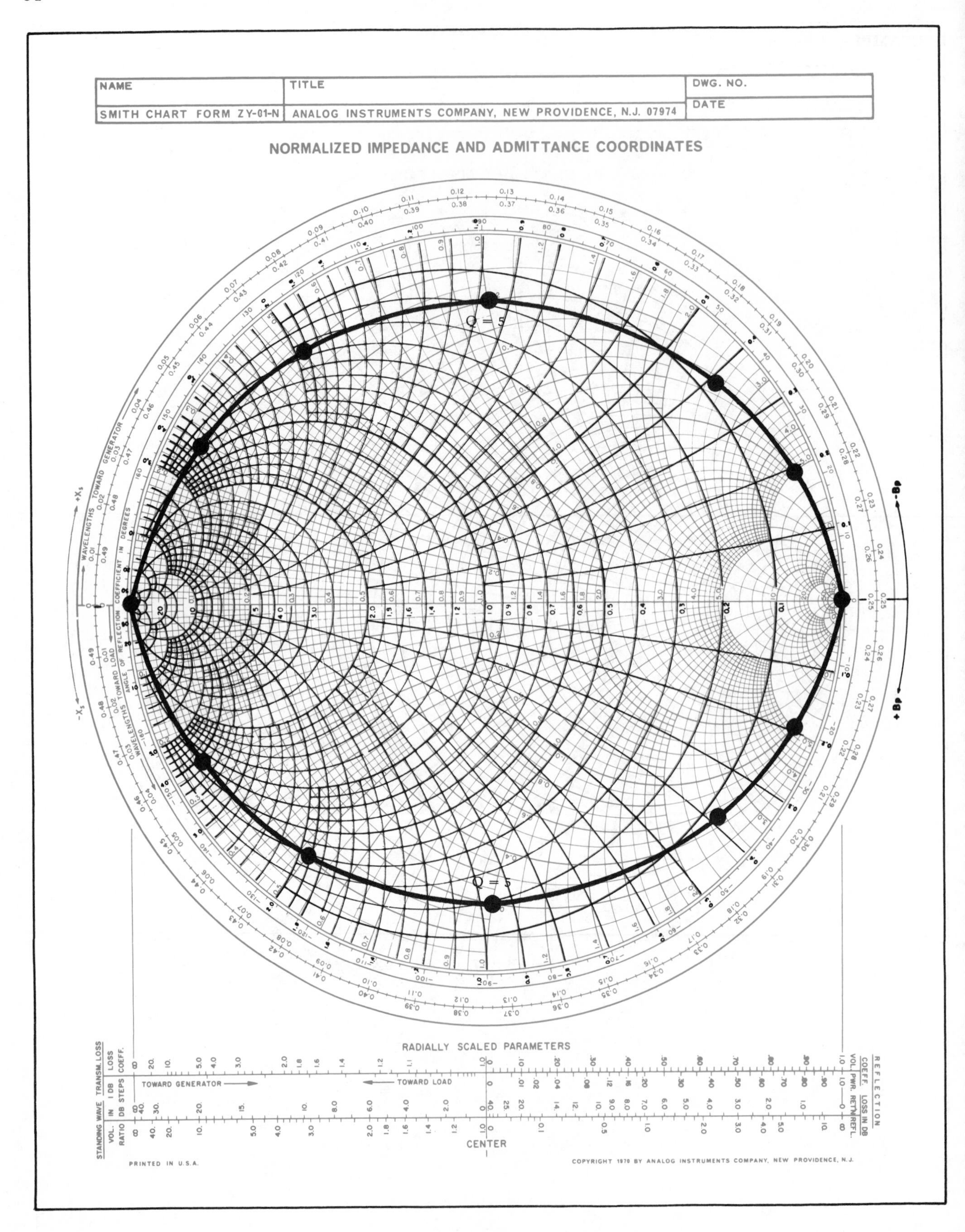

Fig. 4-45. Lines of constant Q.

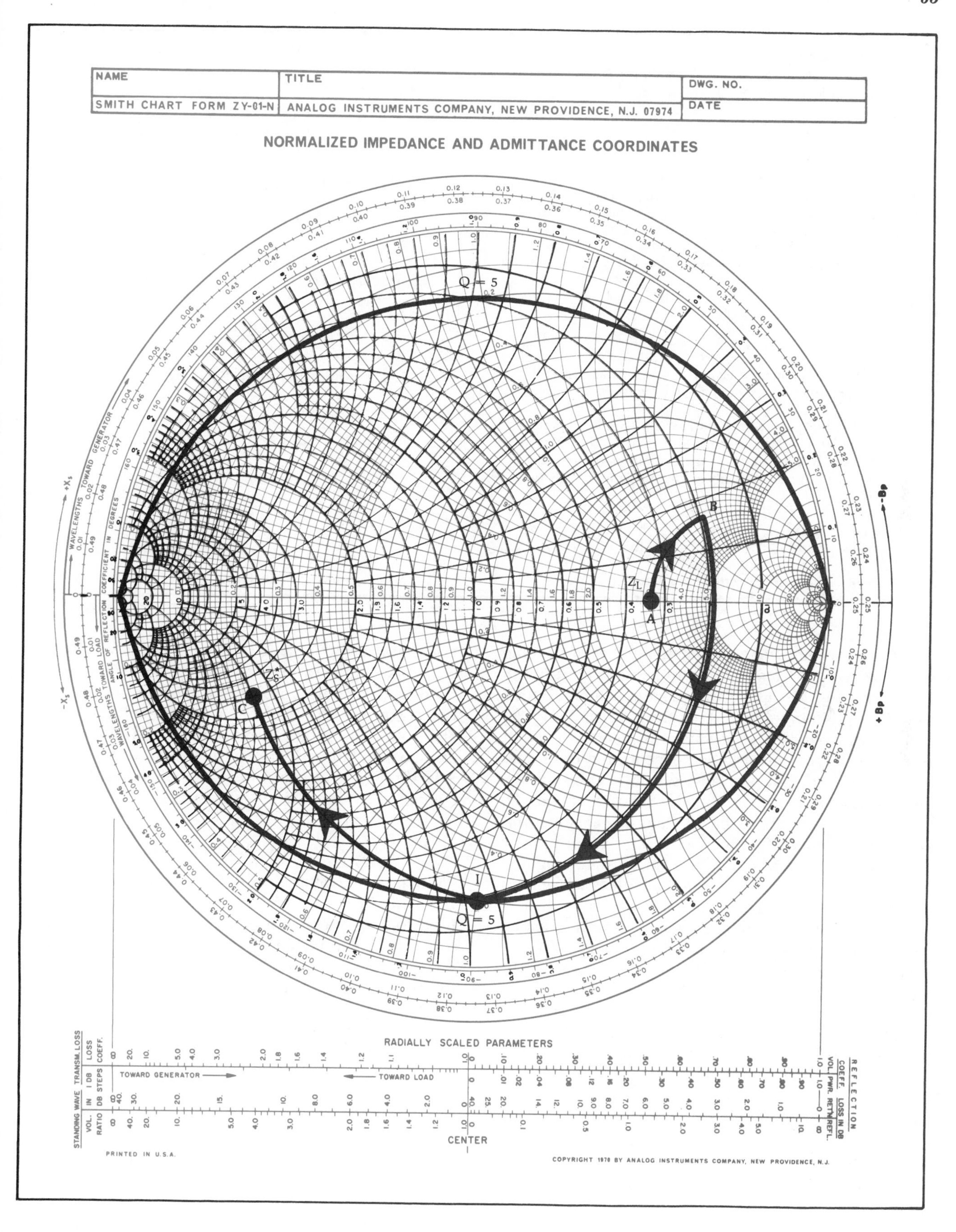

Fig. 4-46. Smith Chart solution for Example 4-8.

EXAMPLE 4-8

Design a T network to match a $Z = 15 + j15$-ohm source to a 225-ohm load at 30 MHz with a loaded Q of 5.

Solution

Following the procedures previously outlined, draw the arcs for $Q = 5$ first and, then, plot the load impedance and the complex conjugate of the source impedance. Obviously, normalization is necessary as the impedances are too large to be located on the chart. Divide by a convenient value (choose $N = 75$) for normalization. Therefore:

$$Z_s^* = 0.2 - j0.2 \text{ ohm}$$

$$Z_L = 3 \text{ ohms}$$

The construction details for the design are shown in Fig. 4-46.

The design statement specifies a T network. Thus, the source termination will determine the network Q because $R_S < R_L$.

Following the procedure for $R_S < R_L$ (Step 4, above), first plot point I, which is the intersection of the $Q = 5$ curve and the $R =$ constant circuit that passes through Z_s^*. Then, move from the load impedance to point I with two elements.

Element 1 = arc AB = series L = j2.5 ohms
Element 2 = arc BI = shunt C = j1.15 mhos

Then, move from point I to Z_s^* along the $R =$ constant circle.

Element 3 = arc IC = series L = j0.8 ohm

Use Equations 4-11 through 4-14 to find the actual element values.

Element 1 = series L:

$$L = \frac{(2.5)75}{2\pi(30 \times 10^6)} = 995 \text{ nH}$$

Element 2 = shunt C:

$$C = \frac{1.15}{2\pi(30 \times 10^6)75} = 81 \text{ pF}$$

Element 3 = series L:

$$L = \frac{(0.8)75}{2\pi(30 \times 10^6)} = 318 \text{ nH}$$

The final network is shown in Fig. 4-47.

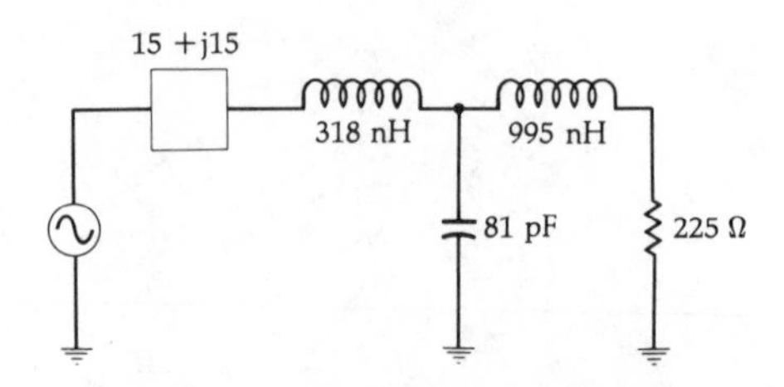

Fig. 4-47. Final circuit for Example 4-8.

treasure trove containing an infinite number of possible solutions. To get from point A to point B on a Smith Chart, there is, of course, an optimum solution. However, the optimum solution is not the only solution. The two-element network gets you from point A to point B with the least number of components and the three-element network can provide a specified Q by following a different route. If you do not care about Q, however, there are 3-, 4-, 5-, 10-, and 20-element (and more) impedance-matching networks that are easily designed on a Smith Chart by simply following the constant-conductance and constant-resistance circles until you eventually arrive at point B, which, in our case, is usually the complex conjugate of the source impedance. Fig. 4-48 illustrates this point. In the lower right-hand corner of the chart is point A. In the upper left-hand corner is point B. Three of the infinite number of possible solutions that can be used to get from point A to point B, by adding series and shunt inductances and capacitances, are shown. Solution 1 starts with a series-L configuration and takes 9 elements to get to point B. Solution 2 starts with a shunt-L procedure and takes 8 elements, while Solution 3 starts with a shunt-C arrangement and takes 5 elements. The element reactances and susceptances can be read directly from the chart, and Equations 4-11 through 4-14 can be used to calculate the actual component values within minutes.

SUMMARY

Impedance matching is not a form of "black magic" but is a step-by-step well-understood process that is used to help transfer maximum power from a source to its load. The impedance-matching networks can be designed either mathematically or graphically with the aid of a Smith Chart. Simpler networks of two and three elements are usually handled best mathematically, while networks of four or more elements are very easily handled using the Smith Chart.

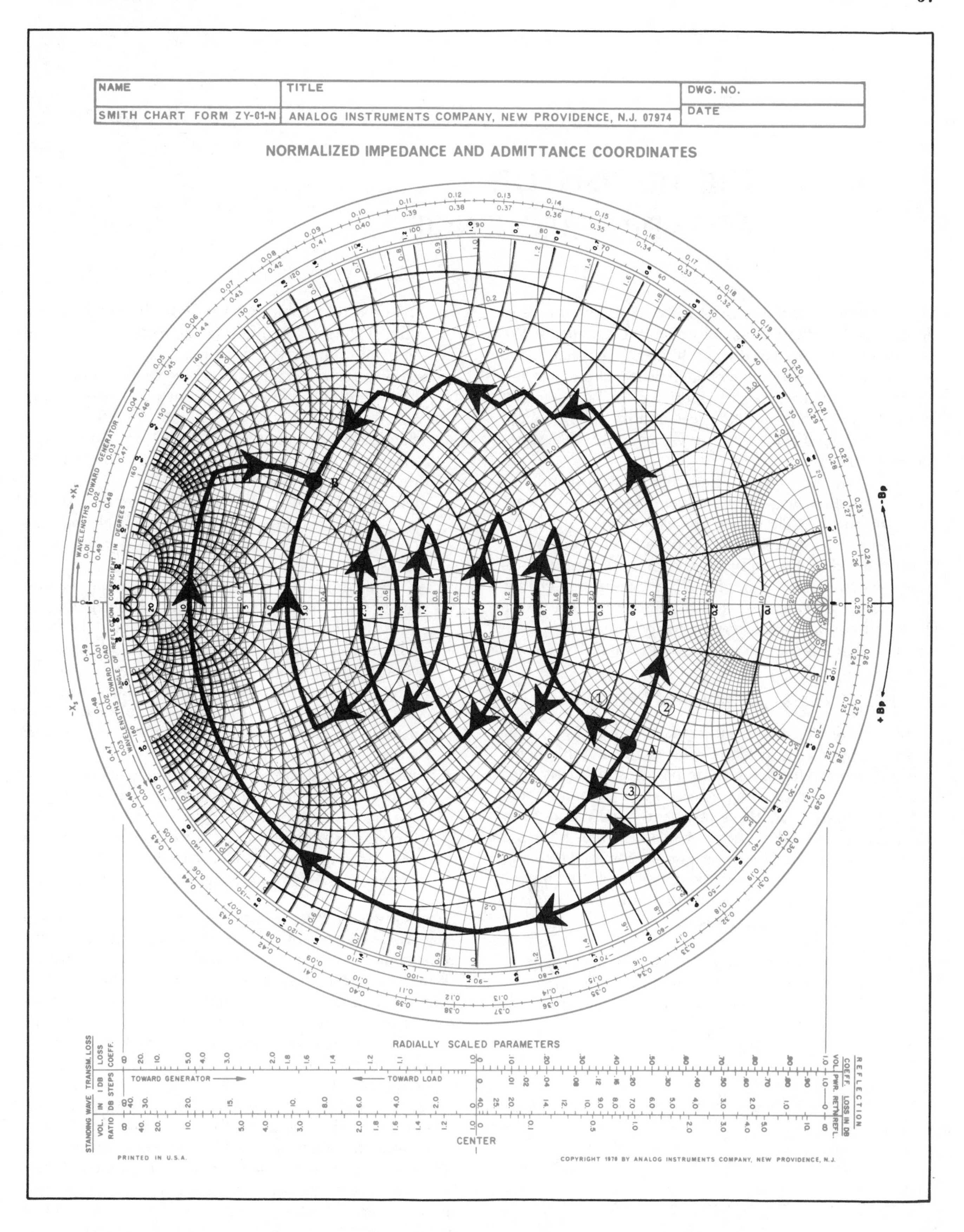

Fig. 4-48. Multielement matching.

THE TRANSISTOR AT RADIO FREQUENCIES

In Chapter 1, we discussed resistors, capacitors, and inductors, and their behavior at radio frequencies. We found that when working at higher frequencies, we could no longer think of a capacitor as just a capacitor, or an inductor as a perfect inductor. In fact, each of these components can be represented by an equivalent circuit that indicates just how imperfect that component really is.

In this chapter, we will find that the transistor, too, is an imperfect device whose characteristics also vary with frequency. Therefore, the equivalent circuit for a typical transistor is introduced and analyzed. Then, with the aid of the equivalent circuit, the input, output, feedback, and gain characteristics are described. We will then examine Y and S parameters and take a look at how manufacturers typically present the transistor's characteristics on their data sheets.

THE TRANSISTOR EQUIVALENT CIRCUIT

Just as resistors, capacitors, and inductors can be modeled by an equivalent circuit at radio frequencies, transistor behavior can also be best described by such a circuit as shown in Fig. 5-1. This is a common-emitter configuration of the equivalent circuit known as the hybrid-π model. At first glance, the hybrid-π model looks to be quite formidable for analysis purposes. After defining each component of the model, however, some simplifying assumptions will be made to aid in the analysis process.

$r_{bb'}$—*Base spreading resistance.* This is an inevitable resistance that occurs at the junction between the base terminal or contact and the semiconductor material that composes the base. Its value is usually in the tens of ohms. Smaller transistors tend to exhibit larger values of $r_{bb'}$.

$r_{b'e}$—*Input resistance.* The resistance that occurs at the base-emitter junction of a forward-biased transistor. Typical values range around 1000 ohms.

$r_{b'c}$—*Feedback resistance.* This is a very large (≈ 5 megohm) resistance appearing from the base to the collector of the transistor.

r_{ce}—*Output resistance.* As the name implies, this is simply the resistance seen looking back into the collector of the transistor. A value for a typical transistor would be about 100K.

C_e—*Emitter diffusion capacitance.* This capacitance is really the sum of the emitter diffusion capacitance and the emitter junction capacitance, both of which are associated with the physics of the semiconductor junction itself and which is beyond the scope of this book. It does exist, however, and since the junction capacitance is so small, C_e is usually called diffusion capacitance with a typical value of 100 pF.

C_c—*Feedback capacitance.* This component is formed at the reverse-biased collector-to-base junction of the transistor. As the frequency of operation for the transistor increases, C_c can begin to have a very pronounced effect on transistor operation. A typical value for this component might be 3 pF.

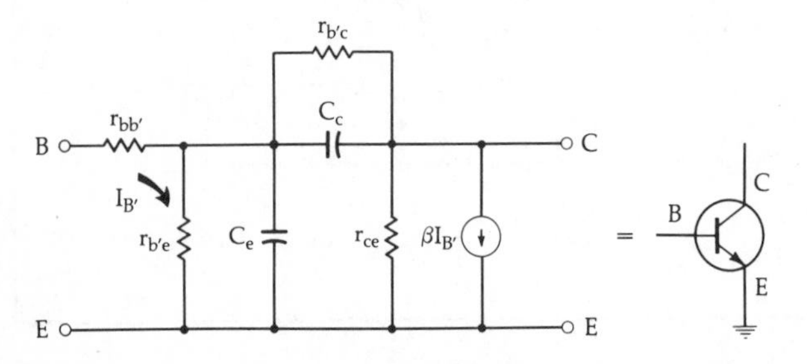

Fig. 5-1. Transistor equivalent circuit—common-emitter configuration.

Also shown in Fig. 5-1 is a *current source* of value $\beta I_{B'}$. Beta (β) is, of course, the small-signal ac current gain of the transistor while I_B' is the current through $r_{b'e}$. The current source can be thought of as simply an indication of current flow in the collector that is dependent upon the current that flows in the base of the transistor. Therefore, the collector current is equal to the base current times the β of the transistor, or $I_c = \beta I_{B'}$.

Keep in mind that Fig. 5-1 depicts only those inherent parasitic elements that are *internal* to the semiconductor material itself. Somehow, however, a connection has to be made from the semiconductor material to the transistor leads. This is done with a minute piece of wire called a bonding wire, which, at high frequencies, adds a bit of inductance to the equivalent circuit. The transistor leads themselves tend to exhibit additional series inductance and the equivalent circuit begins to resemble that of Fig. 5-2, where L_B, L_E, and L_C are the base, emitter, and collector lead and bonding inductance, respectively.

It certainly should be obvious now that the equivalent circuit for a typical transistor is not trivial, but contains numerous components, all of which will affect the device's operation at high frequency *to a certain degree.* If some simplifying assumptions are made, however, we should be able to use the equivalent circuit to determine how the transistor behaves at radio frequencies.

Input Impedance

One of the first simplifications that can be made to the circuit of Fig. 5-2 is to eliminate $r_{b'c}$. Five megohms is, after all, a rather large resistance and, for our purposes, looks like an open circuit. The next step is to use a principle called the *Miller effect* to transpose C_c from its series base-to-collector connection to a position that is in parallel with C_e, with a new value of $(C_c)(1-\beta R_L)$, where R_L is the load resistance. This capacitance is then combined with C_e to form a new total capacitance, C_T. These changes are shown in Fig. 5-3.

The input impedance variation over frequency for a transistor is very easily found by analyzing the circuit of Fig. 5-4. Here we have included only the elements of the equivalent circuit that have an effect on the transistor's input impedance. Notice that the *primary* contributors are $r_{b'e}$ and C_T—neither of which the designer has any control over. The quantity $r_{bb'}$, on the other hand, is a very small resistance while L_B and L_E can vary in size depending on circuit layout. If you are very careful, L_B and L_E can be limited practically to the bonding inductance that was mentioned previously. If this is the case, these elements will have practically no effect on input impedance until well above very high frequencies (vhf).

If we begin our analysis at dc, the circuit of Fig. 5-4 reduces to $r_{bb'}$ in series with $r_{b'e}$ and the input impedance is a pure resistance and is at its maximum value. As the frequency of operation increases, however, C_T begins to play an increasingly important role. Its shunting effect (around $r_{b'e}$) tends to reduce the impedance considerably, until at high frequencies, it effectively eliminates $r_{b'e}$ from the circuit. When this occurs, $r_{bb'}$, L_B, and L_E become the major contributors to the transistor's input impedance.

The impedance looking into the terminals of Fig. 5-4 can be described as follows:

$$Z_{in} = j\omega L_B + r_{bb'} + \frac{\frac{1}{j\omega C_T}(r_{b'e})}{\frac{1}{j\omega C_T} + r_{b'e}} + j\omega L_E$$

$$= j\omega(L_B + L_E) + r_{bb'} + \frac{r_{b'e}}{1 + r_{b'e}\, j\omega C_T}$$

$$= j\omega L_T + r_{bb'} + \frac{r_{b'e}}{1 + j\omega r_{b'e} C_T}$$

This equation is plotted on the Smith Chart shown in Fig. 5-5 with the following values inserted into the equation.

$L_T = 20\ \text{nH}$ $\quad$ $r_{b'e} = 1000$ ohms

$r_{bb'} = 50$ ohms $\quad$ $C_T = 100\ \text{pF}$

Notice that the chart is normalized for convenience. The actual input impedance of the hypothetical transistor is 1050 ohms at dc and 50 ohms at 112 MHz. Therefore, to find the *actual impedance* of this transistor at any frequency, simply multiply the value found on the chart by 100.

The impedance is presented on the Smith Chart for two reasons. First, and most obvious, is for practice and, second, because of the ease with which both impedance and admittance can be read from the chart at a glance. Most manufacturers, as you will see, use *admittance* parameters rather than *impedance* parameters to describe transistor characteristics on their data sheets. This can sometimes be confusing to the

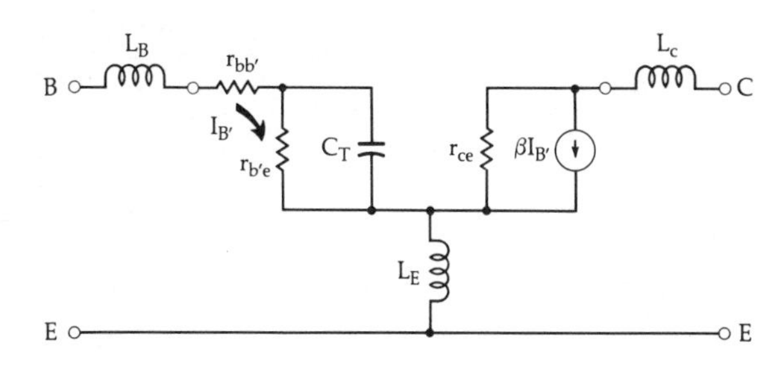

Fig. 5-3. An equivalent circuit using the Miller effect.

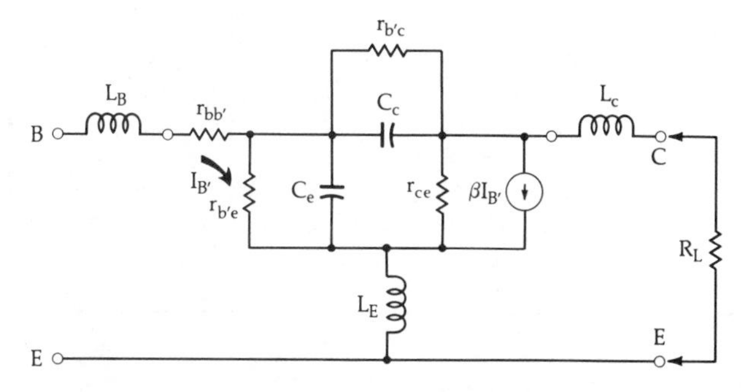

Fig. 5-2. An equivalent circuit including lead inductance.

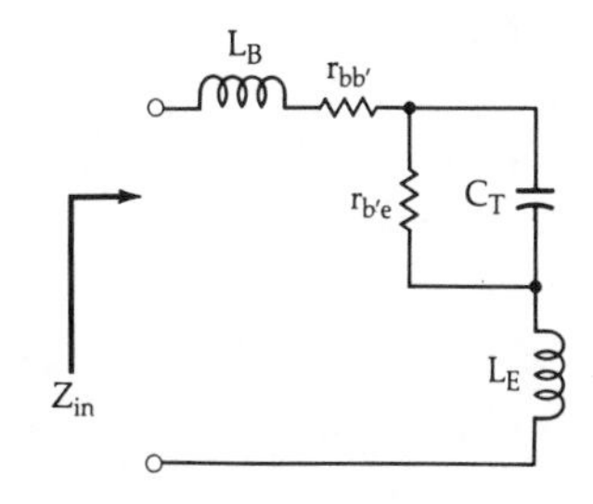

Fig. 5-4. Equivalent input impedance.

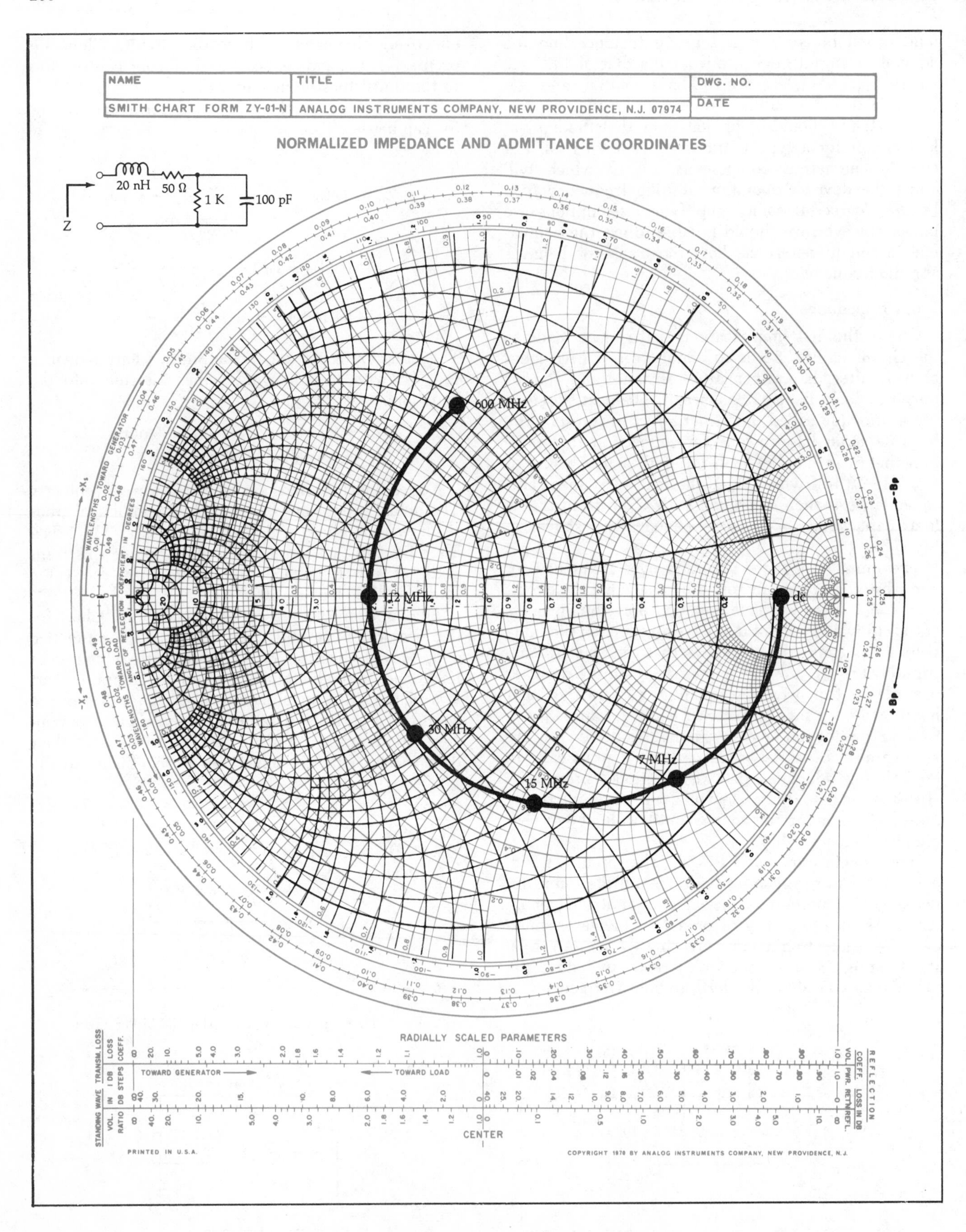

Fig. 5-5. Input impedance vs. frequency.

designer who is not used to working with admittances. However, you will soon be handling both impedance and admittance information equally well.

Output Impedance

The output impedance of a transistor typically decreases with frequency. Let's go back to the original circuit of Fig. 5-1 to see why. We can manipulate Fig. 5-1 in much the same manner as was done in the last section, and can arrive at a convenient circuit that will be useful for an output impedance analysis. Looking into the collector terminal, the first component quantity that we see is r_{ce}, which has a typical value of 100K. This resistance is very large in comparison to the other components in the network and can usually be ignored. The same thing can be said for $r_{b'c}$. This leaves us with the circuit of Fig. 5-6.

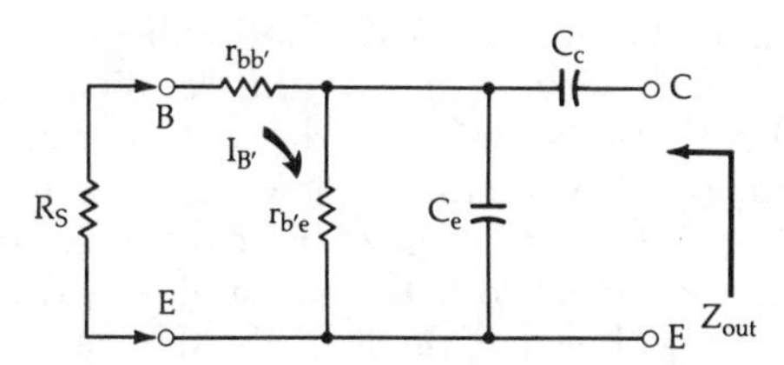

Fig. 5-6. Equivalent output impedance.

The first inclination in an analysis of this circuit would be to assume that C_c and C_e are the determining factors in any output impedance calculation and that they alone cause the output impedance to decrease with frequency. Although C_c and C_e do have an effect on the output impedance of the device, there is another mechanism that is not so obvious that also has quite an effect. This can best be understood if we assume that the transistor is in operation and that some of the collector signal is being fed back to the base through C_c. When this occurs, some of the signal voltage being fed back appears across $r_{b'e}$ causing current to flow in the resistor. This current flow in the base region is amplified by the β of the transistor, thus increasing the collector current. The increase in collector current appears as a decrease in collector impedance. Therefore, even though C_c and C_e act to reduce the output impedance level of the transistor through a decrease in their capacitive reactances, there is also a *hidden* element which tends to further decrease the impedance level beyond that which you would ordinarily expect to find by just looking at the equivalent circuit. Any changes in an external source resistance (R_s) will also change Z_{out}. Increasing R_s decreases Z_{out} because more of the signal current being fed back is forced through $r_{b'e}$.

Feedback Characteristics

The feedback components of the transistor equivalent circuit that is shown in Fig. 5-1 are $r_{b'c}$ and C_c. Of the two, C_c is the most important since it is the element whose value changes with frequency. The quantity $r_{b'c}$, on the other hand, is very large and constant and contributes very little to the feedback characteristics of the device.

As the frequency of operation for a transistor increases, C_c becomes more and more important to the circuit designer because, of course, its reactance is decreasing. Thus, more and more of the collector signal is fed back to the base. At low frequencies, the feedback is usually not much of a problem because C_c, coupled with other stray capacitances located in and around the circuit or circuit-board area, is usually not enough to cause instability. At high frequencies, however, stray reactances coupled with C_c could act to produce a 180° phase shift from collector to base in the fed-back signal. This 180° phase shift, when added to the 180° phase shift that is produced in the normal signal inversion from base-to-collector during amplification, could turn an amplifier into an oscillator very quickly.

Another problem associated with the internal feedback of the transistor is the fact that the collector circuitry is not truly isolated from the base circuitry. Thus, any change in the load resistance of the collector circuitry directly affects the input impedance of the transistor. Or, similarly, any change in the source resistance in the base circuitry directly affects the output impedance of the transistor. This malady is especially important to consider when you are trying to perform an impedance match on both the input and the output of the transistor simultaneously. If, for example, you first match the transistor's input impedance to the source and then match the load to the transistor's output impedance, the output matching network will cause the transistor's input impedance to change from its original value. Therefore, the input matching network is no longer valid and must be redesigned. Once you redesign the input matching network, however, this impedance change will reflect through to the collector causing an output impedance change which invalidates the output matching network. Therefore, if you totally ignore the feedback components in the transistor's equivalent circuit when designing impedance matching networks, you will not obtain a *perfect* match for the transistor. Nevertheless, if C_c is small, the match at both the input and the output might be tolerable in many cases.

It should be pointed out that there is a method for performing a *simultaneous conjugate match* on a transistor while taking into account the effects of C_c. This method is covered in detail in Chapter 6.

Gain

The gain that we are normally interested in for rf transistors is the *power gain* of the device, rather than just the voltage or current gain. It is power gain that is important because of the myriad of impedance levels which abound in rf circuitry. When an impedance level changes in a circuit, voltage and current gains alone no longer mean anything. Even a passive

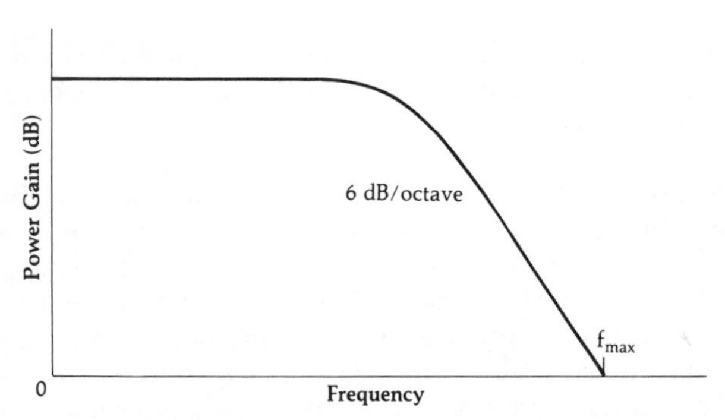

Fig. 5-7. Typical power gain vs. frequency curve.

device can produce a voltage or current gain but it cannot produce both simultaneously. That is what transistors are for—to produce real gain.

The power gain of a transistor typically resembles a curve similar to that shown in Fig. 5-7. This curve is not at all surprising if you again consider the equivalent transistor circuit of Fig. 5-3. Notice that what we have, in effect, is an RC low-pass filter with a gain which must fall off (neglecting lead inductance) at the rate of 6 dB per octave. The maximum frequency at which the transistor provides a power gain is labeled as f_{max} in the diagram. The gain curve passes through f_{max} at 0 dB (Gain = 1), and at the rate of 6 dB per octave.

Gain is usually classified as either unilateralized, neutralized, or unneutralized. *Unilateralized power gain* is defined as the gain available from the transistor when the effects of both feedback components ($r_{b'c}$ and C_c) are negated. Remember that $r_{b'c}$ and C_c are providing negative feedback internal to the transistor and, thus, decrease the gain of the device. Eliminating the negative feedback increases the gain of the transistor. *Neutralized power gain* is that gain which occurs when only the feedback capacitance (C_c) is negated or neutralized. *Unneutralized gain,* on the other hand, occurs when *neither* feedback component is compensated for. Of the three, the unilateralized amplifier produces the most gain and the unneutralized amplifier produces the least. The difference in power gains between the unilateralized case and the neutralized case is usually so small as to be negligible. Thus, neutralization is usually sufficient.

Neutralization is accomplished by providing *external* feedback from the collector to the base of the transistor at just the right amplitude and phase to exactly cancel the internal negative feedback. Further details on this are provided in Chapter 6.

Y PARAMETERS

In Chapter 4, admittance was introduced, with the help of the Smith Chart, as the reciprocal of impedance. It is expressed in the form of $Y = G \pm jB$, where G is conductance or the reciprocal of resistance and B is susceptance or the reciprocal of reactance. Both G and jB are taken to be parallel components as opposed to the series representation ($Z = R \pm jX$) for impedance.

The admittance parameters of a transistor are simply a tool to aid in the unambiguous presentation of the characteristics of the device at a certain frequency and bias point. Or, put another way, they are a method of indicating to a potential user what the transistor "looks like" to something connected to its terminals under certain conditions. Admittance parameters can be used to design impedance-matching networks for the transistor, to determine its maximum available gain, and to determine its stability—or lack thereof. In short, they present a model of the transistor to the designer so that he may best utilize the device in his particular application.

The Transistor as a Two-Port Network

The transistor is obviously a three-terminal device consisting of an emitter, base, and collector. In most applications, however, one of the terminals is common to both the input and the output network as shown in Fig. 5-8. In the common-emitter configuration of Fig. 5-8A, for instance, the emitter is grounded and is thus common to both the input and the output network. So, rather than describe the device as a three-terminal network, it is convenient to describe the transistor in a *black-box* fashion by calling it a two-port network. One port is described as the input port and the other as the output port. This is shown in Fig. 5-8. Once the two-port realization is made, the transistor can be completely characterized by observing its behavior at the two ports.

Two-Port Y Parameters

Two-port admittance parameters can be used to completely characterize the behavior of a transistor at a certain frequency and bias point, and are con-

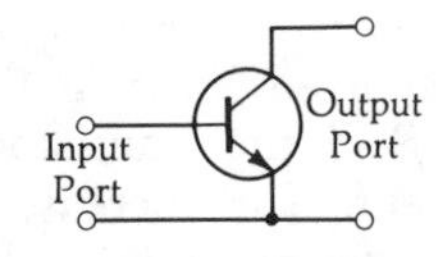

(*A*) *Common emitter.*

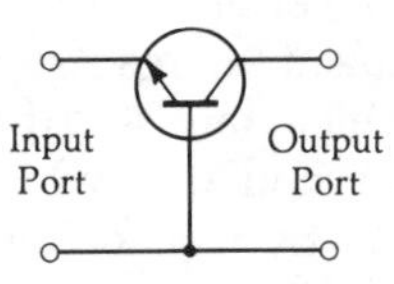

(*B*) *Common base.*

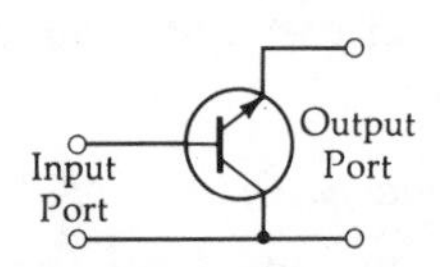

(*C*) *Common collector.*

Fig. 5-8. The three-terminal transistor as a two-port network.

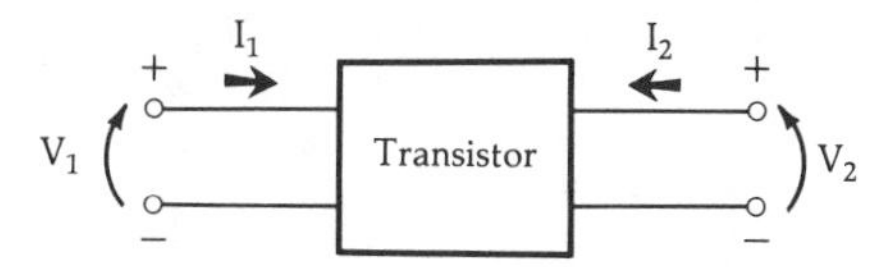

Fig. 5-9. The transistor as a two-port "black box."

sidered to be independent of applied signal level as long as linear operation is maintained. The *black-box* configuration used to create the Y-parameter characterization is shown in Fig. 5-9.

The short-circuit Y parameters for the two-port configuration of Fig. 5-9 are given by:

$$y_i = \frac{I_1}{V_1}\bigg|\ V_2 = 0 \qquad \text{(Eq. 5-1)}$$

$$y_r = \frac{I_1}{V_2}\bigg|\ V_1 = 0 \qquad \text{(Eq. 5-2)}$$

$$y_f = \frac{I_2}{V_1}\bigg|\ V_2 = 0 \qquad \text{(Eq. 5-3)}$$

$$y_o = \frac{I_2}{V_2}\bigg|\ V_1 = 0 \qquad \text{(Eq. 5-4)}$$

where,

y_i = the short-circuit input admittance,
y_r = the short-circuit reverse-transfer admittance,
y_f = the short-circuit forward-transfer admittance,
y_o = the short-circuit output admittance.

The short circuit that is used to make V_1 and V_2 equal to zero is not a dc short circuit, but a short circuit presented at the signal or test frequency. This is usually accomplished by placing a large capacitor across the terminal which requires a short. An examination of Equation 5-1, for instance, reveals that in order to measure y_i in a laboratory, you would first have to connect a large capacitor across the output terminals of the device. This will set V_2 equal to zero. Then, a known signal voltage (V_1) is injected into the input port and a measurement of I_1 is made. The ratio of I_1 to V_1, with their appropriate phase relationships accounted for, is the short-circuit input admittance of the device, which is usually a complex number in the form of $G \pm jB$. Similarly, to measure y_f, simply leave the short circuit in place, inject signal voltage V_1, and measure I_2. The complex ratio of I_2 to V_1 is the short-circuit transfer admittance. Similar methods are used to measure y_o and y_r.

The short circuit is used in the measurement of Y parameters because of the definition of the two-port model. Referring to Fig. 5-9, it is obvious that the current I_1 is dependent upon the voltage (V_1) at the input terminals of the device. But, what might not be so obvious is the fact that I_1 is also dependent upon V_2. This is due to the internal feedback ($R_{b'c}$ and C_c) of the transistor and it must be accounted for. Stated mathematically:

$$I_1 = y_i V_1 + y_r V_2 \qquad \text{(Eq. 5-5)}$$

which simply states that I_1 is dependent upon the input admittance, the reverse-transfer (feedback) admittance, V_1, and V_2. Notice, however, that if we force V_2 equal to zero, I_1 is totally dependent upon V_1 and the input admittance of the device. Or stated another way, the input admittance can be found by injecting V_1 and measuring I_1. A similar argument can be made for y_r if V_1 is forced equal to zero in Equation 5-5. The equation for the output port is

$$I_2 = y_f V_1 + y_o V_2 \qquad \text{(Eq. 5-6)}$$

Equations 5-3 and 5-4 can be derived from Equation 5-6 by alternately setting V_1 and V_2 equal to zero.

Transistor admittance parameter variations with frequency are often published by manufacturers to aid the designer in his design efforts. They are extremely useful, but often very difficult to measure, especially at high frequency. The difficulty arises at high frequencies mainly due to the fact that a good short circuit is difficult to obtain. As we learned in Chapter 1, a capacitor at high frequencies is not a short circuit at all, but presents some reactance at the operating frequency. Obviously, if any reactance creeps into the "short circuit," the voltage at the port in question is no longer zero and our measurement is no longer valid. The higher the impedance at the "shorted" port, the worse our measurement error becomes. There are, of course, other methods besides capacitors for producing short circuits at the test frequency. But they are generally cumbersome, tedious, and time-consuming, and, as such, leave a lot to be desired. Because of the problems associated with finding a true short circuit at high frequencies, the trend in recent years has been to characterize higher frequency transistors in terms of their scattering or S parameters.

S PARAMETERS

Scattering, or S, parameters are another extremely useful design aid that most manufacturers supply for their higher frequency transistors. S parameters are becoming more and more widely used because they are much easier to measure and work with than Y parameters. They are easy to understand, convenient, and provide a wealth of information at a glance.

While Y parameters utilize input and output voltages and currents to characterize the operation of the two-port network, S parameters use normalized incident and reflected traveling waves at each network port. Furthermore, with S parameters, there is no need to present a short circuit to the two-port device. Instead, the network is always terminated in the characteristic impedance of the measuring system. In the majority of measuring systems, this impedance is 50 ohms (purely resistive). The 50-ohm termination requirement is much easier to control than the short-circuit Y-parameter requirement, thus facilitating measurement. In addition, the 50-ohm source and load seen by the two-port network generally forces the device under test, if active, to be stable and not oscillate. This was not always true in a short-circuit measuring system where an active device often does not want to see a short circuit applied to one of its ports. Often such a termination would cause an active device, such as a transistor, to become unstable, thus making measurements impossible. S parameters, therefore, are usually much easier for the manufacturer to measure and, because they are also conceptually easy to understand, are widely used in the design of transistor amplifiers and oscillators.

Transmission Line Background

In order to understand the concept of S parameters, it is necessary to first have a working knowledge of some very simplified transmission line theory. The mathematics have been extensively discussed in the many references cited at the end of the book (Appendix C) and will not be covered here. Instead, you should try to gain an intuitive feel for the incident and reflected traveling waves in a transmission line system.

As shown in Fig. 5-10, voltage, current, or power emanating from a source impedance (Z_s) and delivered to a load (Z_L) can be considered to be in the form of incident and reflected waves traveling in opposite directions along a transmission line of characteristic impedance (Z_o). If the load impedance (Z_L) is exactly equal to Z_o, the incident wave is totally absorbed in the load and there is no reflected wave. If, on the other hand, Z_L differs from Z_o, some of the incident wave is not absorbed in the load but is reflected back toward the source. If the source impedance Z_s were equal to Z_o, the reflected wave from the load would be absorbed in the source and no further reflections would occur. Of course, for a Z_s not equal to Z_o, a portion of the reflected wave from the load is re-reflected from the source back toward the load and the entire process repeats itself perpetually (for a lossless transmission line). The degree of mismatch between Z_o and Z_L, or Z_s, determines the amount of the incident wave that is reflected. The ratio of the reflected wave to the incident wave is known as the *reflection coefficient* and is simply a measure of the quality of the match between the transmission line and the terminating impedances. The reflection coefficient is a complex quantity expressed as a magnitude and an angle in polar form.

$$\Gamma = \text{reflection coefficient}$$
$$= \frac{V_{reflected}}{V_{incident}}$$
$$= \rho \angle\theta \qquad \text{(Eq. 5-7)}$$

As the match between the characteristic impedance of the transmission line and the terminating impedances improves, the reflected wave becomes smaller. Therefore, using Equation 5-7, the reflection coefficient decreases. When a perfect match exists, there is no reflected wave and the reflection coefficient is

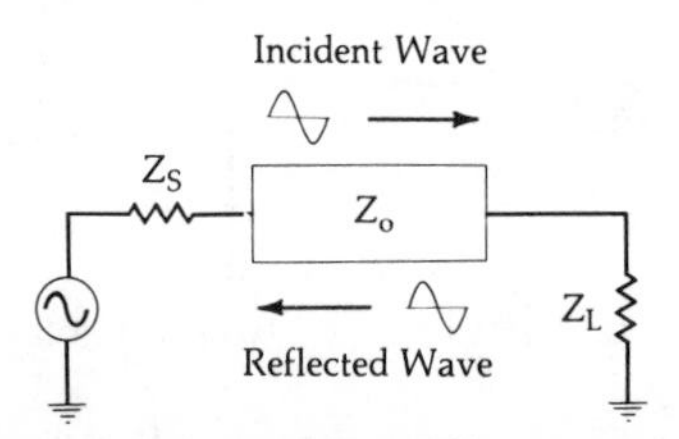

Fig. 5-10. Incident and reflected waves on a transmission line.

zero. If the load impedance, on the other hand, is an open or short circuit, none of the incident power can be absorbed in the load and all of it must be reflected back toward the source. In this case, the reflection coefficient is equal to 1, or a perfect *mismatch*. Thus, the *normal* range of values for the *magnitude* of the reflection coefficient is between zero and one. The reason *normal* is stressed is that in order for the reflection coefficient to be *greater than one*, the magnitude of the reflected wave from a load impedance must be greater than the magnitude of the incident wave to that load. In order for that to occur, it follows that the load in question must be a source of power. This concept is useful in the design of oscillators, but reflection coefficients that are greater than unity, in the input networks of amplifiers, are very bad news.

As we learned in Chapter 4, reflection coefficient can be expressed in terms of the impedances under consideration. For example, the reflection coefficient at the load of the circuit shown in Fig. 5-10 can be expressed as:

$$\Gamma = \frac{Z_L - Z_o}{Z_L + Z_o} \qquad \text{(Eq. 5-8)}$$

Notice that if Z_L is set equal to Z_o in Equation 5-8, the reflection coefficient becomes zero. Conversely, setting Z_L equal to zero (a short circuit), the magnitude of the reflection coefficient goes to unity. Thus, Equation 5-8 holds true for the concepts we have discussed thus far.

Often Equation 5-8 is normalized to the characteristic impedance of the transmission line. Thus, dividing the numerator and denominator of Equation 5-8 by Z_o, we have:

$$\Gamma = \frac{\dfrac{Z_L}{Z_o} - 1}{\dfrac{Z_L}{Z_o} + 1}$$
$$= \frac{Z_n - 1}{Z_n + 1} \qquad \text{(Eq. 5-9)}$$

where,

Z_n is the normalized load impedance.

Equation 5-9 is the same equation that was used in Chapter 4 to develop the Smith Chart. In fact, you will find that reflection coefficients may be plotted directly on the Smith Chart, and the corresponding load impedance read off of the chart immediately—without the need for any calculation using Equations 5-8 or 5-9. The converse is also true. Given a specific characteristic impedance of a transmission line and a load impedance, the reflection coefficient can be read directly from the chart. No calculation is necessary. Example 5-1 illustrates this fact.

The construction performed in Example 5-1 should take less than 30 seconds once you become familiar with the chart. Obviously, an alternate solution would have been to use Equation 5-8 or 5-9 and perform the computation mathematically. Without the aid of a good scientific calculator to perform the complex number manipulation, however, the numerical computation becomes both tedious and time-consuming. That is the reason Mr. Smith developed the chart in the first place—to perform the transformation between impedance and reflection coefficient, and vice-versa, without the need for complex number manipulation.

To find the impedance that gives a certain value of reflection coefficient in a normalized system, you would simply perform the reverse of the construction given in Example 5-1. These procedures are outlined as follows:

1. Draw a line from the center of the chart to the angle of the given reflection coefficient. The normalized impedance is located *somewhere* along that line.
2. From the voltage-reflection coefficient scale located at the bottom of the chart, transfer the value for distance (d), corresponding to the magnitude of the reflection coefficient, to the line drawn in Step 1. Plot this point for a distance (d) from the center of the chart along the line drawn in Step 1.
3. The *normalized* impedance at the point plotted in Step 2 is then read directly from the chart just as any other impedance would be read.

S Parameters and the Two-Port Network

Let us now insert a two-port network between the source and the load in the circuit of Fig. 5-10. This yields the circuit of Fig. 5-13. The following may be said for any traveling wave that originates at the source:

1. A portion of the wave originating from the source and incident upon the two-port device (a_1) will be reflected (b_1) and another portion will be transmitted through the two-port device.
2. A fraction of the *transmitted* signal is then reflected from the load and becomes incident upon the *output* of the two-port device (a_2).
3. A portion of the signal (a_2) is then reflected from the output port back toward the load (b_2), while a fraction is transmitted through the two-port device back to the source.

It is obvious from the above discussion that any traveling wave present in the circuit of Fig. 5-13 is made up of two components. For instance, the total traveling-wave component flowing from the output of the two-port device to the load is actually made up of that portion of a_2 which is reflected from the output of the two-port device, *plus* that portion of a_1 that is transmitted through the two-port device. Similarly, the total traveling wave flowing from the input of the two-port device back toward the source is made up of that portion of a_1 that is reflected from the input port *plus* that fraction of a_2 that is transmitted through the two-port device.

If we set these observations in equation form, just as was done for the Y parameters, we get the following:

$$b_1 = S_{11}a_1 + S_{12}a_2 \qquad \text{(Eq. 5-10)}$$

$$b_2 = S_{21}a_1 + S_{22}a_2 \qquad \text{(Eq. 5-11)}$$

EXAMPLE 5-1

What is the load reflection coefficient for the circuit shown in Fig. 5-11?

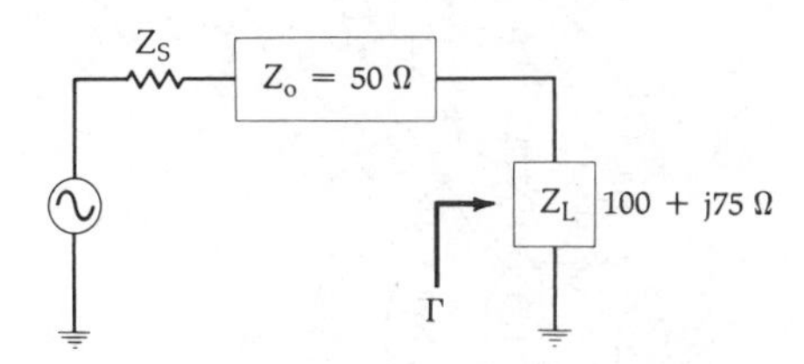

Fig. 5-11. Transmission line circuit for Example 5-1.

Solution

The first step is to normalize the load impedance so that you may plot it on a Smith Chart as was done in Chapter 4. In this case, however, since we are dealing with transmission lines, you must normalize the chart to the characteristic impedance of the line rather than just some convenient number.

$$Z_L = \frac{100 + j75}{50}$$

$$= 2 + j1.5$$

Plot this point on the chart as shown in Fig. 5-12. Draw a line from the center of the chart through the point 2 + j1.5 and extend this line to the outside edge of the chart (which is calibrated in degrees). Note that for clarity, all extraneous scales normally shown around the periphery of the chart have been eliminated. The reflection coefficient can now be read directly from the chart. The distance from the center of the chart to the point Z = 2 + j1.5 is equal to the *magnitude* of the reflection coefficient. To find its numerical value, simply transfer this distance (d) to the *voltage-reflection coefficient* scale located at the bottom of the Smith Chart. This yields a value of 0.54 for the magnitude. To find the angle in degrees, simply read the angle at the intersection of the previously constructed line and the outside edge of the chart. This angle is approximately 29.7°. Thus, the load-reflection coefficient for a load impedance of 100 +j75 ohms in a 50-ohm system is:

$$0.54 \angle 29.7°$$

Continued on next page

EXAMPLE 5-1—Cont.

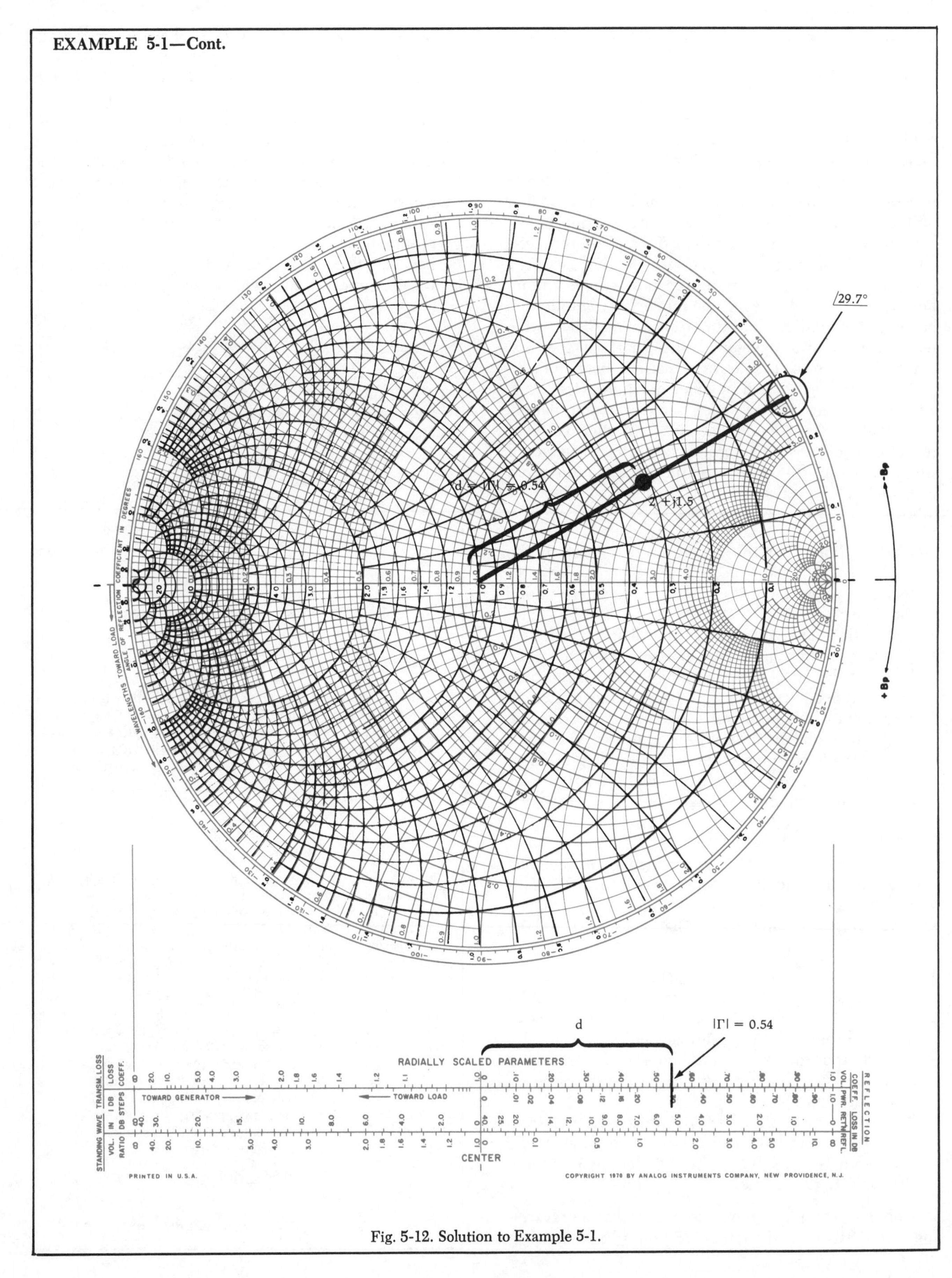

Fig. 5-12. Solution to Example 5-1.

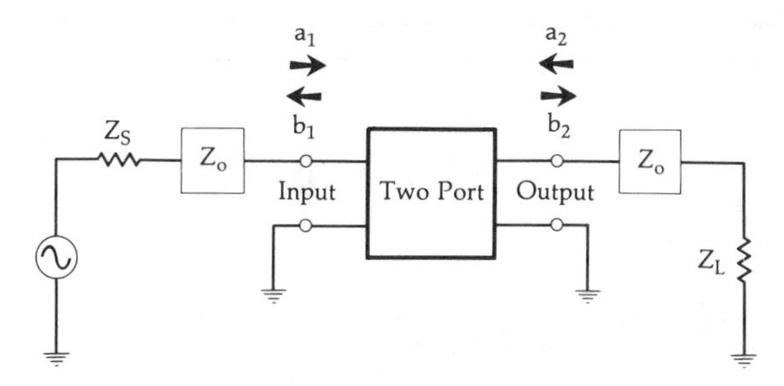

Fig. 5-13. Incident and reflected waves for a two-port device.

where,

S_{11} = the input reflection coefficient,
S_{12} = the reverse transmission coefficient,
S_{21} = the forward transmission coefficient,
S_{22} = the output reflection coefficient.

Notice, in Equation 5-10, that if we set a_2 equal to zero, then,

$$S_{11} = \frac{b_1}{a_1}\bigg|\ a_2 = 0 \qquad \text{(Eq. 5-12)}$$

which is a reflected wave divided by an incident wave and, therefore, by definition, is equal to the input reflection coefficient studied previously. Thus, S_{11} can be plotted on a Smith Chart and the input impedance of the two-port device can be found immediately.

Similarly, using Equation 5-11,

$$S_{22} = \frac{b_2}{a_2}\bigg|\ a_1 = 0 \qquad \text{(Eq. 5-13)}$$

This is also a reflection coefficient and can be plotted on a Smith Chart. Thus, the output impedance of the two-port device can also be found immediately.

The other two S parameters are found as follows:

$$S_{21} = \frac{b_2}{a_1}\bigg|\ a_2 = 0 \qquad \text{(Eq. 5-14)}$$

$$S_{12} = \frac{b_1}{a_2}\bigg|\ a_1 = 0 \qquad \text{(Eq. 5-15)}$$

Notice that Equations 5-12 through 5-15 all require that a_1 or a_2 be set equal to zero in order to measure the individual S parameters. This is easily done by forcing Z_s and Z_L to be equal to the characteristic impedance of the measuring system. Therefore, any wave that is incident upon Z_s or Z_L is totally absorbed and none is reflected back toward the two-port device. For example, let's consider the measurement of the input reflection coefficient, S_{11}. Ideally, we would like to provide an input signal to the two-port device and measure only that fraction of the input signal that is *reflected* back toward the source. In a practical situation, however, some of the incident signal is transmitted through the two-port device, reflected (a_2) from load impedance, and, then, reverse transmitted through the two-port device back to the source. The measured reflected signal is then an aggregate consisting of that portion of a_1, which is reflected, and that portion of a_2, which is transmitted. Obviously, this is not what we need. If Z_L is set equal to Z_o, however, then there is no reflection from the load and the *measured* reflected signal from the input port, divided by the signal incident upon that port, is truly the input reflection coefficient, S_{11}. Similar arguments can be made for the other S parameters to be measured. Therefore, to measure the S parameters of a two-port network, the network is always terminated (source and load) in the characteristic impedance of the measuring system; thus, eliminating all reflections from the terminations.

The significance of S_{21} and S_{12}, as shown in Equations 5-14 and 5-15, is that they are simply the *forward* and *reverse gain* (or *loss*) of the two-port network, respectively, when the two-port device is terminated in the characteristic impedance of the measuring system. These are more meaningful than the equivalent Y parameters y_f and y_r, which were previously studied. Parameter y_f, for instance, is a forward transadmittance and y_r is a reverse transadmittance, neither of which can be intuitively related to an insertion gain or loss for the two-port network.

S parameters, like Y parameters, are simply a convenient method of presenting the characteristics of a device to a potential user. Often, a manufacturer will publish both sets of parameters, along with their variation over frequency, to give the designer the flexibility of working with the parameters with which he feels more comfortable. There will come a time, however, when you will be given only one set of parameters and, as things usually go, it will be the wrong set. If you ever run into this problem, simply refer to the following conversion formulas:

1. $$S_{11} = \frac{(1 - y_i)(1 + y_o) + y_r y_f}{(1 + y_i)(1 + y_o) - y_r y_f}$$

2. $$S_{12} = \frac{-2y_r}{(1 + y_i)(1 + y_o) - y_r y_f}$$

3. $$S_{21} = \frac{-2y_f}{(1 + y_i)(1 + y_o) - y_r y_f}$$

4. $$S_{22} = \frac{(1 + y_i)(1 - y_o) + y_f y_r}{(1 + y_i)(1 + y_o) - y_r y_f}$$

5. $$y_i = \frac{(1 + S_{22})(1 - S_{11}) + S_{12}S_{21}}{(1 + S_{11})(1 + S_{22}) - S_{12}S_{21}} \times \frac{1}{Z_o}$$

6. $$y_r = \frac{-2S_{12}}{(1 + S_{11})(1 + S_{22}) - S_{12}S_{21}} \times \frac{1}{Z_o}$$

7. $$y_f = \frac{-2S_{21}}{(1 + S_{11})(1 + S_{22}) - S_{12}S_{21}} \times \frac{1}{Z_o}$$

8. $$y_o = \frac{(1 + S_{11})(1 - S_{22}) + S_{12}S_{21}}{(1 + S_{22})(1 + S_{11}) - S_{12}S_{21}} \times \frac{1}{Z_o}$$

Notice that if you are converting from Y to S parameters, as in the first four formulas, each individual Y parameter must first be multiplied by Z_o before being substituted into the equations.

2N5179

The RF Line

4.5 dB @ 200 MHz

HIGH FREQUENCY TRANSISTOR

NPN SILICON

NPN SILICON RF HIGH FREQUENCY TRANSISTOR

. . . designed primarily for use in high-gain, low-noise amplifier, oscillator, and mixer applications. Can also be used in UHF converter applications.

- High Current-Gain – Bandwidth Product – f_T = 1.4 GHz (Typ) @ I_C = 10 mAdc
- Low Collector-Base Time Constant – $r_b'C_c$ = 14 ps (Max) @ I_E = 2.0 mAdc
- Characterized with Scattering Parameters
- Low Noise Figure – NF = 4.5 dB (Max) @ f = 200 MHz

*MAXIMUM RATINGS

Rating	Symbol	Value	Unit
Collector-Emitter Voltage Applicable 1.0 to 20 mAdc	V_{CEO}	12	Vdc
Collector-Base Voltage	V_{CB}	20	Vdc
Emitter-Base Voltage	V_{EB}	2.5	Vdc
Collector Current	I_C	50	mAdc
Total Device Dissipation @ T_A = 25°C Derate above 25°C	P_D	200 1.14	mW mW/°C
Total Device Dissipation @ T_C = 25°C Derate above 25°C	P_D	300 1.71	mW mW/°C
Storage Temperature Range	T_{stg}	-65 to +200	°C

*Indicates JEDEC Registered Data.

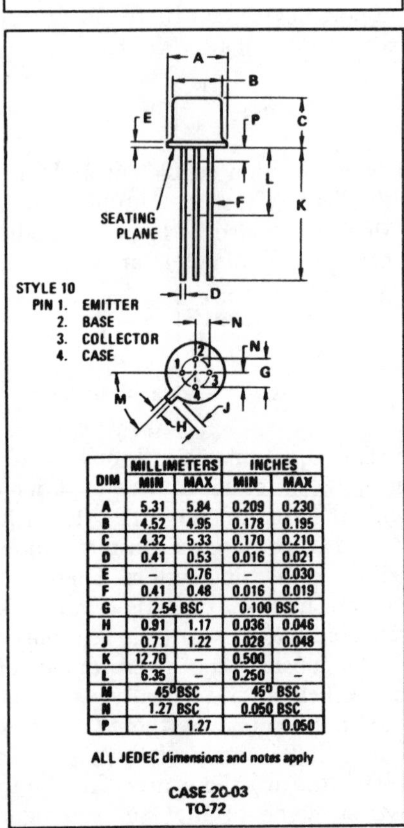

DIM	MILLIMETERS		INCHES	
	MIN	MAX	MIN	MAX
A	5.31	5.84	0.209	0.230
B	4.52	4.95	0.178	0.195
C	4.32	5.33	0.170	0.210
D	0.41	0.53	0.016	0.021
E	–	0.76	–	0.030
F	0.41	0.48	0.016	0.019
G	2.54 BSC		0.100 BSC	
H	0.91	1.17	0.036	0.046
J	0.71	1.22	0.028	0.048
K	12.70	–	0.500	–
L	6.35	–	0.250	–
M	45° BSC		45° BSC	
N	1.27 BSC		0.050 BSC	
P	–	1.27	–	0.050

ALL JEDEC dimensions and notes apply

CASE 20-03
TO-72

Cont. on next page

Fig. 5-14. Data sheet. (*Courtesy Motorola Semiconductor Products Inc.*).

2N5179

*ELECTRICAL CHARACTERISTICS (T_A = 25°C unless otherwise noted)

Characteristic	Symbol	Min	Max	Unit
OFF CHARACTERISTICS				
Collector-Emitter Sustaining Voltage (I_C = 3.0 mAdc, I_B = 0)	$V_{CEO(sus)}$	12	–	Vdc
Collector-Base Breakdown Voltage (I_C = 0.001 mAdc, I_E = 0)	BV_{CBO}	20	–	Vdc
Emitter-Base Breakdown Voltage (I_E = 0.01 mAdc, I_C = 0)	BV_{EBO}	2.5	–	Vdc
Collector Cutoff Current (V_{CB} = 15 Vdc, I_E = 0) (V_{CB} = 15 Vdc, I_E = 0, T_A = 150°C)	I_{CBO}	– –	0.02 1.0	μAdc
ON CHARACTERISTICS				
DC Current Gain (I_C = 3.0 mAdc, V_{CE} = 1.0 Vdc)	h_{FE}	25	250	–
Collector-Emitter Saturation Voltage (I_C = 10 mAdc, I_B = 1.0 mAdc)	$V_{CE(sat)}$	–	0.4	Vdc
Base-Emitter Saturation Voltage (I_C = 10 mAdc, I_B = 1.0 mAdc)	$V_{BE(sat)}$	–	1.0	Vdc
DYNAMIC CHARACTERISTICS				
Current-Gain – Bandwidth Product (1) (I_C = 5.0 mAdc, V_{CE} = 6.0 Vdc, f = 100 MHz)	f_T	900	2000	MHz
Collector-Base Capacitance (V_{CB} = 10 Vdc, I_E = 0, f = 0.1 to 1.0 MHz)	C_{cb}	–	1.0	pF
Small-Signal Current Gain (I_C = 2.0 mAdc, V_{CE} = 6.0 Vdc, f = 1.0 kHz)	h_{fe}	25	300	–
Collector-Base Time Constant (I_E = 2.0 mAdc, V_{CB} = 6.0 Vdc, f = 31.9 MHz)	$r_b'C_c$	3.0	14	ps
Noise Figure (See Figure 1) (I_C = 1.5 mAdc, V_{CE} = 6.0 Vdc, R_S = 50 ohms, f = 200 MHz)	NF	–	4.5	dB
FUNCTIONAL TEST				
Common-Emitter Amplifier Power Gain (See Figure 1) (V_{CE} = 6.0 Vdc, I_C = 5.0 mAdc, f = 200 MHz)	G_{pe}	15	–	dB
Power Output (See Figure 2) (V_{CB} = 10 Vdc, I_E = 12 mAdc, f≥500 MHz)	P_{out}	20	–	mW

*Indicates JEDEC Registered Values.
(1) f_T is defined as the frequency at which $|h_{fe}|$ extrapolates to unity.

MOTOROLA Semiconductor Products Inc.

Cont. on next page

Fig. 5-14.—Cont. Data sheet. (*Courtesy Motorola Semiconductor Products Inc.*).

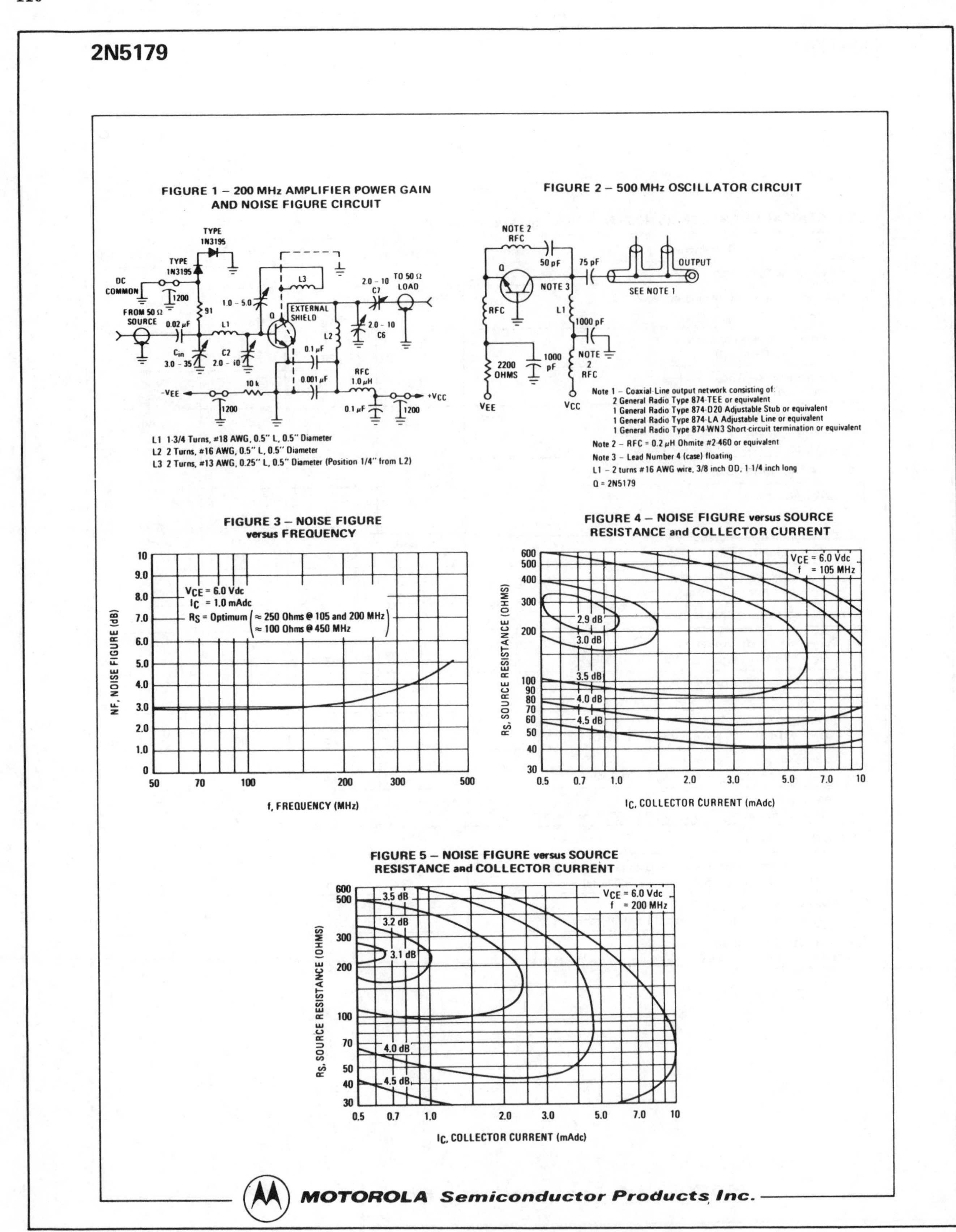

Cont. on next page

Fig. 5-14.—Cont. Data sheet. (*Courtesy Motorola Semiconductor Products Inc.*).

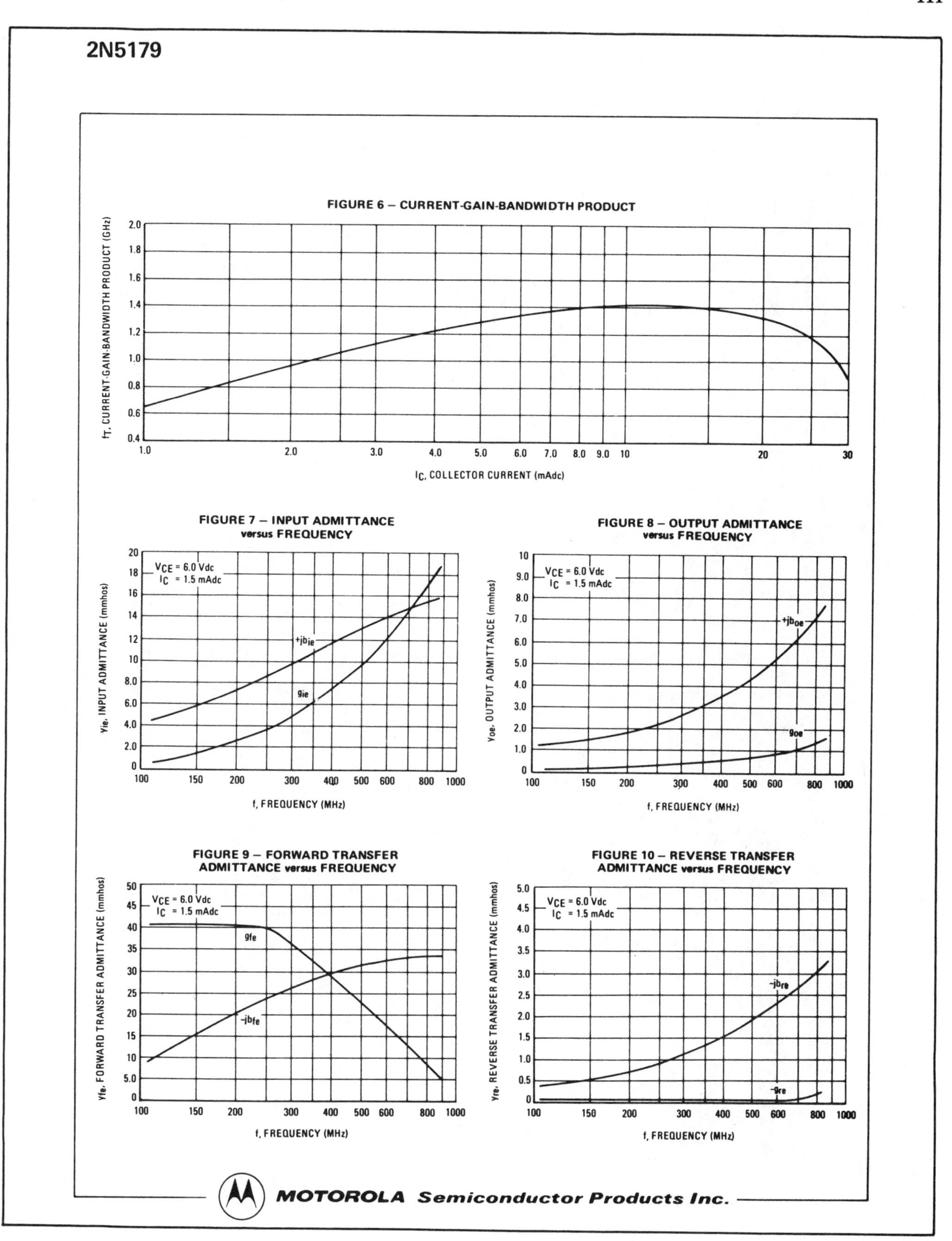

Cont. on next page

Fig. 5-14.—Cont. Data sheet. (*Courtesy Motorola Semiconductor Products Inc.*).

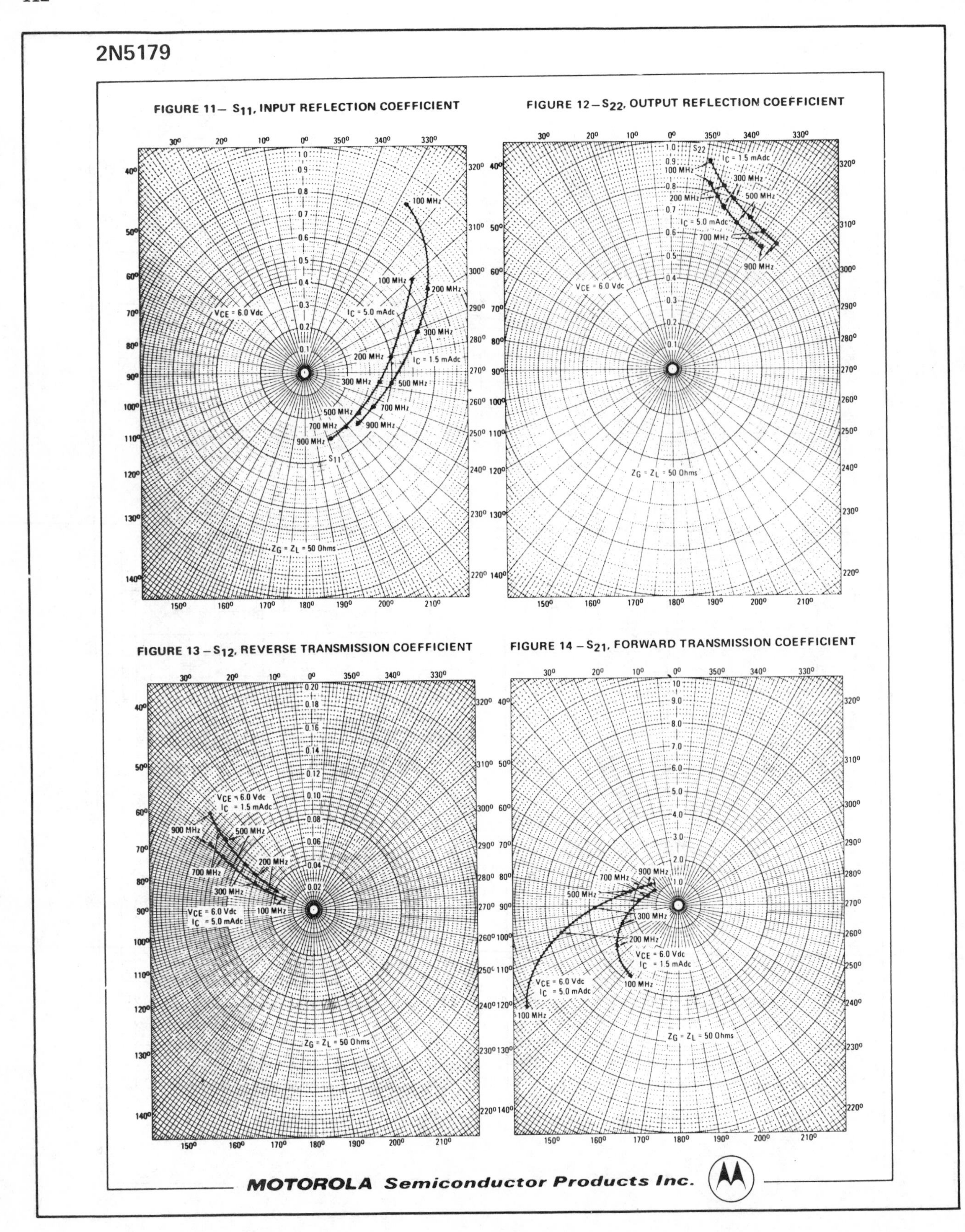

Cont. on next page

Fig. 5-14.—Cont. Data sheet. (*Courtesy Motorola Semiconductor Products Inc.*).

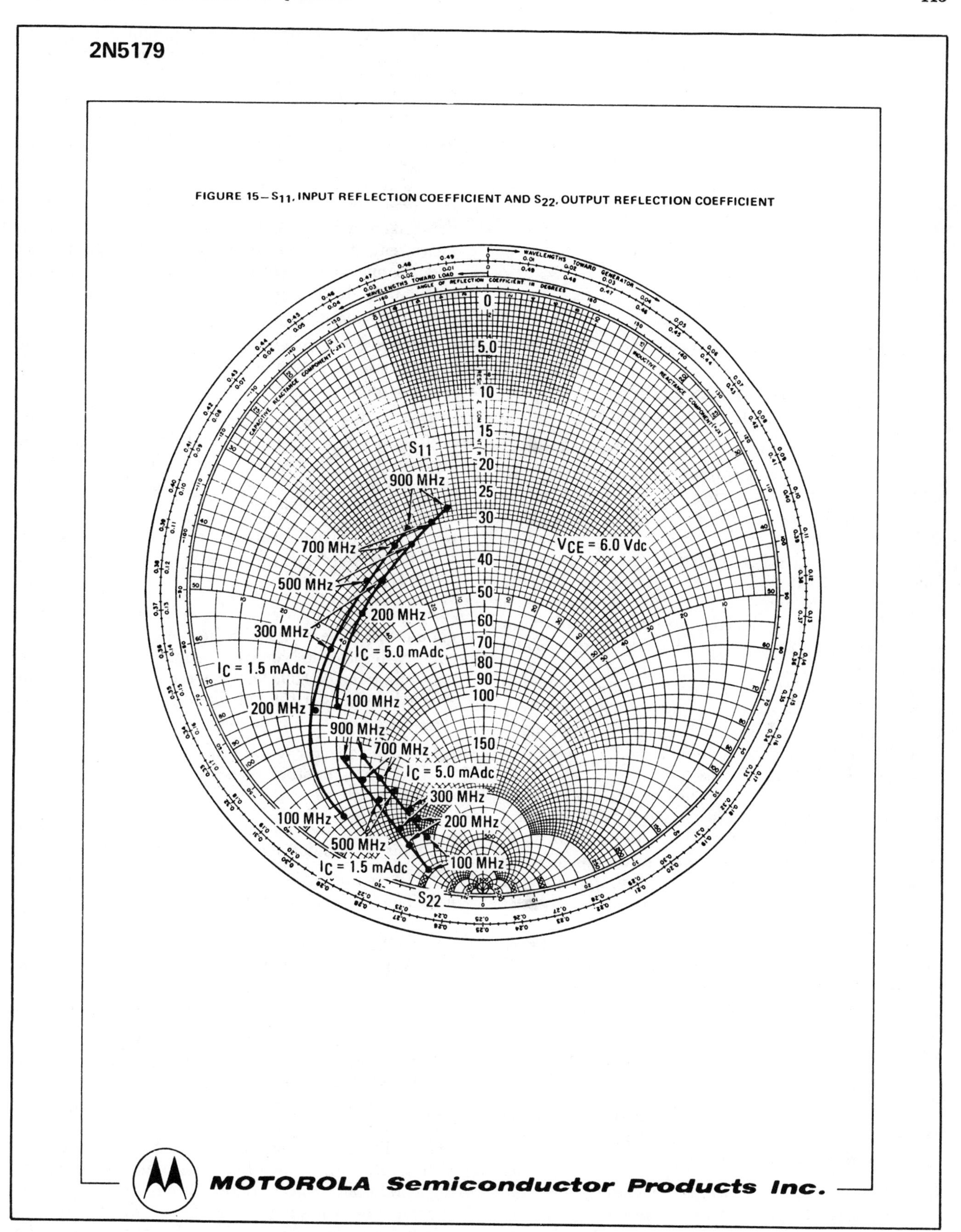

Fig. 5-14.—Cont. Data sheet. (*Courtesy Motorola Semiconductor Products Inc.*).

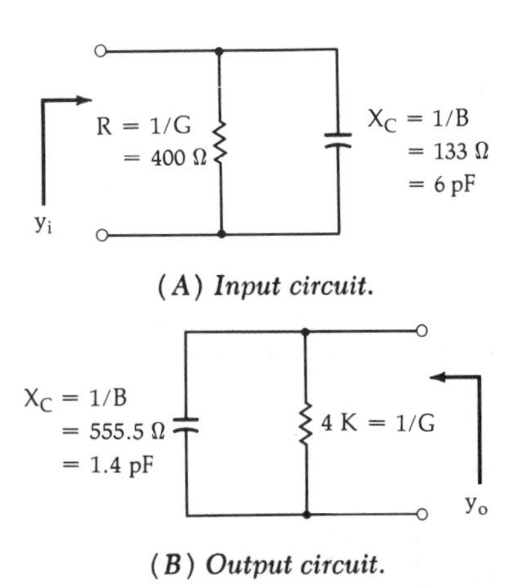

(*A*) *Input circuit.*

(*B*) *Output circuit.*

Fig. 5-15. Equivalent circuit for a 2N5179 (at 200 MHz).

UNDERSTANDING RF TRANSISTOR DATA SHEETS

The rf transistor data sheet is only a bit more complex than that of its low-frequency counterpart. In addition to all of the low-frequency information normally provided on a transistor data sheet, the transistor's rf characteristics, in the form of Y parameters and S parameters and other related information, are included as well. It is assumed that you are already familiar with the low-frequency portion of the data sheet. Therefore, we will concern ourselves only with that information that is typically added specifically for rf transistors.

Fig. 5-14 is a data sheet for the Motorola 2N5179 npn silicon rf high-frequency transistor. This particular transistor was chosen simply because the data sheet provides both Y and S parameters and is, therefore, very good for instructional use.

One of the first things you might notice about this particular transistor is that it has four leads! The extra lead is not connected internally to the device itself, but is connected to the case which just happens to be a metal can. In normal circuit operation, the extra pin is grounded, thus providing a shield around the device to help reduce unwanted stray fields.

The first page of the data sheet is fairly straightforward and provides the *never-exceed* ratings for the transistor. This is a common practice even for low-frequency transistors and is nothing new. Notice that this manufacturer does list those applications in which he feels the transistor might be useful. This particular device was designed for high-gain, low-noise amplifier, oscillator, and mixer applications.

On page 2 of the data sheet, under the heading *Dynamic Characteristics*, several parameters of interest to the rf designer are listed.

f_T—This is called the *transition frequency,* or more commonly, the *gain bandwidth product* of the device. f_T is the theoretical frequency at which the common-emitter current gain (h_{fe}) of the transistor is unity or 0 dB.

Very rarely is f_T used in the rf amplifier-design process except to verify how close you might be to the transistor's upper frequency limit. Keep in mind that f_T is only an indication of the frequency at which the *current* gain of the device drops to 0 dB. Power gain *may* still be possible depending upon the available voltage gain from the device at the frequency in question. Usually f_T is not measured directly for very high-frequency transistors, but is usually extrapolated from data taken at lower frequencies. The accuracy of the measurement is, therefore, somewhat questionable and, as one manufacturer has stated, f_T is provided on the data sheets simply for historical reasons.

C_{cb}—This is the *collector-to-base capacitance* of the transistor as measured at 1 MHz with a collector-to-base voltage of 10 volts and the emitter open-circuited.

The smaller this capacitance is, the better off you will be, if you are using the transistor in an amplifier configuration. This capacitance can be equated to C_c in the transistor equivalent circuit of Fig. 5-1 at the beginning of this chapter.

h_{fe}—This is the *common-emitter current gain* or *beta* of the transistor at the specified low frequency of 1 kHz.

For an rf circuit design, h_{fe} will not do you much good either. The *dc beta* of the transistor (h_{FE}), however, will provide you with needed information in controlling the dc collector or bias current. This parameter is listed under the *On Characteristics* heading for the device (on the second page).

$r_{b'}C_c$—The *collector-to-base time constant* for the transistor is another measure of its feedback characteristics.

The smaller this number is, the better off you will be. This is another bit of information that is often ignored.

NF—The *noise figure* of the transistor is simply a measure of how much noise the transistor adds to the signal during the amplification process (see Appendix B). Notice that for these data sheets, a maximum noise figure of 4.5 dB was measured for the device under a very rigid set of conditions.

Figure 1, on page 3 of the data sheet, was used for the noise figure measurement with the transistor biased at a V_{CE} of 6 volts, $I_C = 1.5$ mA, and the source resistance set equal to 50 ohms. This method of presenting the NF for a transistor, as you can well imagine, is practically useless. Very rarely will the circuit designer ever see the transistor under this exact set of operating conditions. Any variation from these conditions changes the measured noise figure drastically. For this reason, the manufacturer often provides a few noise figure contours which present NF graphically under a wide variety of operating conditions. These contours are shown in Figures 3, 4, and 5 of the data sheet. Figure 3 is a graph of noise figure versus frequency. The NF is measured at various frequencies under the same *bias* conditions. Notice, however, that this measurement was taken with a variable source resistance, where R_S was made equal to its optimum value for a minimum noise figure. The concept of an optimum source resistance for a minimum noise figure is presented in Chapter 6 of this book. Notice that the minimum noise figure increases as the frequency is increased. This is typical of rf transistors.

Figure 4 on the data sheet is a plot of noise figure versus both collector current and source resistance for the transistor at a V_{CE} of 6 volts and an operating frequency of 105 MHz. It is obvious from the diagram that there are an infinite number of R_S, I_C combinations which will provide you with a specified noise figure. For example, the following combinations of R_S and I_C will provide you with a 3.5-dB noise figure:

I_C (mA)	R_S (ohms)
0.5	105 or 600
1.0	90 or 500
1.5	85 or 430
2.0	82 or 390
3.0	81 or 320
5.0	94 or 250

Notice that for each value of collector current there are two values of source resistance that will provide the specified noise figure. Obviously, any variation from the intended bias current or source resistance could change the noise figure drastically.

Figure 5 is simply another set of contours measured at the same bias levels but at a different frequency (200 MHz). If you intended to use the transistor at 300 MHz and wanted to know what bias current and source resistance to use for a specific noise figure, you would be out of luck. There are no noise contours provided for that frequency.

Figure 6, on page 4 of the data sheet, is a graph of f_T versus collector current. Optimum f_T is obtained at the peak of the curve which occurs at approximately 12 mA of collector current. This graph becomes more important at frequencies close to f_T, when you are trying to squeeze every last bit of gain out of the device that you possibly can. It will indicate the optimum collector current at which to operate the device. Once the value of collector current is defined, a sample device could then be biased accordingly and its Y or S parameters measured so that the design could proceed.

Figures 7, 8, 9, and 10 are a graphical presentation of the Y parameters versus frequency for the 2N5179. Measurements were plotted at $V_{CE} = 6$ volts and $I_c = 1.5$ mA. If you prefer a different set of bias conditions, you will have to measure your own Y parameters as no other data is provided.

The vertical axis of each diagram is calibrated in millimhos (mmhos). Therefore, the input admittance (Figure 7) of the 2N5179 at 200 MHz is approximately $y_i = 2.5 + j7.5$ mmhos, which can be represented by the circuit of Fig. 5-15A. Remember that positive susceptance (+jB) indicates a shunt capacitor while negative susceptance (−jB) indicates an inductor. Similarly, the *output* admittance of the transistor at 200 MHz is read in Figure 8 as $y_o = 0.25 + j1.8$ mmhos. The equivalent circuit for this output admittance is shown in Fig. 5-15B. The forward and reverse transfer admittances for the transistor are given in Figures 9 and 10, respectively, on the data sheet.

In Chapter 6, you will learn how to apply the given Y parameter information from data sheets to the design of rf small-signal amplifiers.

Figures 11, 12, 13, 14, and 15 on the data sheet are plots of the transistor's S parameters versus frequency at two different bias levels: $V_{CE} = 6.0$ volts, $I_c = 1.5$ mA and $V_{CE} = 6.0$ volts, $I_c = 5.0$ mA. The four plots shown on page 5 of the data sheet provide S-parameter data in polar form. The radial distance outward from the center of the chart is equal to the magnitude; and the angle is read along the perimeter of the chart. For example, the S parameters for the 2N5179 with a bias of $V_{CE} = 6$ volts, $I_c = 5$ mA, at 100 MHz, are:

$$S_{11} = 0.65 \angle 309^\circ$$

$$S_{22} = 0.84 \angle 348^\circ$$

$$S_{12} = 0.03 \angle 70^\circ$$

$$S_{21} = 8.2 \angle 123^\circ$$

Parameters S_{21} and S_{12} are the forward and reverse gain of the device in magnitude form. To find the gain in dB, simply take the logarithm of the number and multiply it by 20.

$$S_{12}(\text{dB}) = 20 \log_{10} 0.03 = -30.5 \text{ dB}$$

$$S_{21}(\text{dB}) = 20 \log_{10} 8.2 = 18.3 \text{ dB}$$

From the preceding calculations, we can deduce that the output port to input port isolation (S_{12}) of the transistor is very good at −30.5 dB. Also, the gain of the transistor (S_{21}), when driven with a 50-ohm source and terminated in a 50-ohm load (even without impedance matching), is better than 18 dB. Notice that each gain was calculated as a voltage gain. In actuality, voltage and power gains are identical in this instance because the input and the output impedance levels are the same (50 ohms).

Figure 15 on the last page of the data sheet is another plot of the input and output reflection coefficients of the transistor. This time, however, a Smith Chart is used. As was stated previously in this chapter, S_{11} and S_{22} are simply reflection coefficients and can be plotted just like any other reflection coefficient. Once the information is plotted, the input and output impedance of the device can be read directly from the chart.

The chart shown in Figure 15 is a bit different from the charts we have used thus far in the book. It has been normalized to 50 ohms rather than the usual 1 ohm. Thus, the center of the chart now represents $50 \pm j0$ ohms rather than $1 \pm j0$. This type of chart normalization is often used when the impedances that the designer is working with tend to concentrate around a certain value—in this case, 50 ohms.

The input impedance of the 2N5179 as read from the Smith Chart, at 100 MHz with a collector current of 5 mA and $V_{CE} = 6$ volts, is:

$$Z_{in} = 48 - j79 \text{ ohms}$$

This agrees, within reading accuracy, with the polar plot of S_{11} (Figure 11) under the same operating conditions and can be verified numerically by plugging S_{11} in Equation 5-8 for Γ and solving for Z_L.

SUMMARY

The transistor is no different from any other component when it comes to misbehaving at radio frequencies. Like other components, the transistor, too, has stray inductance and capacitance which tends to limit its high-frequency performance. Y and S parameters were devised as a means of presenting this complex transistor behavior over frequency with a minimum of effort. Manufacturers typically present the Y- and S-parameter information in the form of data sheets, which should be considered only as a starting point when used in any rf design task. Manufacturers cannot possibly hope to provide Y- and S-parameter information at every conceivable bias point and in every possible circuit configuration. Instead, they usually try to provide a set of *typical* operating conditions for the device in question. Inevitably, however, the day will come when the data sheet provided with a device is of no use to you whatsoever, and you will find yourself measuring your own parameters and creating your own data sheets in order to complete a design task.

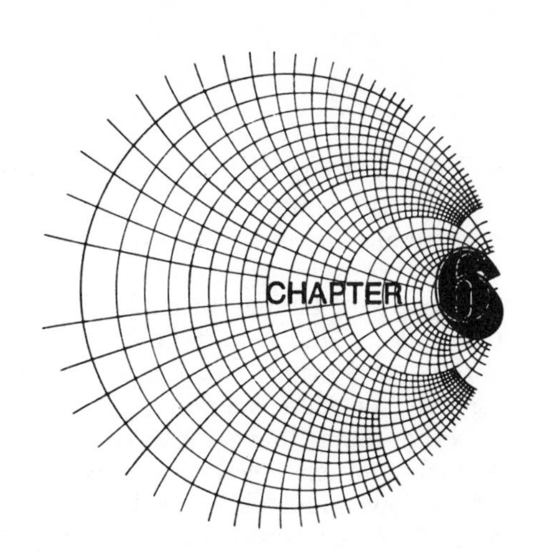

SMALL-SIGNAL RF AMPLIFIER DESIGN

The design of rf small-signal amplifiers is a step-by-step logical procedure with an exact solution for each problem. There are many books available on the market today that offer schematics (complete with parts values) which are "adaptable to any of your circuit needs." That is, a circuit which the author may have designed for a specific set of operating conditions is offered and it may or may not meet our needs. Nonetheless, the design is presented without any design procedure attached, and the reader is left out in the cold when he tries to adapt the circuit to his particular set of operating conditions.

The chapter presented here, however, takes the opposite approach. Detailed step-by-step procedures are followed in the design process so that you can choose the transistor you want and use it under any (realistic) operating conditions that you desire. You will no longer have to adapt someone else's schematic to your needs. Rather you will create your own *homemade* rf amplifiers and optimize them for your personal application.

We will begin our discussion with a very brief overview of transistor biasing. We will discuss both the bipolar and the field-effect transistor (FET). As was shown in the last chapter, the quiescent bias point of a transistor has a great effect on its Y and S parameters. Biasing a transistor is, therefore, serious business and should not be taken lightly.

Next, we'll jump head first into the rf aspect of amplifiers by examining stability (tendency for oscillation), gain, impedance matching, and general amplifier design, with emphasis on the use of Y and S parameters as a design tool.

TRANSISTOR BIASING

In most rf amplifier designs, unfortunately, very little thought is ever given to the design of bias networks for the individual transistors involved. Often, the lack of interest in bias networks may be justified. If, for instance, the amplifier is to be operated only at room temperature, there would be little need to spend much time developing an extremely temperature-stable dc operating point. If, on the other hand, the amplifier must operate reliably and maintain certain specifications (gain, noise figure, etc.) over large temperature extremes, the dc bias network *must* be carefully considered. Consider, for example, the 2N5179 data sheet presented in the last chapter. A quick look at the Y- and S-parameter curves for the device will reveal that a change in the transistor's bias point does in fact change all of its rf operating characteristics. It only stands to reason, then, that the dc operating point must remain stable under your specified operating conditions or the rf characteristics may change drastically.

It has been shown that there are two basic internal transistor characteristics that have a profound effect upon the transistor's dc operating point over temperature; they are ΔV_{BE} and $\Delta\beta$. The object of a good temperature-stable bias design (see Fig. 6-1) is to minimize the effects of these parameters.

As the temperature *increases,* the base-to-emitter voltage (V_{BE}) of a transistor *decreases* at the rate of about 2.5 mV/°C from its nominal room-temperature value of 0.7 volt (for a silicon device). As V_{BE} decreases, more base current is allowed to flow which, in turn, produces more collector current and that is exactly what we would like to prevent. The total change in V_{BE} for a given temperature change is called ΔV_{BE}. The primary external circuit factor that the circuit designer has control over, and which tends to minimize the effects of ΔV_{BE}, is the emitter voltage (V_E) of the transistor. This is shown in Fig. 6-1. Here, a decrease in V_{BE} with temperature would cause an increase in emitter current and, hence, an increase in V_E. The increase in V_E is a form of negative feedback that tends to reverse bias the base-emitter junction and, therefore, *decrease* the collector current. A decrease in V_{BE}, therefore, tends to be counteracted by the increase in V_E, and the collector current does not increase as much with temperature. If these observations were put into equation form, we would have:

$$\Delta I_C \approx -\frac{\Delta V_{BE} I_C}{V_E} \qquad \text{(Eq. 6-1)}$$

where,

ΔI_C = the change in collector current,
I_C = the quiescent collector current,
ΔV_{BE} = the change in base-to-emitter voltage,
V_E = the quiescent emitter voltage.

Thus, if V_E were made equal to 20 times ΔV_{BE}, the collector current would change only 5% over temperature due to ΔV_{BE}. It is important to note that it is the value of the emitter voltage (V_E) and not the value of the emitter resistor (R_E) that is the important bias-design criteria.

Equation 6-1 tends to imply that the higher V_E is, the better. This would be exactly true if we had nothing to worry about except biasing the transistor for the specified operating point. Obviously, there are other things that must be considered in the design. A high emitter voltage, for instance, does tend to waste power and decrease the ac signal gain. A bypass capacitor across R_E at the signal frequency is usually used to prevent the loss in gain, but the wasted power may still be a problem.

If we assume that the amplifier is to operate over a change in temperature of no more than ±50 °C, then an emitter voltage of 2.5 volts will provide a ±5% variation in I_C due to ΔV_{BE}. In fact, you will find that the majority of the transistor bias networks that are similar to Fig. 6-1 will provide a value of V_E from two to four volts depending upon the values of V_{CC} and V_C chosen. Higher values are, of course, possible depending upon the degree of stability you need.

The change in a transistor's dc current gain, or β, over temperature, is also of importance to the circuit designer. Any variation in β will produce a corresponding change in quiescent collector current and will, therefore, disrupt the transistor's designed operating point. The β of a silicon transistor typically *increases* with temperature at the rate of about 0.5% per °C. Thus, for a ±50 °C temperature variation you can expect the β of the transistor and, hence, its collector current to vary as much as ±25%.

Not only does β vary with *temperature*, but the manufacturing tolerance for β among transistors of the same part number is typically very poor. It is not uncommon, for instance, for a manufacturer to specify a 10 to 1 range for β on the data sheet (such as 50 to 500). This, of course, makes it extremely difficult to design a bias network for the device in question when it is to be used in a production environment. Thus, a stable operating point with respect to β is difficult to obtain from a production standpoint as well as from a temperature standpoint.

The change in collector current for a corresponding change in β can be approximated as:

$$\Delta I_C = I_{C1}\left(\frac{\Delta\beta}{\beta_1\beta_2}\right)\left(1+\frac{R_B}{R_E}\right) \qquad \text{(Eq. 6-2)}$$

where,

I_{C1} = the collector current at $\beta = \beta_1$,
β_1 = the lowest value of β,
β_2 = the highest value of β,
$\Delta\beta = \beta_2 - \beta_1$
R_B = the parallel combination of R_1 and R_2 (in Fig. 6-1),
R_E = the emitter resistor.

This equation indicates that once a transistor is specified, the only control that the designer has over the

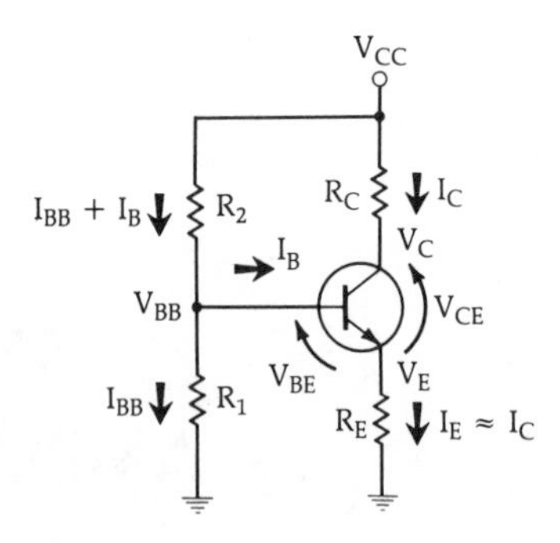

1. Choose the operating point for the transistor.

$$I_C = 10\text{ mA},\ V_C = 10\text{ V},\ V_{CC} = 20\text{ V},\ \beta = 50$$

2. Assume a value for V_E that considers bias stability:

$$V_E = 2.5\text{ volts}$$

3. Assume $I_E \approx I_C$ for high-beta transistors.
4. Knowing I_E and V_E, calculate R_E.

$$R_E = \frac{V_E}{I_E} = \frac{2.5}{10 \times 10^{-3}} = 250\text{ ohms}$$

5. Knowing V_{CC}, V_C, and I_C, calculate R_C.

$$R_C = \frac{V_{CC} - V_C}{I_C} = \frac{20 - 10}{10 \times 10^{-3}} = 1000\text{ ohms}$$

6. Knowing I_C and β, calculate I_B.

$$I_B = \frac{I_c}{\beta} = 0.2\text{ mA}$$

7. Knowing V_E and V_{BE}, calculate V_{BB}.

$$V_{BB} = V_E + V_{BE} = 2.5 + 0.7 = 3.2\text{ volts}$$

8. Assume a value for I_{BB}, the larger the better (see text):

$$I_{BB} = 1.5\text{ mA}$$

9. Knowing I_{BB} and V_{BB}, calculate R_1.

$$R_1 = \frac{V_{BB}}{I_{BB}} = \frac{3.2}{1.5 \times 10^{-3}}\text{ ohms} = 2133\text{ ohms}$$

10. Knowing V_{CC}, V_{BB}, I_{BB}, and I_B, calculate R_2.

$$R_2 = \frac{V_{CC} - V_{BB}}{I_{BB} + I_B} = \frac{20 - 3.2}{1.7 \times 10^{-3}} = 9882\text{ ohms}$$

Fig. 6-1. Bias network design 1.

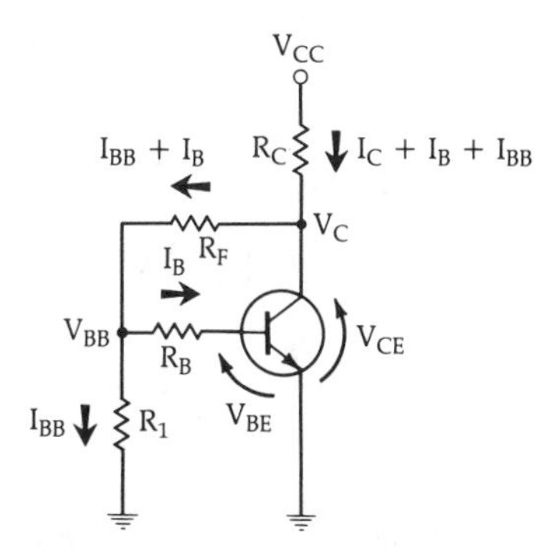

1. Choose the operating point for the transistor.

$$I_C = 10 \text{ mA}, V_C = 10 \text{ V}, V_{CC} = 20 \text{ V}, \beta = 50$$

2. Assume values for V_{BB} and I_{BB} to supply a constant current, I_B.

$$V_{BB} = 2 \text{ volts}$$
$$I_{BB} = 1 \text{ mA}$$

3. Knowing I_C and β, calculate I_B.

$$I_B = \frac{I_C}{\beta} = 0.2 \text{ mA}$$

4. Knowing V_{BB}, $V_{BE} = 0.7$ V, and I_B, calculate R_B.

$$R_B = \frac{V_{BB} - V_{BE}}{I_B} = \frac{2 - 0.7}{0.2 \times 10^{-3}} = 6500 \text{ ohms}$$

5. Knowing V_{BB} and I_{BB}, calculate R_1.

$$R_1 = \frac{V_{BB}}{I_{BB}} = \frac{2}{1 \times 10^{-3}} = 2000 \text{ ohms}$$

6. Knowing V_{BB}, I_{BB}, I_B, and V_C, calculate R_F.

$$R_F = \frac{V_C - V_{BB}}{I_{BB} + I_B} = \frac{10 - 2}{1.2 \times 10^{-3}} = 6667 \text{ ohms}$$

7. Knowing V_{CC}, V_C, I_C, I_B, and I_{BB}, calculate R_C.

$$R_C = \frac{V_{CC} - V_C}{I_C + I_B + I_{BB}} = \frac{20 - 10}{11.2 \times 10^{-3}} = 893 \text{ ohms}$$

Fig. 6-2. Bias network design 2.

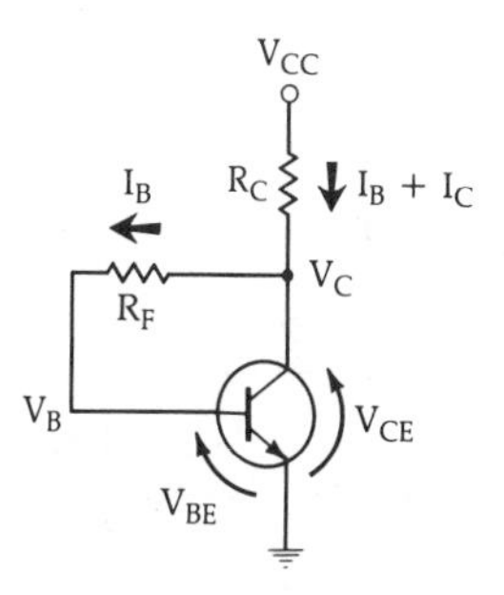

1. Choose the operating point for the transistor (V_C, I_C).

$$I_C = 10 \text{ mA}, V_C = 10 \text{ V}, V_{CC} = 20 \text{ V}, \beta = 50$$

2. Knowing I_C and β, calculate I_B.

$$I_B = \frac{I_c}{\beta} = 0.2 \text{ mA}$$

3. Knowing V_C, $V_B = V_{BE} = 0.7$ V, and I_B, calculate R_F.

$$R_F = \frac{V_C - V_B}{I_B} = \frac{10 - 0.7}{200 \times 10^{-6}} = 46.5\text{K}$$

4. Knowing I_B, I_C, V_{CC}, and V_C, calculate R_C.

$$R_C = \frac{V_{CC} - V_C}{I_B + I_C} = \frac{20 - 10}{10.2 \times 10^{-3}} = 980 \text{ ohms}$$

Fig. 6-3. Bias network design 3.

effect of β changes on collector current is through the resistance ratio R_B/R_E. *The smaller this ratio, the less the collector current varies.* Again, however, some compromise is necessary. As you decrease the ratio R_B/R_E, you also produce the undesirable effect of decreasing the current gain of the amplifier. Also, as the ratio approaches unity, the improvement in operating-point stability rapidly decreases. As a practical rule of thumb for stable designs, the ratio R_B/R_E should be less than 10.

Figs. 6-1, 6-2, and 6-3 indicate three possible bias configurations for bipolar transistors—in order of *decreasing* bias stability. Complete step-by-step design instructions using a typical example are included with each circuit-configuration sketch. Note that the bias networks of Figs. 6-2 and 6-3 do not contain the emitter resistor (R_E) which provides the negative feedback needed to counteract collector-current variations over temperature. Instead, resistor R_F is connected from the collector to the base of the transistor to provide the negative feedback. Obviously, for these two designs, the designer has control over neither the ratio R_B/R_E nor the voltage V_E of Fig. 6-1. The designs are, therefore, of the "pot-luck" variety as far as dc stability is concerned. You basically take what you get. Surprisingly, however, R_F works quite well in minimizing the effects of transistor-parameter variations over temperature.

Figs. 6-4 and 6-5 show similar bias arrangements and design procedures for a field-effect transistor (FET). These are based on the well-known formula:

$$I_D = I_{DSS}\left(1 - \frac{V_{GS}}{V_p}\right)^2 \qquad \text{(Eq. 6-3)}$$

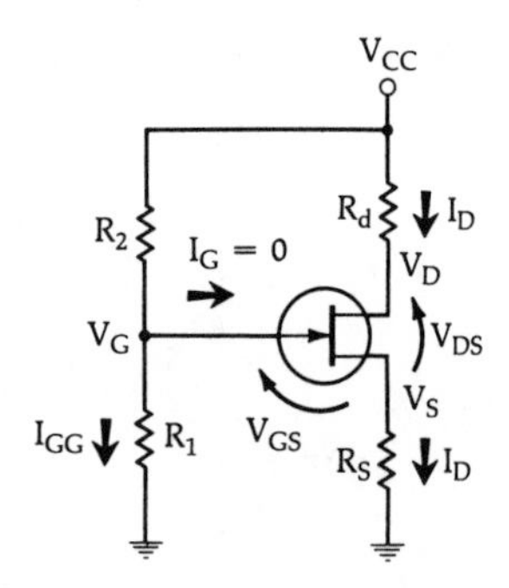

1. Choose the operating point for the transistor.

$$I_D = 10 \text{ mA}, V_D = 10 \text{ V}, V_{CC} = 20 \text{ V}$$

2. Knowing V_{CC}, V_D, and I_D, calculate R_d.

$$R_d = \frac{V_{CC} - V_D}{I_D} = \frac{10 \text{ V}}{10 \text{ mA}} = 1000 \text{ ohms}$$

3. Determine V_p and I_{DSS} from the data sheet.

$$V_p = -6 \text{ volts}$$
$$I_{DSS} = 5 \text{ mA}$$

4. Knowing I_D, I_{DSS}, and V_p, calculate V_{GS}.

$$V_{GS} = V_p\left(1 - \sqrt{\frac{I_D}{I_{DSS}}}\right) = -6\left(1 - \sqrt{\frac{10 \times 10^{-3}}{5 \times 10^{-3}}}\right) = 2.48 \text{ volts}$$

5. Assume a value for V_S in the 2- to 3-volt range.

$$V_S = 2.5 \text{ volts}$$

6. Knowing V_S and I_D, calculate R_S.

$$R_S = \frac{V_S}{I_D} = \frac{2.5}{10 \times 10^{-3}} = 250 \text{ ohms}$$

7. Knowing V_S and V_{GS}, calculate V_G.

$$V_G = V_{GS} + V_S = 2.48 + 2.5 = 4.98 \text{ volts}$$

8. Assume a value for R_1 based upon dc input resistance needs.

$$R_1 = 220\text{K}$$

9. Knowing R_1, V_G, and V_{CC}, calculate R_2.

$$R_2 = \frac{R_1(V_{CC} - V_G)}{V_G} = \frac{220 \times 10^3(20 - 4.98)}{4.98} = 664\text{K}$$

Fig. 6-4. Bias network design 4.

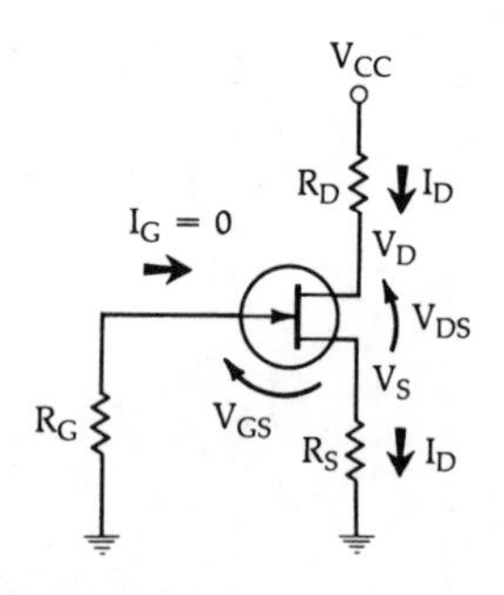

1. Choose an operating point for the transistor.

$$I_D = 10 \text{ mA}, V_D = 10 \text{ V}, V_{CC} = 20 \text{ V}$$

2. Knowing V_{CC}, V_D, and I_D, calculate R_D.

$$R_D = \frac{V_{CC} - V_D}{I_D} = \frac{20 - 10}{10 \times 10^{-3}} = 1000 \text{ ohms}$$

3. Determine V_p and I_{DSS} from the transistor data sheet.

$$V_p = -6 \text{ volts}$$
$$I_{DSS} = 5 \text{ mA}$$

4. Knowing I_D, I_{DSS}, and V_p, calculate V_{GS}.

$$V_{GS} = V_p\left(1 - \sqrt{\frac{I_D}{I_{DSS}}}\right) = -6\left(1 - \sqrt{\frac{10 \times 10^{-3}}{5 \times 10^{-3}}}\right) = 2.48 \text{ volts}$$

5. Knowing $I_G = 0$, $V_{GS} = V_S$, and I_D, calculate R_S.

$$R_S = \frac{V_S}{I_D} = \frac{V_{GS}}{I_D} = \frac{2.48}{10 \times 10^{-3}} = 248 \text{ ohms}$$

6. Since $I_G = 0$, R_G can be chosen to be any large value of resistor—approximately 1 megohm.

Fig. 6-5. Bias network design 5.

where,

I_D = the drain current,
I_{DSS} = the drain current with $V_{GS} = 0$,
V_{GS} = the gate-to-source voltage,
V_p = the pinch-off voltage.

I_D is usually a value chosen by the user as part of the bias specifications, and I_{DSS} and V_p can be found on the data sheet for the transistor. Once these three values are known, Equation 6-3 can be used to solve for V_{GS}, and a suitable bias circuit can then be found.

DESIGN USING Y PARAMETERS

The rf small-signal performance of a transistor can be completely characterized by its two-port admittance parameters. Based on these parameters, equations can be written to aid you both in finding a transistor to suit your needs and in completing the design once the transistor is selected.

One of the first requirements in any amplifier design is to choose the transistor which is best suited for the job. Many rf amplifier designs are doomed from the beginning simply because the active device chosen for the job should have never been considered. Spend a little time shopping for the right device for your application. The more time you spend shopping prior to the start of the actual design, the less hair-pulling there will be later. Two of the most important considerations, in choosing a transistor for use in any amplifier design, are its stability and its maximum available gain (MAG). Stability, as it is used here, is a measure of the transistor's tendency toward oscillation. MAG is a type of figure-of-merit for the transistor, which indicates the maximum theoretical power gain you can expect to obtain from the device when it is *conjugately* matched to its source and load impedance. The MAG is never actually reached in practice; nevertheless, it is quite useful in gauging the capabilities of a transistor.

Stability Calculations

It has been said that one of the easiest methods of building an oscillator is to design an amplifier. Although experience has found this to be true, it really need not be the case. A bit of prior planning and basic *apriori* knowledge about the transistor that is to be used can go a long way toward preventing oscillations in any amplifier design.

It is possible to predict the degree of stability (or lack thereof) of a transistor before you actually place the device in a circuit. This is done through a calculation of the Linvill stability factor, C.

$$C = \frac{|y_r y_f|}{2g_i g_o - \mathrm{Re}(y_r y_f)} \qquad \text{(Eq. 6-4)}$$

where,

$|\ |$ = the magnitude of the product in brackets,
y_r = the reverse-transfer admittance,
y_f = the forward-transfer admittance,
g_i = the input conductance,
g_o = the output conductance,
Re = the real part of the product in parentheses.

When C is *less than 1,* the transistor is *unconditionally stable at the bias point you have chosen.* This means that you could choose any possible combination of source and load impedance for the device, and the amplifier would remain stable *providing* that no external feedback paths exist which have not been accounted for.

If C is *greater than 1,* the transistor is *potentially unstable* and will oscillate for certain values of source and load impedance. A C-factor greater than 1 does not indicate, however, that the transistor cannot be used as an amplifier. It merely indicates that you must exercise extreme care in choosing your source and load impedances or oscillations may occur.

The Linvill stability factor is useful in predicting a *potential* stability problem. It does not indicate the actual impedance values between which the transistor will go unstable. Obviously, if a transistor is chosen for a particular design problem, and the transistor's C-factor is less than 1 (unconditionally stable), it will be much easier to work with than a transistor which is potentially unstable. Keep in mind also that if C is less than *but very close to* 1 for any transistor, then any change in the bias point due to temperature variation could cause the transistor to become potentially unstable and most likely oscillate at some frequency. This is because Y parameters are specified *at a particular bias point* which varies with temperature. This is a very important concept to remember. The smaller C is, the better.

Y parameters can also be used to predict the stability of an amplifier given certain values of load and source impedance. This is called the Stern stability factor and is given by

$$K = \frac{2(g_i + G_S)(g_o + G_L)}{|y_r y_f| + \mathrm{Re}(y_r y_f)} \qquad \text{(Eq. 6-5)}$$

where,

G_S = the source conductance,
G_L = the load conductance.

In this case, if K is *greater than* 1, the circuit will be stable for that value of source and load impedance. If K is *less than* 1, the circuit is potentially unstable and will most likely oscillate at some frequency. Note that the K-factor is a more definitive calculation for stability in that it predicts stability for a particular circuit. The C-factor, on the other hand, predicts a kind of nebulous possibility for instability without giving you an indication as to where the instability may occur.

The Linvill stability factor is, therefore, useful in finding stable transistors while the Stern stability factor predicts possible stability problems with circuits.

Maximum Available Gain

The MAG of a transistor can be found by using the following equation:

$$\mathrm{MAG} = \frac{|y_f|^2}{4g_i g_o} \qquad \text{(Eq. 6-6)}$$

MAG is a useful calculation in the initial search for a transistor for any particular application. It will give you a good indication as to whether or not the transistor can provide enough gain for the task.

The maximum available gain for a transistor occurs when $y_r = 0$, and when Y_L and Y_S are the complex conjugates of y_o and y_i, respectively. The condition that y_r must equal zero for maximum gain to occur is due to the fact that under normal conditions, y_r acts as a *negative* feedback path internal to the transistor. With

$y_r = 0$, no feedback is allowed and the gain is at a maximum.

In practical situations, it is physically impossible to reduce y_r to zero and, as a result, MAG can never truly be attained. It is possible, however, to *very nearly achieve* the MAG calculated in Equation 6-6 through a *simultaneous conjugate match* of the input and output admittance of the transistor. Thus, Equation 6-6 remains a valid tool in the search for a suitable transistor as long as you understand its limitations. For example, if an amplifier design calls for a minimum power gain of 18 dB at 200 MHz, don't choose a transistor with a calculated MAG of 19 dB. Allow yourself a small margin to cover for realistic values of y_r, component losses in the matching network, and variation in the bias point over temperature.

Simultaneous Conjugate Matching (Unconditionally Stable Transistors)

Optimum power gain is obtained from a transistor when y_i and y_o are conjugately matched to Y_S and Y_L, respectively. As was discussed in Chapter 5, however, the reverse-transfer admittance (y_r) associated with each transistor tends to reflect any impedance changes made at one port back toward the other port, causing a change in that port's impedance characteristics. This makes it very difficult to design good matching networks for a transistor while using only its input and output admittances and totally ignoring the contribution that y_r makes to the transistor's impedance characteristics. Even though Y_L affects the input admittance of the transistor and Y_S affects its output admittance, it is still possible to provide the transistor with a simultaneous conjugate match for maximum power transfer (from source to load) by using the following design equations:

$$G_S = \frac{\sqrt{[2g_ig_o - \text{Re}(y_fy_r)]^2 - |y_fy_r|^2}}{2g_o} \quad \text{(Eq. 6-7)}$$

$$B_S = -jb_i + \frac{\text{Im}(y_fy_r)}{2g_o} \quad \text{(Eq. 6-8)}$$

$$G_L = \frac{\sqrt{[2g_ig_o - \text{Re}(y_fy_r)]^2 - |y_fy_r|^2}}{2g_i} \quad \text{(Eq. 6-9)}$$

$$= \frac{G_Sg_o}{g_i} \quad \text{(Eq. 6-10)}$$

$$B_L = -jb_o + \frac{\text{Im}(y_fy_r)}{2g_i} \quad \text{(Eq. 6-11)}$$

where,

G_S = the source conductance,
B_S = the source susceptance,
G_L = the load conductance,
B_L = the load susceptance,
Im = the imaginary part of the product in parentheses.

The above equations may look formidable but actually they are not—once you have used them a few times. Let's try an example of a simultaneous conjugate match for clarification (Example 6-1).

EXAMPLE 6-1

A transistor has the following Y parameters at 100 MHz, with V_{CE} = 10 volts and I_C = 5 mA.

$$y_i = 8 + j5.7 \text{ mmhos}$$
$$y_o = 0.4 + j1.5 \text{ mmhos}$$
$$y_f = 52 - j20 \text{ mmhos}$$
$$y_r = 0.01 - j0.1 \text{ mmho}$$

Design an amplifier which will provide maximum power gain between a 50-ohm source and a 50-ohm load at 100 MHz.

Solution

First, calculate the Linvill stability factor using Equation 6-4.

$$C = \frac{|y_fy_r|}{2g_ig_o - \text{Re}(y_fy_r)}$$

$$= \frac{|(52 - j20)(0.01 - j0.1)|}{2(8)(0.4) - \text{Re}[(52 - j20)(0.01 - j0.1)]}$$

$$= \frac{5.57}{6.4 - (-1.47)}$$

$$= 0.71$$

Since C is less than 1, the device is *unconditionally stable* and we may proceed with the design. Had C been greater than 1, however, we would have had to be extremely careful in matching the transistor to the source and load as instability could occur.

The MAG of this transistor is computed with Equation 6-6:

$$\text{MAG} = \frac{|y_f|^2}{4g_ig_o}$$

$$= \frac{|52 - j20|^2}{4(8)(0.4)}$$

$$= 242.5$$

$$= 23.8 \text{ dB}$$

The actual gain we can achieve will be somewhat less than this due to y_r and component losses.

Using Equations 6-7 through 6-11, calculate the source and load admittances for a simultaneous conjugate match. For the source, using Equation 6-7:

$$G_S = \frac{\sqrt{[2g_ig_o - \text{Re}(y_fy_r)]^2 - |y_fy_r|^2}}{2g_o}$$

$$= \frac{\sqrt{[6.4 + 1.47]^2 - (5.57)^2}}{2(.4)}$$

$$= 6.95 \text{ mmhos}$$

And, with Equation 6-8:

$$B_s = -jb_i + \frac{\text{Im}(y_fy_r)}{2g_o}$$

$$= -j5.7 + j\frac{-5.37}{2(.4)}$$

$$= -j12.41 \text{ mmhos}$$

Continued on next page

EXAMPLE 6-1—Cont.

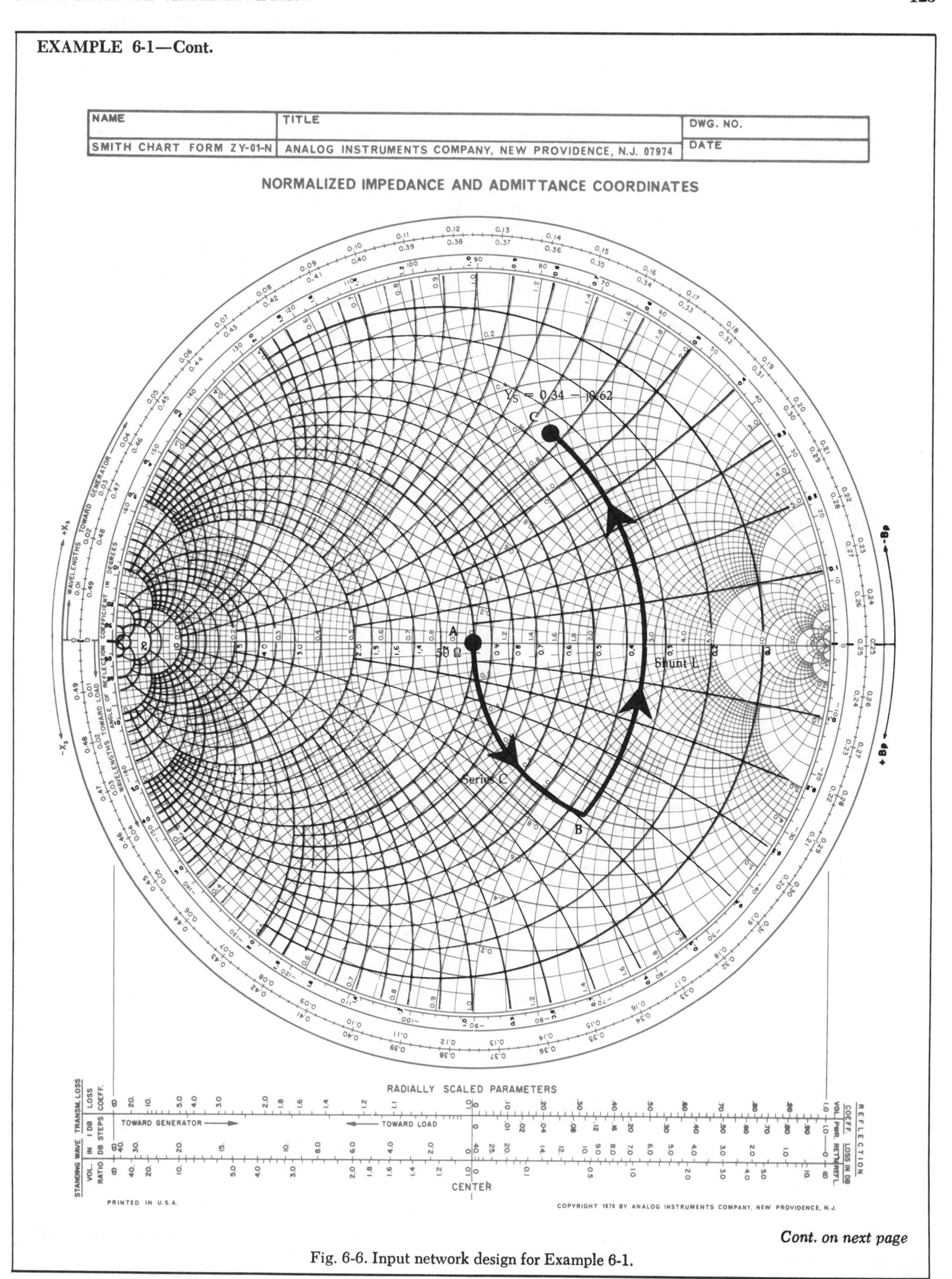

Cont. on next page

Fig. 6-6. Input network design for Example 6-1.

EXAMPLE 6-1—Cont.

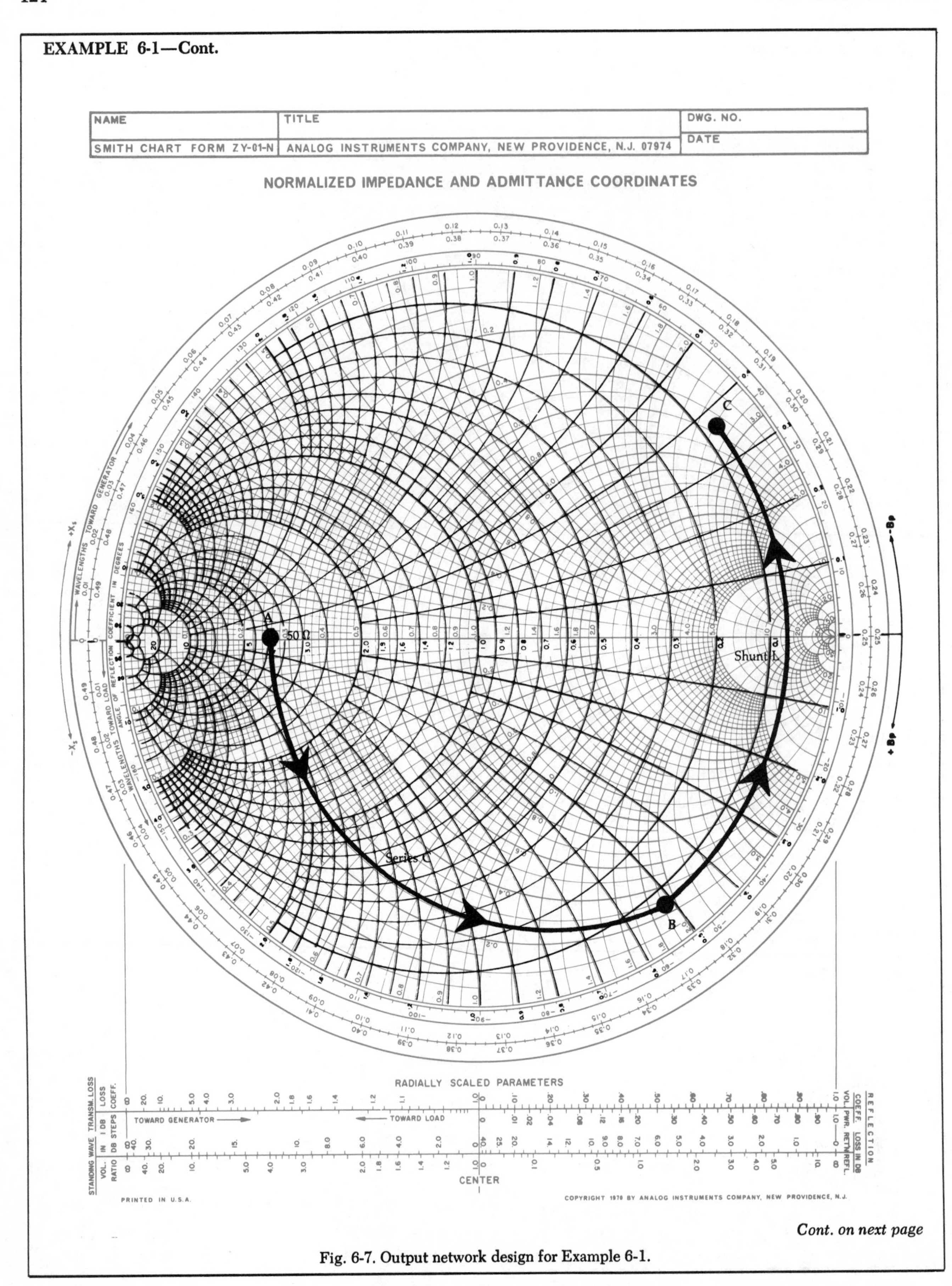

Cont. on next page

Fig. 6-7. Output network design for Example 6-1.

EXAMPLE 6-1—Cont.

Therefore, the source admittance that the transistor must "see" for optimum power transfer is 6.95 − j12.41 mmhos. The transistor's actual input admittance is the conjugate of this number, or 6.95 + j12.41 mmhos. For the load, using Equation 6-10:

$$G_L = \frac{G_S g_o}{g_i}$$
$$= \frac{(6.95)(0.4)}{8}$$
$$= 0.347 \text{ mmho}$$

And, with Equation 6-11:

$$B_L = -jb_o + \frac{\text{Im}(y_f y_r)}{2g_i}$$
$$= -j1.5 + j\frac{-5.37}{2(8)}$$
$$= -j1.84 \text{ mmhos}$$

Thus, for optimum power transfer, the load admittance must be 0.347 − j1.84 mmhos. The actual output admittance of the transistor is the conjugate of the load admittance, or 0.347 + j1.84 mmhos.

The next step is to design the input and output impedance-matching networks that will transform the 50-ohm source and load to the impedance which the transistor would like to see for optimum power transfer. The input matching design is shown on the Smith Chart of Fig. 6-6. This chart is normalized so that the center of the chart represents 50 ohms or 20 mmhos. Thus, the point, Y_S = 6.95 − j12.42 mmhos, is normalized to:

$$Y_S = 50(6.95 - j12.41) \text{ mmhos}$$
$$= 0.34 - j0.62 \text{ mho}$$

This normalized admittance is shown plotted in Fig. 6-6. Note that its corresponding impedance can be read directly from the chart as Z_S = 0.69 + j1.2 ohms. The input matching network must transform the 50-ohm source impedance to the impedance represented by this point. As was discussed in Chapter 4, there are numerous impedance-matching networks available to do the trick. The two-element L network was chosen here for simplicity and convenience.

$$\text{Arc AB} = \text{series C} = -j1.3 \text{ ohms}$$
$$\text{Arc BC} = \text{shunt L} = -j1.1 \text{ mhos}$$

The output circuit is designed and plotted in Fig. 6-7. Because the admittance values needed in the output network are so small, this chart had to be normalized to 200 ohms (5 mmhos). Thus, the normalized admittance plotted on the chart is:

$$Y_L = 200(0.347 - j1.84) \text{ mmhos}$$
$$= 0.069 - j0.368 \text{ mho}$$

or,

$$Z_L = 0.495 + j2.62 \text{ ohms}$$

The normalized 50-ohm load must be transformed to this impedance for maximum transfer of power. Again, the two-element L network was chosen to perform the match.

$$\text{Arc AB} = \text{series C} = -j1.9 \text{ ohms}$$
$$\text{Arc BC} = \text{shunt L} = -j0.89 \text{ mho}$$

The input and output matching networks are shown in Fig. 6-8. For clarity, the bias circuitry is not shown.

Actual component values are found using Equations 4-11 through 4-14. For the input network:

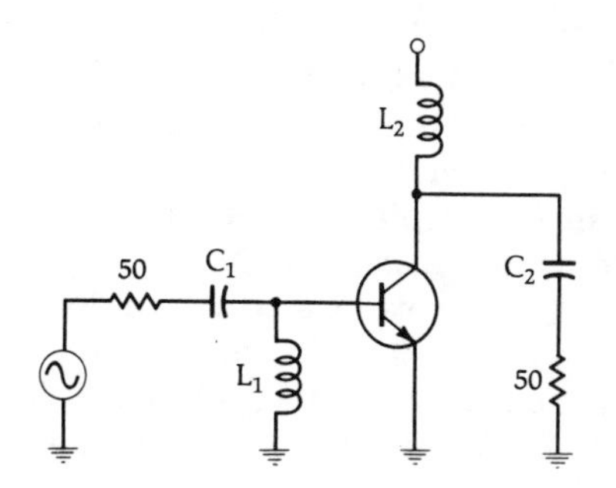

Fig. 6-8. Circuit topology for Example 6-1.

$$C_1 = \frac{1}{\omega XN}$$
$$= \frac{1}{2\pi(100 \times 10^6)(1.3)(50)}$$
$$= 24.5 \text{ pF}$$

and,

$$L_1 = \frac{N}{\omega B}$$
$$= \frac{50}{2\pi(100 \times 10^6)(1.1)}$$
$$= 72 \text{ nH}$$

Similarly, for the output network:

$$C_2 = \frac{1}{2\pi(100 \times 10^6)(1.9)(200)}$$
$$= 4.18 \text{ pF}$$

and,

$$L_2 = \frac{200}{2\pi(100 \times 10^6)(0.89)}$$
$$= 358 \text{ nH}$$

The final circuit, including the bias network, might appear as shown in Fig. 6-9. The 0.1-μF capacitors provide rf bypass at 100 MHz.

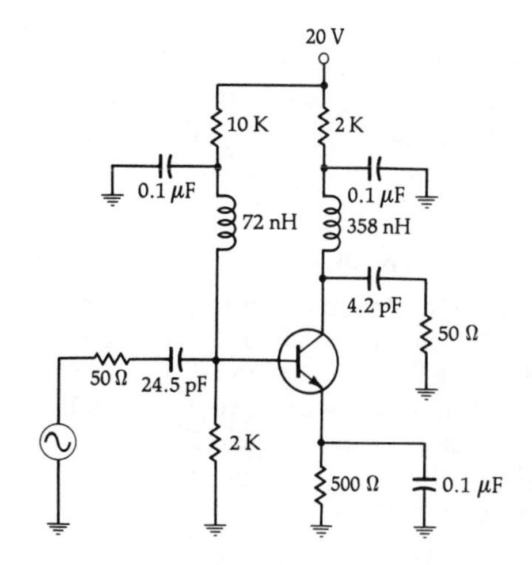

Fig. 6-9. Final circuit for Example 6-1.

Transducer gain

Transducer gain is defined as the output power that is delivered to a load by a source, divided by the maximum power available from the source. This is the gain term most often referenced in rf amplifier design work. Transducer gain includes the effects of input and output impedance matching as well as the contribution that the transistor makes to the overall gain of the amplifier stage. Component resistive losses are neglected.

Given the source admittance (Y_S) and load admittance (Y_L) as seen by the transistor, the transducer gain is given by:

$$G_T = \frac{4G_SG_L|y_f|^2}{|(y_i + Y_S)(y_o + Y_L) - y_fy_r|^2} \quad \text{(Eq. 6-12)}$$

EXAMPLE 6-2

Find the gain of the circuit that was designed in Example 6-1. Disregard any component losses.

Solution

The transducer gain for the amplifier is determined by substituting the values given in Example 6-1 into the Equation 6-12:

$$G_T = \frac{4(6.95)(0.347)}{|(8 + j5.7 + 6.95 - j12.41)(0.4 + j1.5 + 0.347 - j1.84)} \frac{|52 - j20|^2}{- (52 - j20)(0.01 - j0.1)|^2}$$

$$= \frac{29943}{|8.88 - j10.1 + 1.47 + j5.37|^2}$$

$= 231.2$

$= 23.64$ dB

The transducer gain calculated in Example 6-2 is very close to the MAG that was calculated in Example 6-1. Therefore, in this case, the reverse-transfer admittance (y_r) of the transistor has very little effect on the overall gain of the stage. In many instances, however, y_r can take an appreciable toll on gain. For this reason, it is best to calculate G_T once the transistor's load and source admittances are determined. The calculation will provide you with a very good estimate of what the *actual* gain of the amplifier will be.

Designing with Potentially Unstable Transistors

If the Linvill stability factor (C) calculated with Equation 6-4 is greater than 1, the transistor you have chosen is *potentially* unstable and may oscillate under certain conditions of source and load impedance. If this is the case, there are several options available that will enable you to use the transistor in a stable amplifier configuration:

1. Select a new bias point for the transistor.
2. Unilateralize or neutralize the transistor.
3. Selectivity mismatch the input and output impedance of the transistor to reduce the gain of the stage.

The simplest solution to a stability problem is very often Option 1. This is especially true if C calculates to be very close to, but greater than, 1. Remember, any change in a transistor's operating point has a direct effect on its rf characteristics. Therefore, by simply changing the dc bias point, it is possible to change the Y parameters of the transistor and, hence, its stability. Of course, if this approach is taken, it is absolutely critical that the bias point be temperature-stable over the range of temperatures that the device must operate.

Since instability is generally caused by the feedback path, which consists of the reverse-transfer admittance (y_r) of the transistor, unilateralization or neutralization will often stabilize a design. *Unilateralization* consists of providing an external feedback path (Y_f) from the output to the input, such that $Y_f = -y_r$. Thus, Y_f cancels y_r leaving a composite reverse-transfer admittance (y_{rc}) equal to zero. With y_{rc} equal to zero, the device is unconditionally stable. This can be verified by substituting $y_{rc} = 0$ for y_r in Equation 6-4. The Linvill stability factor in this case becomes zero, thus, indicating unconditional stability.

Often, when y_r is a complex admittance consisting of $g_r \pm jb_r$, it becomes very difficult to provide the correct *external* reverse admittance needed to totally eliminate the effect of y_r. In such cases, neutralization is often used. *Neutralization* is similar to unilateralization except that only the imaginary component of y_r is counteracted. An external feedback path is constructed from output to input such that $B_f = -b_r$. Thus, the composite reverse-transfer susceptance (b_{rc}) is equal to zero. Neutralization also tends to tame wild amplifiers because, in most transistors, g_r is negligible when compared to b_r. Thus, the elimination of b_r very nearly eliminates y_r. For this reason, neutralization is generally preferred over unilateralization. Two types of neutralizing circuits are shown in Fig. 6-10. In Fig. 6-10A, the series inductor and capacitor can be tuned to provide the correct amount of negative susceptance (inductance) necessary to cancel a positive reverse-transfer susceptance internal to the transistor. The circuit of Fig. 6-10B can be used to provide the correct amount of external positive susceptance necessary to cancel any $-jb$ that is internal to the transistor.

The addition of external components, in order to neutralize an amplifier, tends to increase the cost and complexity of the circuit. Also, most neutralization circuits tend to neutralize the amplifier at the operating frequency only, and may cause problems (instability) at other frequencies. It is possible, however, to stabilize an amplifier without any form of external feedback. Another look at the *Stern stability factor* (K) in Equation 6-5 will reveal how.

If G_S and G_L are made sufficiently large enough to force K to be *greater than 1,* then the amplifier will

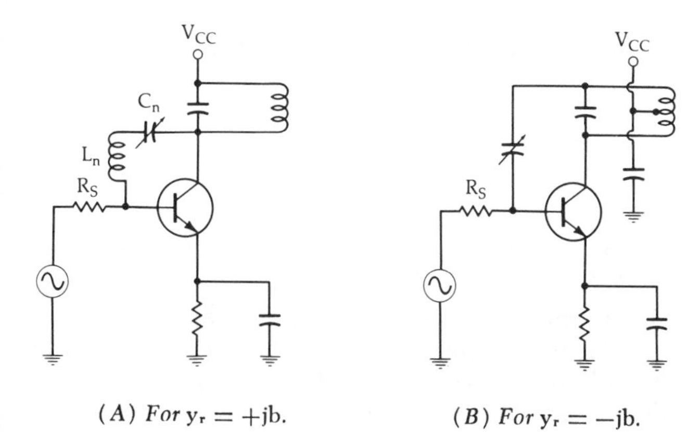

(*A*) *For* $y_r = +jb$. (*B*) *For* $y_r = -jb$.

Fig. 6-10. Neutralization circuits.

remain stable for those terminations. This suggests *selectively mismatching* the transistor to achieve stability. Thus, the gain of the amplifier must be less than that which would be possible with a simultaneous conjugate match. The procedure for a design using unstable devices is as follows:

1. Choose G_S based on the optimum noise-figure information in the transistor's data sheet. Alternately, choose G_S based on some other criteria, such as convenience or input-network Q.
2. Select a value of K that will assure you of a stable amplifier ($K > 1$).
3. Substitute the above values for K and G_S into Equation 6-5 and solve for G_L.
4. Now that G_S and G_L are known, all that remains is to find B_S and B_L. Choose a value of B_L equal to the $-b_o$ of the transistor. The corresponding Y_L, which results, will then be *very close to* the true Y_L that is theoretically needed to complete the design.
5. Next, calculate the transistor's input admittance (Y_{in}) using the load chosen in Step 4 and the formula in Equation 6-13.

$$Y_{in} = y_i - \frac{y_r y_f}{y_o + Y_L} \qquad \text{(Eq. 6-13)}$$

where,

$Y_L = G_L \pm jB_L$ (found in Steps 3 and 4).

6. Once Y_{in} is known, set B_S equal to the *negative* of the imaginary part of Y_{in}, or:

$$B_S = -B_{in}$$

7. Calculate the gain of the stage using Equation 6-12.

From this point forward, it is only necessary to provide input and output networks that will present the calculated Y_S and Y_L to the transistor. Example 6-3 illustrates the procedure.

DESIGN USING S PARAMETERS

As we discussed in Chapter 5, transistors can also be completely characterized by their scattering or S

EXAMPLE 6-3

A 2N5179 transistor has the following Y parameters at 200 MHz:

$$y_i = 2.25 + j7.2$$
$$y_o = 0.4 + j1.9$$
$$y_f = 40 - j20$$
$$y_r = 0.05 - j0.7$$

All of the above parameters are in mmhos. Find source and load admittances that will assure you of a stable design. Find the gain of the amplifier.

Solution

The *Linvill stability factor* (C) for the transistor is equal to 2.27 as calculated using Equation 6-4. Therefore, the device is potentially unstable and you must exercise extreme caution in choosing a source and load admittance for the transistor. Proceed as previously outlined in Steps 1 through 7.

The data sheet for the 2N5179 transistor states that the optimum source resistance for the best noise figure is 250 ohms. Thus, $G_S = 1/R_S = 4$ mmhos. Choose a Stern stability factor of $K = 3$ for an adequate safety margin.

Substitute G_S and K into Equation 6-5 and solve for G_L.

$$K = \frac{2(g_i + G_S)(g_o + G_L)}{|y_r y_f| + \text{Re}(y_r y_f)}$$

$$3 = \frac{2(2.25 + 4)(0.4 + G_L)}{31.35 + (-12)}$$

and,

$$G_L = 4.24 \text{ mmhos}$$

Set B_L equal to $-b_o$ of the transistor,

$$B_L = -j1.9 \text{ mmhos}$$

The load admittance is now defined.

$$Y_L = 4.24 - j1.9 \text{ mmhos}$$

Calculate the input admittance of the transistor using Equation 6-13 and Y_L.

$$Y_{in} = y_i - \frac{y_r y_f}{y_o + Y_L}$$

$$= 2.25 + j7.2 - \frac{(0.701\angle -85.9^\circ)(44.72\angle -26.6^\circ)}{0.4 + j1.9 + 4.24 - j1.9}$$

$$= 4.84 + j13.44 \text{ mmhos}$$

Set B_S equal to the negative of the imaginary part of Y_{in}.

$$B_s = -j13.44 \text{ mmhos}$$

The source admittance needed for the design is now defined as:

$$Y_S = 4.84 - j13.44 \text{ mmhos}$$

Now that Y_S and Y_L are known, you can calculate the expected gain of the amplifier using Equation 6-12.

$$G_T = \frac{4\,(4.84)\,(4.24)\,|\,(44.72)\,|^2}{|\,(7.08 - j6.24)(4.64) - (-12 - j28.96)\,|^2}$$

$$= \frac{135{,}671.7}{2011}$$

$$= 67.61$$

$$= 18.3 \text{ dB}$$

Therefore, even though the transistor is not conjugately matched, you can still realize a respectable amount of gain while maintaining a perfectly stable amplifier. Component values can be found by following the procedures outlined in Example 6-1.

parameters. With these parameters, it is possible to calculate potential instabilities (tendency toward oscillation), maximum available gain, input and output impedances, and transducer gain. It is also possible to calculate optimum source and load impedances either for simultaneous conjugate matching or simply to help you choose specific source and load impedances for a specified transducer gain.

Like Y parameters, S parameters vary with frequency and bias level. Therefore, you must first choose a transistor, select a stable operating point, and determine its S parameters at that operating point (either by measurement or from a data sheet) before following the procedures given in the following sections.

Stability

The tendency of a transistor toward oscillation can be gauged by its S-parameter data in much the same manner as was done in an earlier section with Y parameters. The calculation can be made even before an amplifier is built and, thus, it serves as a useful tool in finding a suitable transistor for your application.

To calculate the stability of a transistor with S parameters, you must first calculate the intermediate quantity D_s:

$$D_s = S_{11}S_{22} - S_{12}S_{21} \qquad \text{(Eq. 6-14)}$$

The Rollett Stability Factor (K) is then calculated as:

$$K = \frac{1 + |D_s|^2 - |S_{11}|^2 - |S_{22}|^2}{2 \cdot |S_{21}| \cdot |S_{12}|} \qquad \text{(Eq. 6-15)}$$

If K is *greater than 1*, then the device will be *unconditionally stable* for any combination of source and load impedance. If, on the other hand, K calculates to be *less than* 1, the device is *potentially unstable* and will most likely oscillate with certain combinations of source and load impedance. With K less than 1, you must be extremely careful in choosing source and load impedances for the transistor. It does not mean that the transistor cannot be used for your application, it merely indicates that the transistor will be *more difficult* to use.

If K calculates to be less than 1, there are several approaches that you can take to complete the design:

1. Select another bias point for the transistor.
2. Choose a different transistor.
3. Follow the procedures outlined later in this chapter.

Maximum Available Gain

The maximum gain you could ever hope to achieve from a transistor under conjugately matched conditions is called the Maximum Available Gain (MAG). To calculate MAG, first calculate the intermediate quantity B_1:

$$B_1 = 1 + |S_{11}|^2 - |S_{22}|^2 - |D_s|^2 \qquad \text{(Eq. 6-16)}$$

where D_s is the quantity calculated using Equation 6-14.

The MAG is then calculated:

$$MAG = 10 \log \frac{|S_{21}|}{|S_{12}|} + 10 \log \left| K \pm \sqrt{K^2 - 1} \right| \qquad \text{(Eq. 6-17)}$$

where,

MAG is in dB,

K is the stability factor calculated using Equation 6-15.

The reason B_1 had to be calculated first is because its polarity determines which sign (±) to use before the radical in Equation 6-17. If B_1 is negative, use the plus sign. If B_1 is positive, use the minus sign.

Note that K must be greater than 1 (unconditionally stable) or Equation 6-17 will be undefined. That is, for a K less than 1, the radical in the equation will produce an *imaginary* number and the MAG calculation is no longer valid. Thus, MAG is undefined for unstable transistors.

Simultaneous Conjugate Match (Unconditionally Stable Transistors)

Once a suitable stable transistor has been found, and its gain capabilities have been found to match your requirements, you can proceed with the design.

The following design procedures will result in load and source reflection coefficients which will provide a conjugate match for the *actual* output and input impedances, respectively, of the transistor. Remember that the actual *output* impedance of a transistor is dependent upon the *source* impedance that the transistor "sees." Conversely, the actual *input* impedance of the transistor is dependent upon the *load* impedance that the transistor "sees." This dependency is, of course, caused by the reverse gain of the transistor (S_{12}). If S_{12} were equal to zero, then, the load and source impedances would have no effect on the transistor's input and output impedances.

To find the desired *load reflection coefficient* for a conjugate match, perform the following calculations:

$$C_2 = S_{22} - (D_s S_{11}^*) \qquad \text{(Eq. 6-18)}$$

where, the asterisk indicates the complex conjugate of S_{11} (same magnitude, but angle has the opposite sign). The quantity D_s is the intermediate quantity as calculated in Equation 6-14.

Next, calculate B_2.

$$B_2 = 1 + |S_{22}|^2 - |S_{11}|^2 - |D_s|^2 \qquad \text{(Eq. 6-19)}$$

The *magnitude* of the reflection coefficient is then found from the equation:

$$|\Gamma_L| = \frac{B_2 \pm \sqrt{B_2^2 - 4|C_2|^2}}{2|C_2|} \qquad \text{(Eq. 6-20)}$$

The sign preceding the radical is the opposite of the sign of B_2 (which was previously calculated in Equation 6-19). The *angle* of the load-reflection coefficient is simply the negative of the angle of C_2 (found in Equation 6-18).

Once the desired load-reflection coefficient is found, it can be plotted on a Smith Chart, and the corresponding load impedance can be found directly. Or, if you prefer, you can substitute Γ_L into Equation 5-8, and solve for Z_L mathematically.

With the desired load-reflection coefficient specified, you can now calculate the source-reflection coefficient that is needed to properly terminate the transistor's input.

$$\Gamma_S = \left[S_{11} + \frac{S_{12}S_{21}\Gamma_L}{1 - (\Gamma_L \bullet S_{22})} \right]^* \qquad \text{(Eq. 6-21)}$$

The asterisk again indicates that you should take the conjugate of the quantity in brackets (same magnitude, but opposite sign for the angle). In other words, once you complete the calculation (within the brackets) of Equation 6-21, the magnitude of the result will be correct, but the angle will have the wrong sign. Simply change the sign of the angle.

Once Γ_S is found, it can either be plotted on a Smith Chart or substituted into Equation 5-8 to find the corresponding source impedance. An example should help clarify matters (Example 6-4).

EXAMPLE 6-4

A transistor has the following S parameters at 200 MHz, with a $V_{CE} = 10$ V and an $I_C = 10$ mA:

$$S_{11} = 0.4 \angle 162^\circ$$
$$S_{22} = 0.35 \angle -39^\circ$$
$$S_{12} = 0.04 \angle 60^\circ$$
$$S_{21} = 5.2 \angle 63^\circ$$

The amplifier must operate between 50-ohm terminations. Design input and output matching networks to simultaneously conjugate match the transistor for maximum gain.

Solution

First use Equations 6-14 and 6-15 to see if the transistor is stable at the operating frequency and bias point:

$$D_s = (0.4 \angle 162^\circ)(0.35 \angle -39^\circ) - (0.04 \angle 60^\circ)(5.2 \angle 63^\circ)$$
$$= 0.14 \angle 123^\circ - 0.208 \angle 123^\circ$$
$$= 0.068 \angle -57^\circ$$

Use the magnitude of D_s to calculate K.

$$K = \frac{1 + (0.068)^2 - (0.4)^2 - (0.35)^2}{2(5.2)(0.04)}$$
$$= 1.74$$

Since K is greater than 1, the transistor is unconditionally stable and we may proceed with the design.

Next, calculate B_1 using Equation 6-16.

$$B_1 = 1 + (0.4)^2 - (0.35)^2 - (0.068)^2$$
$$= 1.03$$

The Maximum Available Gain is then given by Equation 6-17:

$$MAG = 10 \log \frac{5.2}{0.04} + 10 \log | 1.74 - \sqrt{(1.74)^2 - 1} |$$
$$= 21.14 + (-5)$$
$$= 16.1 \text{ dB}$$

The negative sign shown in front of the radical in the above equation results from B_1 being positive.

If the design specification had called out a minimum gain greater than 16.1 dB, a different transistor would be needed. We will consider 16.1 dB adequate for our purposes.

The next step is to find the load-reflection coefficient needed for a conjugate match. The two intermediate quantities (C_2 and B_2) must first be found. From Equation 6-18:

$$C_2 = 0.35 \angle -39^\circ - [(0.068 \angle -57^\circ)(0.4 \angle -162^\circ)]$$
$$= 0.272 - j0.22 - [-0.021 + j0.017]$$
$$= 0.377 \angle -39^\circ$$

and, from Equation 6-19:

$$B_2 = 1 + (0.35)^2 - (0.4)^2 - (0.068)^2$$
$$= 0.958$$

Therefore, the magnitude of the load-reflection coefficient can now be found using Equation 6-20.

$$|\Gamma_L| = \frac{0.958 - \sqrt{(0.958)^2 - 4(0.377)^2}}{2(0.377)}$$
$$= 0.487$$

The angle of the load-reflection coefficient is simply equal to the negative of the angle of C_2, or $+39^\circ$. Thus,

$$\Gamma_L = 0.487 \angle 39^\circ$$

Using Γ_L, calculate Γ_S using Equation 6-21.

$$\Gamma_S = \left[0.4 \angle 162^\circ + \frac{(0.04 \angle 60^\circ)(5.2 \angle 63^\circ)(0.487 \angle 39^\circ)}{1 - (0.487 \angle 39^\circ)(0.35 \angle -39^\circ)} \right]^*$$
$$= [0.522 \angle 162^\circ]^*$$
$$= 0.522 \angle -162^\circ$$

Once the desired Γ_S and Γ_L are known, all that remains is to surround the transistor with components that provide it with source and load impedances which "look like" Γ_S and Γ_L.

The input matching-network design is shown on the Smith Chart of Fig. 6-11. The object of the design is to force the 50-ohm source to present a reflection coefficient of $0.522 \angle -162^\circ$. With Γ_S plotted as shown, the corresponding desired and normalized impedance is read directly from the chart as $Z_S = 0.32 - j0.14$ ohm. Remember, this is a ***normalized*** impedance because the chart has been normalized to 50 ohms. The actual impedance represented by Γ_S is equal to $50(0.32 - j0.14) = 16 - j7$ ohms. To force the 50-ohm source to actually appear as a $16 - j7$ ohm impedance to the transistor, we merely add a shunt and a series reactive component as shown on the chart of Fig. 6-11. Proceeding from the source, we have:

Arc AB = Shunt C = j1.45 mhos
Arc BC = Series L = j0.33 ohm

The actual component values are found using Equations 4-12 and 4-13.

$$C_1 = \frac{1.45}{2\pi(200 \times 10^6)50}$$
$$= 23 \text{ pF}$$

$$L_1 = \frac{(0.33)(50)}{2\pi(200 \times 10^6)}$$
$$= 13 \text{ nH}$$

Continued on next page

EXAMPLE 6-4—Cont.

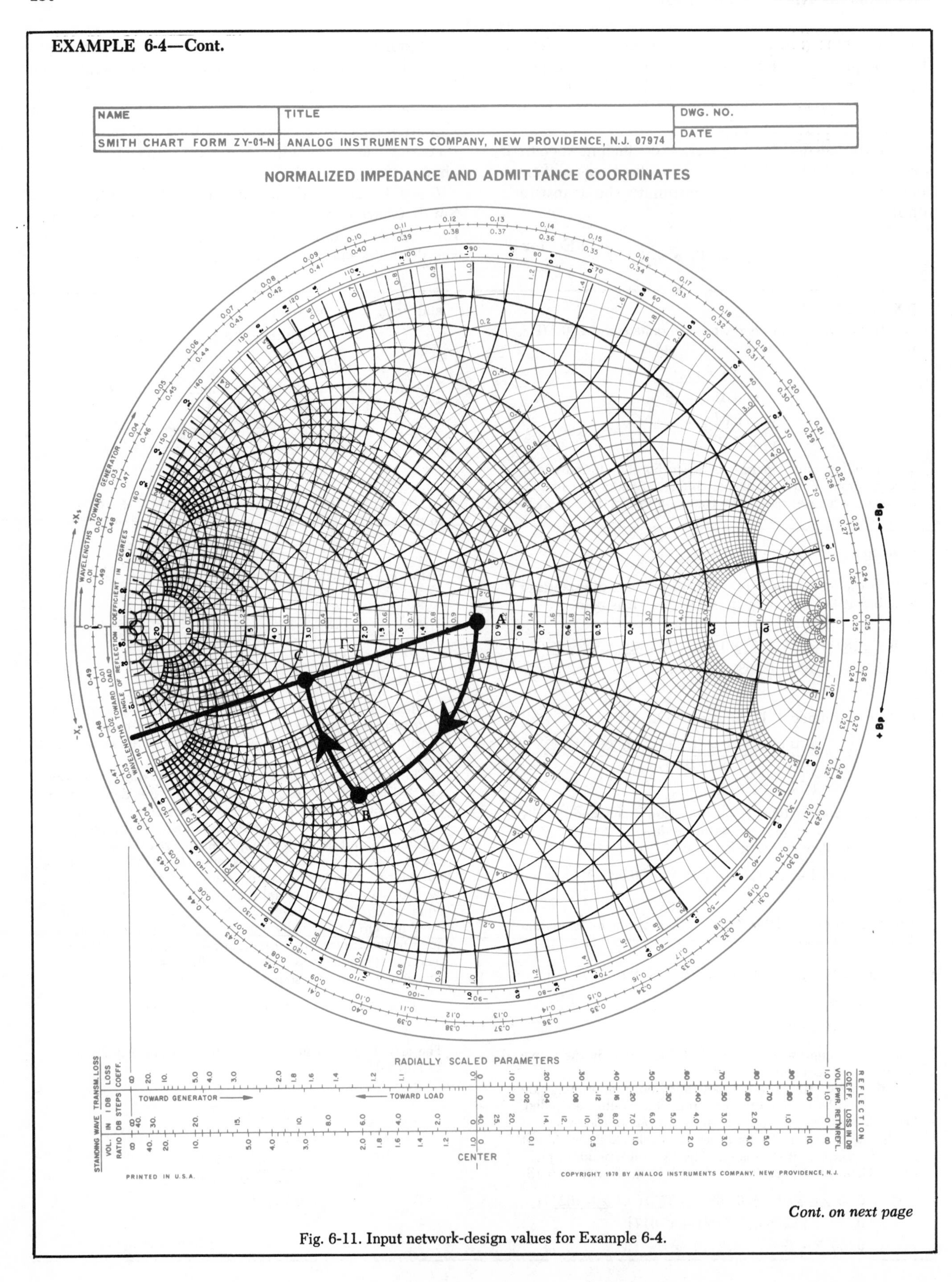

Cont. on next page

Fig. 6-11. Input network-design values for Example 6-4.

EXAMPLE 6-4—Cont.

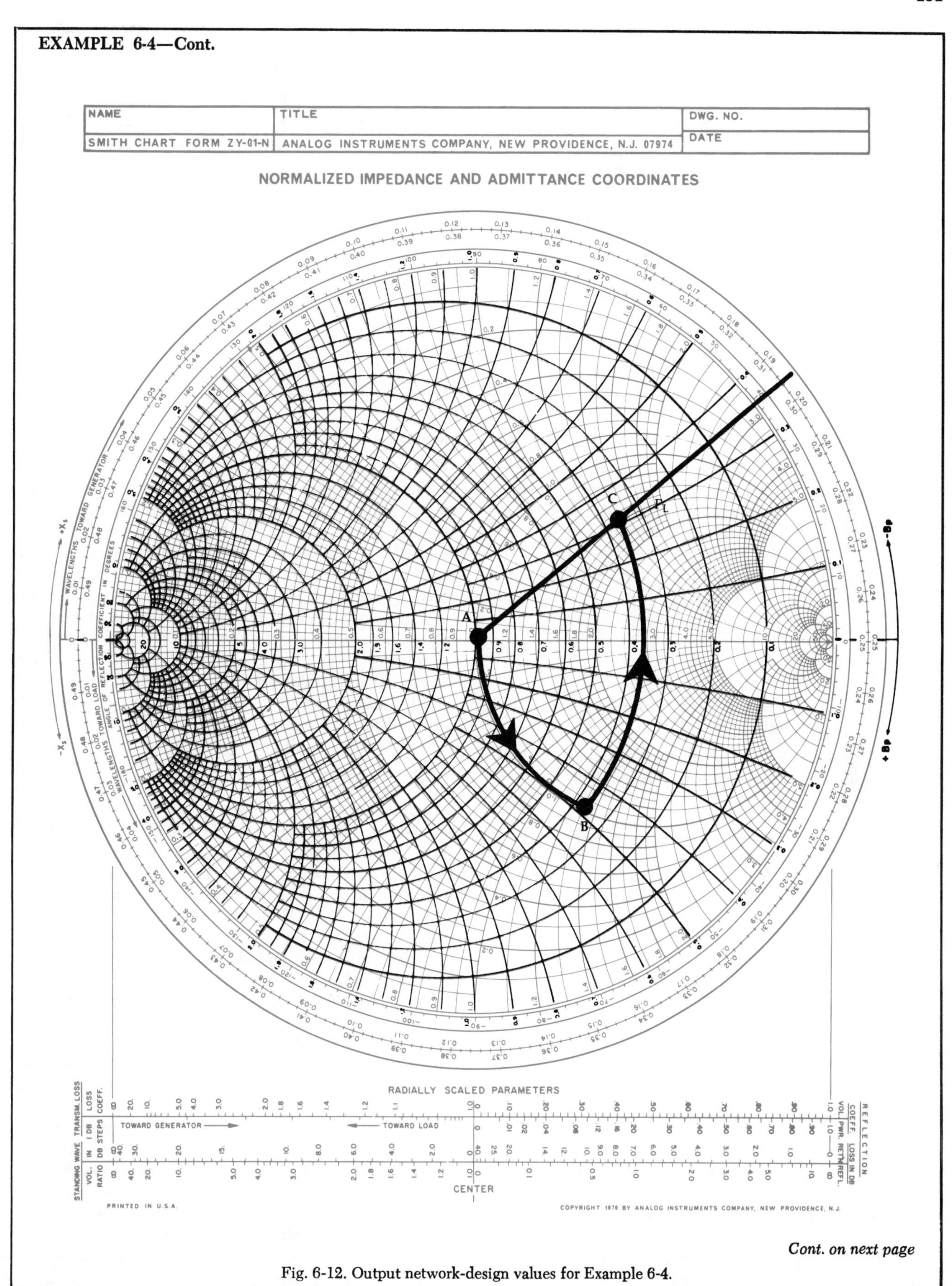

Cont. on next page

Fig. 6-12. Output network-design values for Example 6-4.

EXAMPLE 6-4—Cont.

This completes the input matching network.

The load-reflection coefficient is plotted in Fig. 6-12 and represents a desired load impedance (as read from the chart) of $Z_L = 50\ (1.6 + j1.28)$ ohms, or $80 + j64$ ohms. The matching network is designed as follows. Proceeding from the load:

Arc AB = Series C = −j1.3 ohms
Arc BC = Shunt L = −j0.78 mho

Component values are now found using Equations 4-11 and 4-14.

$$C_2 = \frac{1}{2\pi(200 \times 10^6)(1.3)(50)}$$
$$= 12 \text{ pF}$$

$$L_2 = \frac{50}{2\pi(200 \times 10^6)(0.78)}$$
$$= 51 \text{ nH}$$

The final design, excluding bias circuitry, is shown in Fig. 6-13.

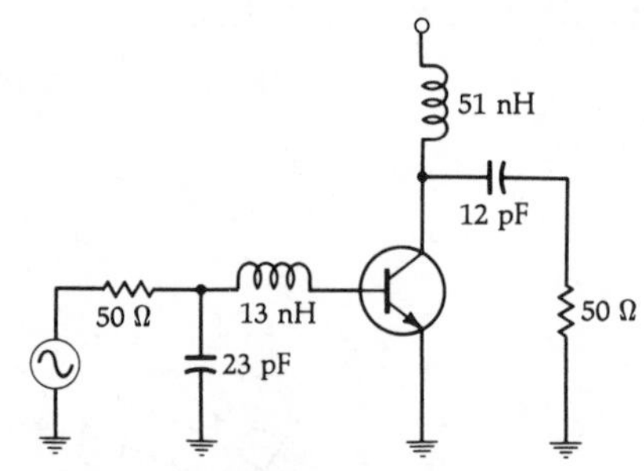

Fig. 6-13. Final circuit for Example 6-4.

Transducer Gain

The transducer gain, as defined earlier in this chapter, is the actual gain of an amplifier stage including the effects of input and output matching and device gain. It does not include losses attributed to power dissipation in imperfect components.

Transducer gain is found by

$$G_T = \frac{|S_{21}|^2(1 - |\Gamma_S|^2)(1 - |\Gamma_L|^2)}{|(1 - S_{11}\Gamma_S)(1 - S_{22}\Gamma_L) - S_{12}S_{21}\Gamma_L\Gamma_S|^2} \quad \text{(Eq. 6-22)}$$

where,

Γ_S and Γ_L are the source- and load-reflection coefficients, respectively.

Calculation of G_T is a useful method of checking the power gain of an amplifier *before* it is built. This is shown by Example 6-5.

EXAMPLE 6-5

Calculate the transducer gain of the amplifier that was designed in Example 6-4.

Solution

Using Equation 6-22, we have:

$$G_T = \frac{(5.2)^2 (1 - (0.522)^2)(1 - (0.487)^2)}{|(1 - 0.2088)(1 - 0.170) - (0.04\angle 60^\circ)(5.2\angle 63^\circ)(0.487\angle 39^\circ)(0.522\angle -162^\circ)|^2}$$
$$= 41.15$$
$$= 16.1 \text{ dB}$$

Notice, again, that the transducer gain calculates to be very close to the MAG. If you carry the calculation out to several decimal places, you will find that G_T is still less than the MAG by a few hundredths of a dB. This is due to the fact that S_{12} is not equal to zero and is, therefore, providing a slight amount of negative feedback internal to the transistor.

Design for a Specified Gain

Often, when designing amplifiers, it is required that a single stage provide a certain amount of gain—no more and no less. In a situation such as this, a simultaneous conjugate match for the transistor would probably provide *too much* gain for the stage and would probably overdrive its load (or the succeeding stage). Obviously, if you so desired, you could search through mountains of manufacturer's literature hoping to find a transistor that, when conjugately matched, would provide exactly the amount of gain desired. This approach could take weeks or even months. Even if you did find a transistor with exactly the gain needed, you are now at the mercy of the manufacturer and are subject to any and all gain variations among transistors of the same type. There is a better way, however, and it alleviates the above problems very easily. It is called *selective mismatching*.

Selective mismatching is simply a controlled manageable way of decreasing gain by not matching the transistor to its load. This may sound like heresy to some, but it is a practical, logical, and well-accepted design procedure. There are still those who believe that at rf frequencies, a transistor *must* be matched to its source and load impedance. This is just not true. A transistor is simultaneously conjugate matched to its source and load only if maximum gain is desired, without regard for any other parameter, such as noise figure and bandwidth.

One of the easiest methods of selectively mismatching a transistor is through the use of a *constant-gain circle* as plotted on a Smith Chart. A constant-gain circle is simply a circle, the circumference of which represents a locus of points (load impedances) that will force the amplifier gain to a specified value. For instance, any of the infinite number of impedances located on the circumference of a 10-dB constant-gain circle would force the amplifier stage gain to 10 dB. Once the circle is drawn on a Smith Chart, you can see the load impedances that will provide a desired gain.

A constant-gain circle is plotted on a Smith Chart by performing a few calculations to determine:

1. Where the center of the circle is located.
2. The radius of the circle.

This information is calculated as follows:

1. Calculate D_s using Equation 6-14.
2. Calculate D_2.

$$D_2 = |S_{22}|^2 - |D_s|^2 \qquad \text{(Eq. 6-23)}$$

3. Calculate C_2.

$$C_2 = S_{22} - D_s S_{11}^* \qquad \text{(Eq. 6-24)}$$

4. Calculate G.

$$G = \frac{\text{Gain desired (absolute)}}{|S_{21}|^2} \qquad \text{(Eq. 6-25)}$$

Note that the numerator in Equation 6-25 must be an absolute gain and not a gain in dB.

5. Calculate the location of the center of the circle.

$$r_o = \frac{GC_2^*}{1 + D_2 G} \qquad \text{(Eq. 6-26)}$$

6. Calculate the radius of the circle.

$$p_o = \frac{\sqrt{1 - 2K\,|S_{12}S_{21}|\,G + |S_{12}S_{21}|^2 G^2}}{1 + D_2 G} \qquad \text{(Eq. 6-27)}$$

Equation 6-26 produces a complex number in magnitude-angle format similar to a reflection coefficient. This number is plotted on the chart exactly as you would plot a value of reflection coefficient.

The radius of the circle that is calculated with Equation 6-27 is simply a fractional number between 0 and 1 which represents the size of that circle in relation to a Smith Chart. A circle with a radius of 1 has the same radius as a Smith Chart, a radius of 0.5 represents half the radius of a Smith Chart, and so on.

Once you choose the load-reflection coefficient and, hence, the load impedance that you will use, the next step is to determine the value of source-reflection coefficient that is needed to complete the design without producing any further decrease in gain. This value of source-reflection coefficient is the conjugate of the *actual input reflection coefficient* of the transistor with the specified load and is given by Equation 6-21. Example 6-6 outlines the procedure to follow.

Stability Circles

When the Rollett stability factor, as calculated with Equation 6-15, indicates a potential instability with the transistor, the chances are that with some combination of source and load impedance, the transistor will oscillate. Therefore, when K calculates to be less than 1, it is extremely important to choose source and load impedances very carefully. One of the best

EXAMPLE 6-6

A transistor has the following S parameters at 250 MHz, with a $V_{CE} = 5$ V and $I_C = 5$ mA.

$$S_{11} = 0.277 \angle -59^\circ$$
$$S_{22} = 0.848 \angle -31^\circ$$
$$S_{12} = 0.078 \angle 93^\circ$$
$$S_{21} = 1.92 \angle 64^\circ$$

Design an amplifier to provide 9 dB of gain at 250 MHz. The source impedance is $Z_S = 35 - j60$ ohms and the load impedance is $Z_L = 50 - j50$ ohms. The transistor is unconditionally stable with K = 1.033.

Solution

Using Equation 6-14 and Equations 6-23 through 6-27, and proceeding "by the numbers," we have:

$$D_s = S_{11}S_{22} - S_{12}S_{21}$$
$$= (0.277 \angle -59^\circ)(0.848 \angle -31^\circ) - (0.078 \angle 93^\circ)(1.92 \angle 64^\circ)$$
$$= 0.324 \angle -64.8^\circ$$

$$D_2 = (0.848)^2 - (0.324)^2$$
$$= 0.614$$

$$C_2 = 0.848 \angle -31^\circ - (0.324 \angle -64.8^\circ)(0.277 \angle 59^\circ)$$
$$= 0.768 \angle -33.9^\circ$$

$$G = \frac{7.94}{(1.92)^2}$$
$$= 2.15$$

The center of the circle is then located at the point:

$$r_o = \frac{2.15(0.768 \angle 33.9^\circ)}{1 + (0.614)(2.15)}$$
$$= 0.712 \angle 33.9^\circ$$

This point can now be plotted on the Smith Chart.

The radius of the 9-dB gain circle is calculated as:

$$p_o = \frac{\sqrt{1 - 2(1.033)(0.078)(1.92)(2.15) + (0.150)^2(2.15)^2}}{1 + (0.614)(2.15)}$$
$$= 0.285$$

The Smith Chart construction is shown in Fig. 6-14. Note that any load impedance located along the circumference of the circle will produce an amplifier gain of 9 dB *if the input impedance of the transistor is conjugately matched.*

The actual load impedance we have to work with is 50 − j50 ohms, as given in the problem statement. Its normalized value (1 − j1) is shown in Fig. 6-14 (point A). The transistor's output network must transform the actual load impedance into a value that falls on the constant-gain circle. Obviously, there are numerous circuit configurations that will do the trick. The configuration shown was chosen for convenience. Proceeding *from* the load:

Arc AB = Series C = −j2 ohms
Arc BC = Shunt L = −j0.425 mho

Again, using Equations 4-11 through 4-14, the actual component values are:

Continued on next page

EXAMPLE 6-6—Cont.

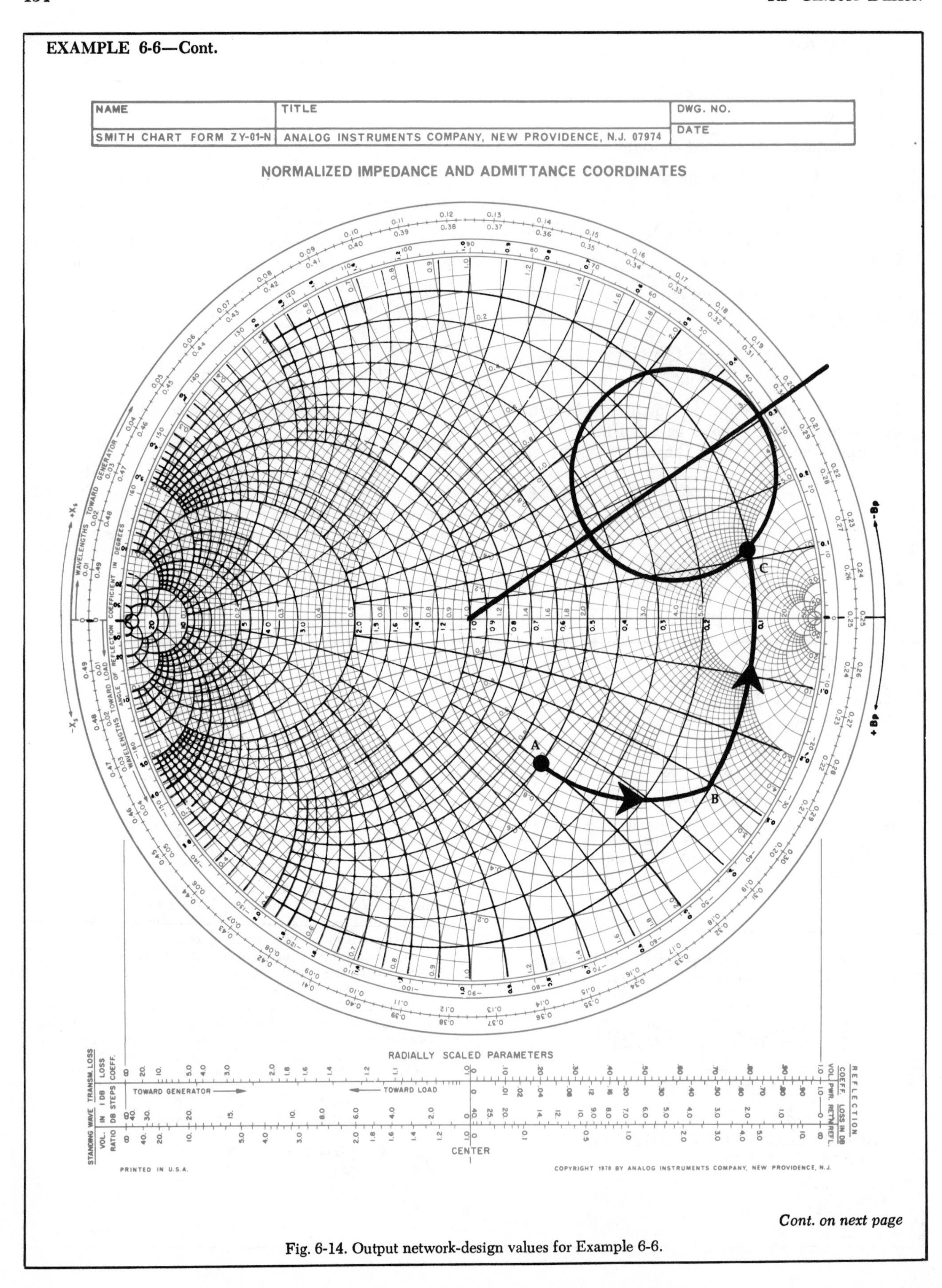

Cont. on next page

Fig. 6-14. Output network-design values for Example 6-6.

EXAMPLE 6-6—Cont.

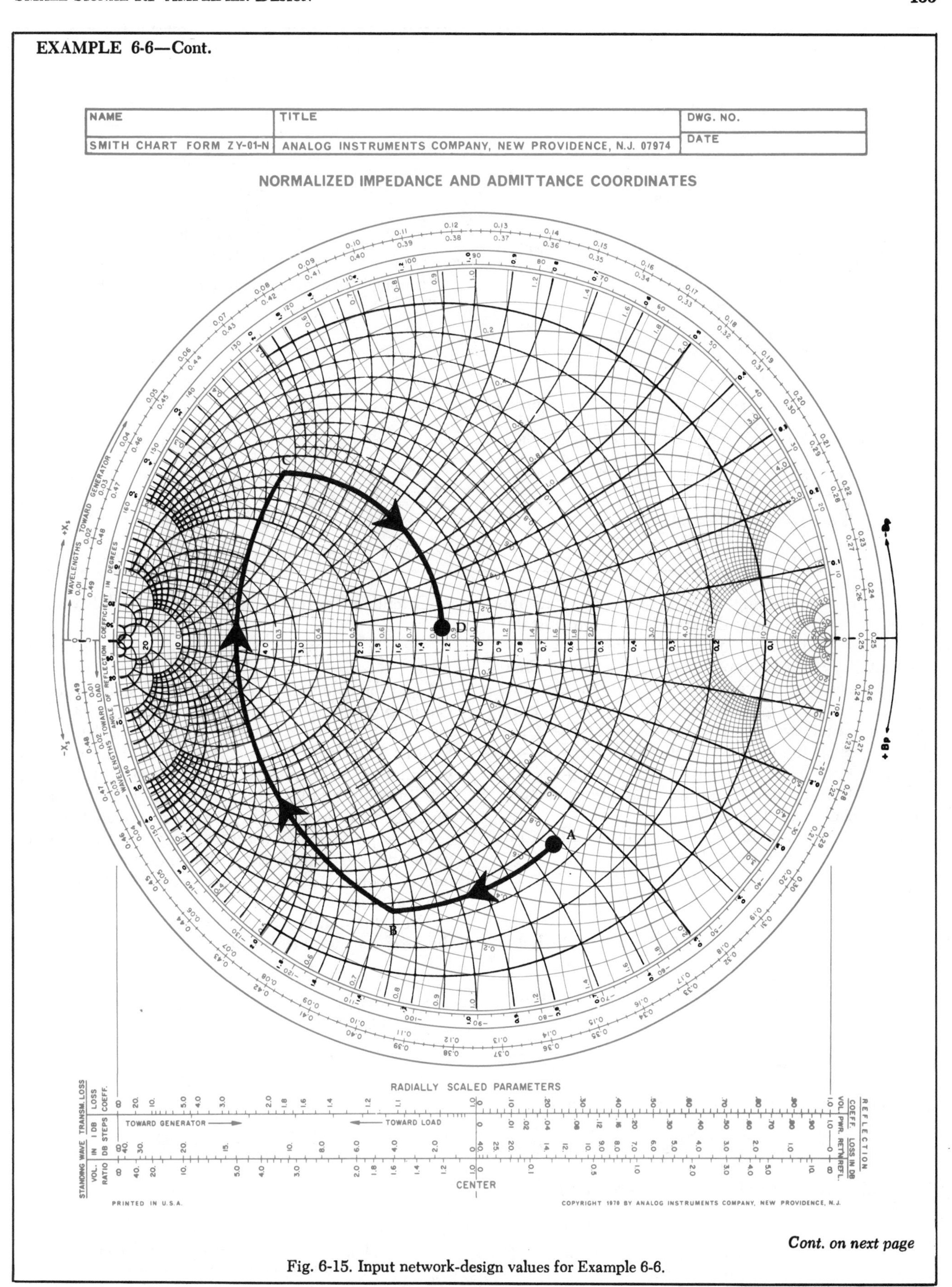

Cont. on next page

Fig. 6-15. Input network-design values for Example 6-6.

EXAMPLE 6-6—Cont.

$$C_1 = \frac{1}{2\pi(250 \times 10^6)(2)(50)}$$
$$= 6.4 \text{ pF}$$

and,

$$L_1 = \frac{50}{2\pi(250 \times 10^6)(0.425)}$$
$$= 75 \text{ nH}$$

For a conjugate match at the input to the transistor with $\Gamma_L = 0.82 \angle 14.2°$ (point C), the desired source-reflection coefficient must be (using Equation 6-21):

$$\Gamma_S = \left[0.277 \angle -59° + \frac{(0.078 \angle 93°)(1.92 \angle 64°)(0.82 \angle 14.2°)}{1 - (0.82 \angle 14.2°)(0.848 \angle -31°)}\right]^*$$
$$= 0.105 \angle 160°$$

This point is plotted as point D in Fig. 6-15. The actual normalized source impedance is plotted at point A (0.7 − j1.2 ohms). Thus, the input network must transform the actual impedance at point A to the desired impedance at point D. For practice, this was done with a three-element design as shown.

Arc AB = Shunt C_2 = j0.62 mho
Arc BC = Series L_2 = j1.09 ohms
Arc CD = Shunt C_3 = j2.1 mhos

From Equations 4-11 through 4-14:

$$C_2 = \frac{(0.62)}{2\pi(250 \times 10^6)(50)}$$
$$= 7.9 \text{ pF}$$

$$C_3 = \frac{2.1}{2\pi(250 \times 10^6)50}$$
$$= 27 \text{ pF}$$

$$L_2 = \frac{(1.09)(50)}{2\pi(250 \times 10^6)}$$
$$= 34.7 \text{ nH}$$

The completed design, excluding the bias network, is shown in Fig. 6-16.

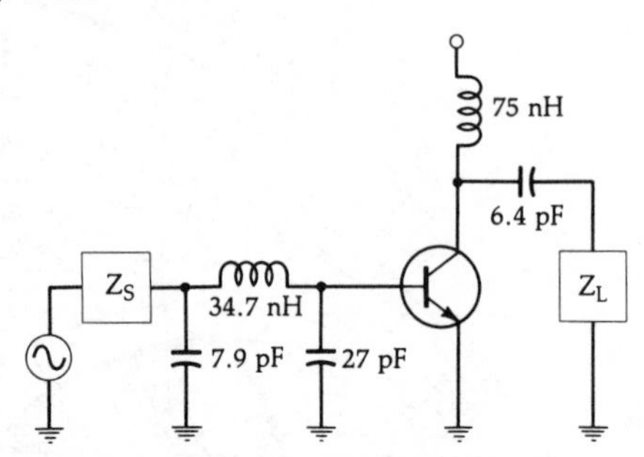

Fig. 6-16. Final circuit for Example 6-6.

methods of determining those source and load impedances that will cause the transistor to go unstable is to plot *stability circles* on a Smith Chart.

A stability circle is simply a circle on a Smith Chart which represents the boundary between those values of source or load impedance that cause instability and those that do not. The perimeter of the circle thus represents the locus of points which forces K = 1. Either the inside *or* the outside of the circle may represent the unstable region and that determination must be made after the circles are drawn.

The locations and radaii of the input and output stability circles are found as follows:

1. Calculate D_s using Equation 6-14.
2. Calculate C_1.

$$C_1 = S_{11} - D_s S_{22}^* \qquad \text{(Eq. 6-28)}$$

3. Calculate C_2 using Equation 6-18.
4. Calculate the *center location* of the *input* stability circle.

$$r_{s1} = \frac{C_1^*}{|S_{11}|^2 - |D_s|^2} \qquad \text{(Eq. 6-29)}$$

5. Calculate the *radius* of the *input* stability circle.

$$p_{s1} = \left|\frac{S_{12}S_{21}}{|S_{11}|^2 - |D_s|^2}\right| \qquad \text{(Eq. 6-30)}$$

6. Calculate the *center location* of the *output* stability circle.

$$r_{s2} = \frac{C_2^*}{|S_{22}|^2 - |D_s|^2} \qquad \text{(Eq. 6-31)}$$

7. Calculate the *radius* of the *output* stability circle.

$$p_{s2} = \left|\frac{S_{12}S_{21}}{|S_{22}|^2 - |D_s|^2}\right| \qquad \text{(Eq. 6-32)}$$

Once the calculations are made, the stability circles can be plotted directly on the Smith Chart. Note, however, that if you try to plot stability circles on the Smith Chart for an unconditionally stable transistor, you may never find them. This is because for an unconditionally stable amplifier the entire chart represents a stable operating region, as shown in Fig. 6-17.

For a *potentially unstable* transistor, the stability circles might resemble those shown in Fig. 6-18. Often, only a portion of the stability circle intersects the chart as shown.

After the stability circles are plotted on the chart, the next step is to determine which side of the circle

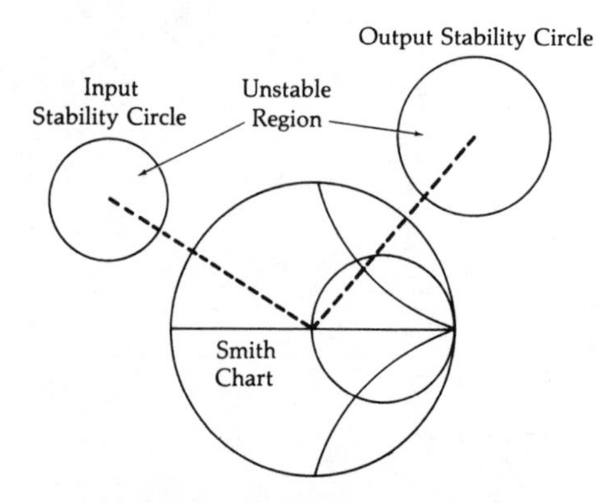

Fig. 6-17. Typical stability circles for an unconditionally stable amplifier.

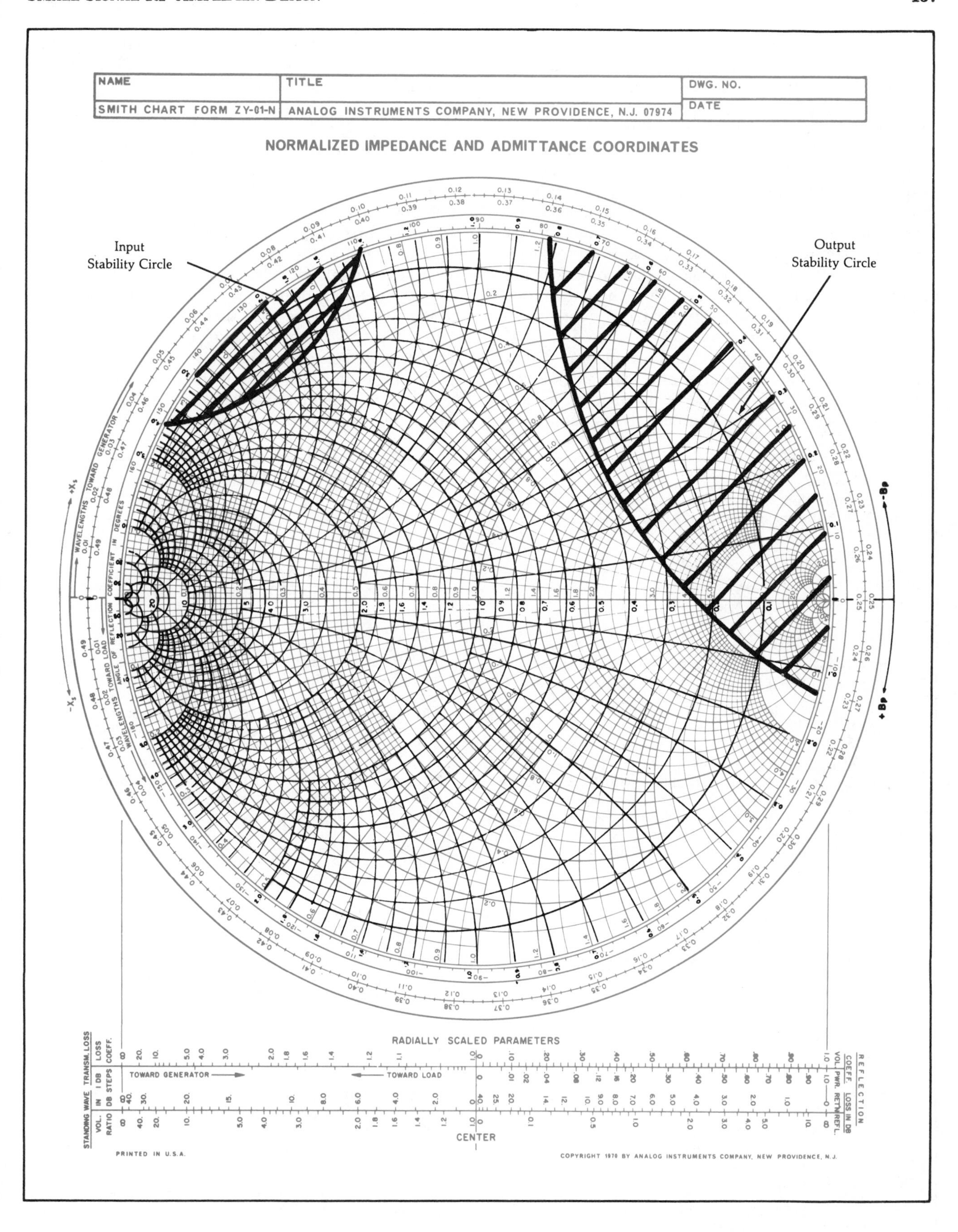

Fig. 6-18. Typical stability circles for a potentially unstable transistor.

(inside or outside) represents the *stable* region. This is very easily done if S_{11} and S_{22} for the transistor are less than 1. Since the S parameters were measured with a 50-ohm source and load, and since the transistor remained stable under these conditions (S_{11} or S_{22} would be greater than 1 for an unstable transistor), *then the center of the normalized Smith Chart must be part of the stable region* as described by the stability circles. Therefore, in this case, if one of the circles surrounds the center of the chart, the inside of that circle must represent the region of *stable* impedances for that port. If, on the other hand, the circle does not surround the center of the chart, then the entire area outside of that circle must represent the *stable* operating region for that port.

It is very rare that you will find a transistor that is unstable with a 50-ohm source and load and, if you do, it would probably be wise to try another device. Therefore, the procedure outlined above should be considered to be the most direct method of locating the stable operating regions on a Smith Chart. Example 6-7 diagrams the procedure.

Design for Optimum Noise Figure

The noise figure of any two-port network gives a measure of the amount of noise that is added to a signal that is transmitted through the network. For any practical circuit, the signal-to-noise ratio at its output will be worse (smaller) than that at its input. In most circuit-design applications, however, it is possible to minimize the noise contribution of each two-port network through a judicious choice of operating point and source resistance.

In Chapter 5, it was briefly mentioned that for each transistor, indeed for each two-port network, there exists an optimum source resistance necessary to establish a minimum noise figure (see also Appendix B). Many manufacturers specify an optimum source

EXAMPLE 6-7

The S parameters for a 2N5179 transistor at 200 MHz, with a $V_{CE} = 6$ volts and an $I_C = 5$ mA, are (see the data sheet in Chapter 5):

$$S_{11} = 0.4 \angle 280°$$
$$S_{22} = 0.78 \angle 345°$$
$$S_{12} = 0.048 \angle 65°$$
$$S_{21} = 5.4 \angle 103°$$

Choose a stable load- and source-reflection coefficient that will provide a power gain of 12 dB at 200 MHz.

Solution

A calculation of Rollett's stability factor (K) for the transistor indicates a potential instability with K = 0.802. Therefore, you must exercise extreme caution in choosing source and load impedances for the device or it may oscillate. To find the stable operating regions on the Smith Chart, plot the input and output stability circles. Proceeding with Step 1, above, we have:

$$D_s = (0.4 \angle 280°)(0.75 \angle 345°) - (0.048 \angle 65°)(5.4 \angle 103°)$$
$$= 0.429 \angle -58.18°$$
$$C_1 = 0.4 \angle 280° - (0.429 \angle -58.2°)(0.78 \angle -345°)$$
$$= 0.241 \angle -136.6°$$
$$C_2 = 0.78 \angle 345° - (0.429 \angle -58.2°)(0.4 \angle -280°)$$
$$= 0.65 \angle -24°$$

Then, the center of the input stability circle is located at the point:

$$r_{s1} = \frac{0.241 \angle 136.6°}{(0.4)^2 - (0.429)^2}$$
$$= 10 \angle 136.6°$$

The radius of the circle is calculated as:

$$p_{s1} = \left| \frac{(0.048 \angle 65°)(5.4 \angle 103°)}{(0.4)^2 - (0.429)^2} \right|$$
$$= 10.78$$

Similarly, for the output stability circle:

$$r_{s2} = \frac{0.65 \angle 24°}{(0.78)^2 - (0.429)^2}$$
$$= 1.53 \angle 24°$$

$$p_{s2} = \left| \frac{(0.048 \angle 65°)(5.4 \angle 103°)}{(0.78)^2 - (0.429)^2} \right|$$
$$= 0.610$$

These circles are shown in Fig. 6-19. Note that the input stability circle is actually drawn as a straight line because the radius of the circle is so large. Since S_{11} and S_{22} are both less than 1, we can deduce that the *inside* of the *input* stability circle represents the region of *stable* source impedances while the *outside* of the *output* stability circle represents the region of *stable* load impedances for the device.

The 12-dB gain circle is also shown plotted in Fig. 6-19. It is found using Equation 6-14 and Equations 6-23 through 6-27. Note that D_s and C_2 have already been calculated. The center location of the circle is found to be:

$$r_o = 0.287 \angle 24°$$

with a radius of:

$$p_o = 0.724$$

The only *load* impedances that we may not select for the transistor are located inside of the input stability circle. Any other load impedance located on the 12-dB gain circle will provide the needed gain as long as the input of the device is conjugately matched and as long as the impedance required for a conjugate match falls inside of the input stability circle.

Choose Γ_L equal to a convenient value on the 12-dB gain circle.

$$\Gamma_L = 0.89 \angle 70°$$

Using Equation 6-21, calculate the source-reflection coefficient needed for a conjugate match and plot this point on the Smith Chart.

$$\Gamma_S = 0.678 \angle 79.4°$$

Notice that Γ_S falls within the stable region of the input stability circle and, therefore, represents a stable termination for the transistor.

Continued on next page

EXAMPLE 6-7—Cont.

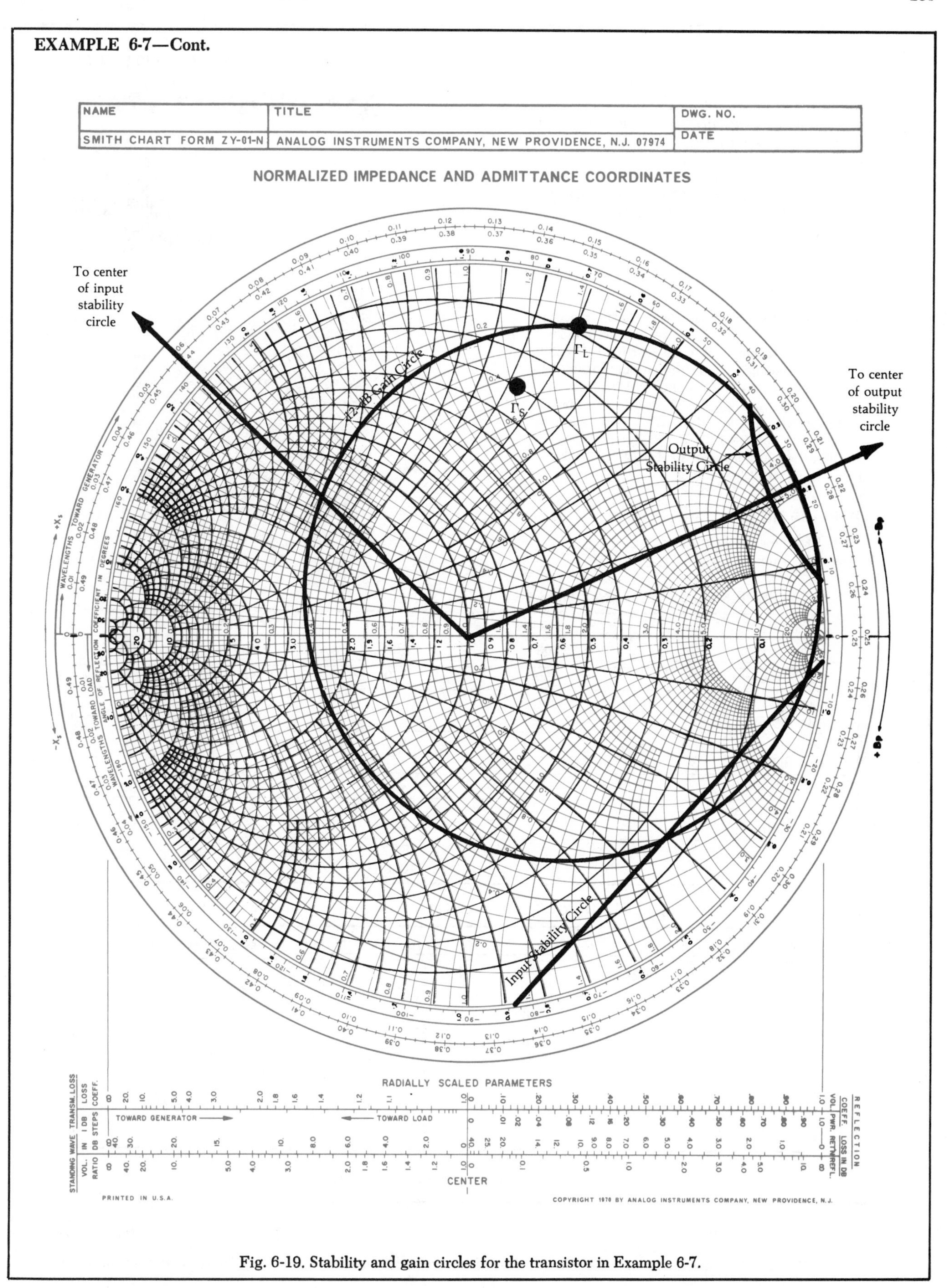

Fig. 6-19. Stability and gain circles for the transistor in Example 6-7.

MA-42120 SERIES

specification sheet npn silicon planar transistors

FEATURES

- HIGH GAIN BANDWIDTH PRODUCT (1.5 GHz)
- HIGH f_{max} (4.2 GHz @ I_C = 20 mA)
- GOLD METALLIZATION
- HIGH RELIABILITY
- DIRECT INTERCHANGABILITY WITH THE FMT 1060 SERIES OF FAIRCHILD TRANSISTORS

DESCRIPTION

This series of NPN Epitaxial Silicon Planar Transistors is designed for VHF/UHF service. The performance of this series is comparable to the Fairchild FMT-1060 series. The high gain bandwidth products make the MA-42122 and MA-42123 useful to 1.0 GHz while the MA-42120 and MA-42121 have a maximum frequency of oscillation of 4.0 GHz. Two packages are offered, the TO-46 (ODS-508), for low power oscillator applications and the TO-72 (ODS-509) for small signal UHF amplifiers.

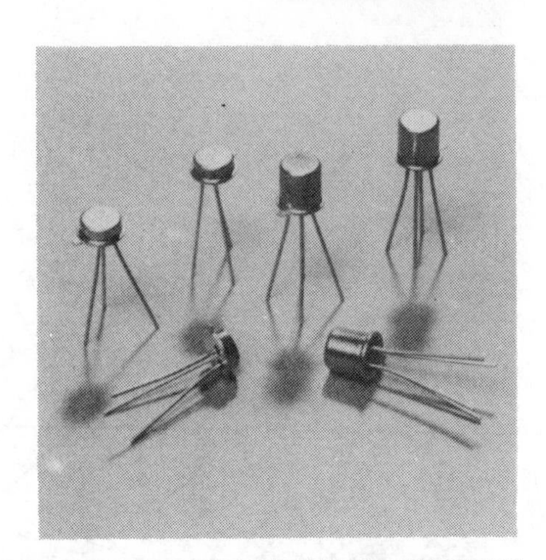

APPLICATIONS

VHF, UHF Low Level Oscillators
IF and RF Amplifiers

MA-42120 SERIES R.F. SPECIFICATIONS

MODEL NO.	MA-42120	MA-42121	MA-42122	MA-42123
CASE STYLE	508	508	509	509
Test Frequency (MHz)	450	450	450	450
Max Noise Fig. @ I_C (dB)	– –	– –	3.5	3.0
Gu (max) Typ. (dB)	13	13	14	14
I_C (mA)	– –	– –	1.5	1.5
1 dB Compression Point (dBm)	– –	– –	–12	–12
Fairchild Equivalent	FMT 1060	FMT 1060A	FMT 1061	FMT 1061A

MA-42120 SERIES HIGH FREQUENCY SPECIFICATIONS (25°C Ambient Temperature Unless Otherwise Noted)

Symbol	Characteristics	Type	Min.	Typ.	Max.	Units	Test Conditions
f_t	Gain Bandwidth Product	MA42121	1.3	1.5		GHz	V_{CE} = 10V, I_C = 20 mA
		MA42120	1.0	1.3		GHz	f = 500 MHz
		MA42123	1.3	1.5		GHz	
		MA42122	1.0	1.3		GHz	
$G_{A\,max}$	Maximum Available Gain	MA42121		12.8		dB	V_{CE} = 10V, I_C = 20 mA
		MA42123		13.8			f = 1.0 GHz
NF	Noise Figure	MA42123		2.3	3.0	dB	V_{CE} = 10V, I_C = 1.5 mA
		MA42122		2.7	3.5	dB	f = 450 MHz, R = 50 ohms
G_{pe}	Neutralized Power Gain	MA42123		17.0		dB	V_{CE} = 10V, I_E = 1.5 mA
							f = 450 MHz, R = 50 ohms
f_{max}	Maximum Frequency of Oscillation[1]	MA42121		4.2		GHz	V_{CE} = 10V, I_C = 20 mA
		MA42120		3.8		GHz	V_{CE} = 10V, I_C = 20 mA

NOTE:
1. Calculated from S-Parameters, f_{max} is the frequency at which the extrapolated GA_{max} is 0 dB.

Cont. on next page

Fig. 6-20. Data sheet for Microwave Associates' MA-42120 series of transistors. (*Courtesy Microwave Associates*)

MA-42120 SERIES

specification sheet

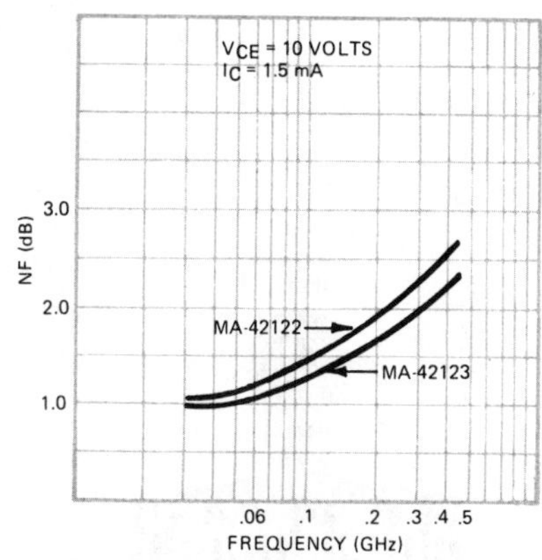

TYPICAL NOISE FIGURE VS FREQUENCY

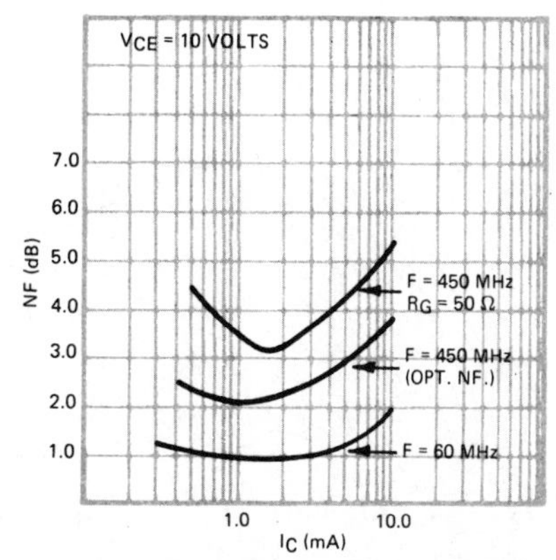

MA-42123-509 TYPICAL OPTIMUM N.F. VS COLLECTOR CURRENT

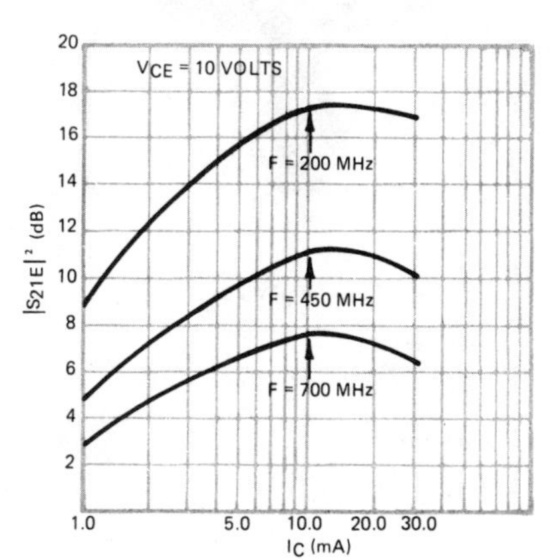

MA-42123-509 TYPICAL $|S_{21E}|^2$ VS COLLECTOR CURRENT

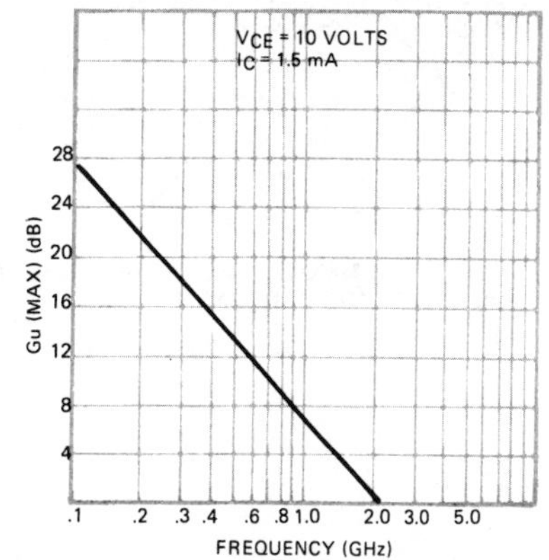

MA-42123-509 TYPICAL Gu (MAX) VS FREQUENCY

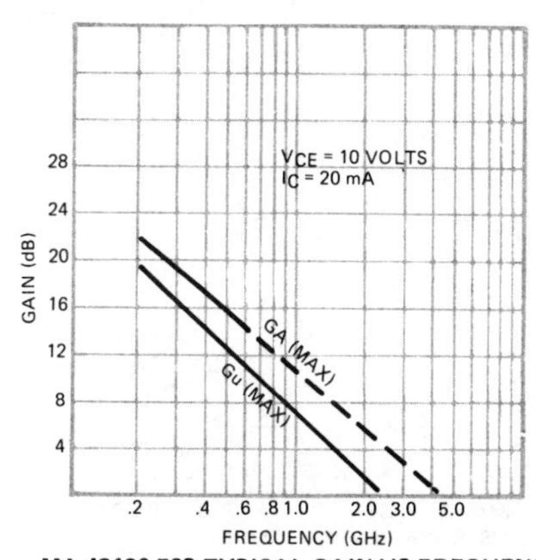

MA-42120-508 TYPICAL GAIN VS FREQUENCY

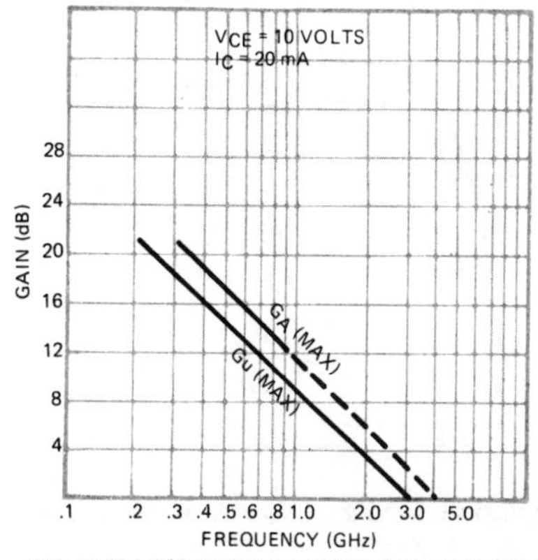

MA-42121-508 TYPICAL GAIN VS FREQUENCY

Cont. on next page

Fig. 6-20.—Cont. Data sheet for Microwave Associates' MA-42120 series of transistors. (*Courtesy Microwave Associates*)

MA-42120 SERIES

specification sheet

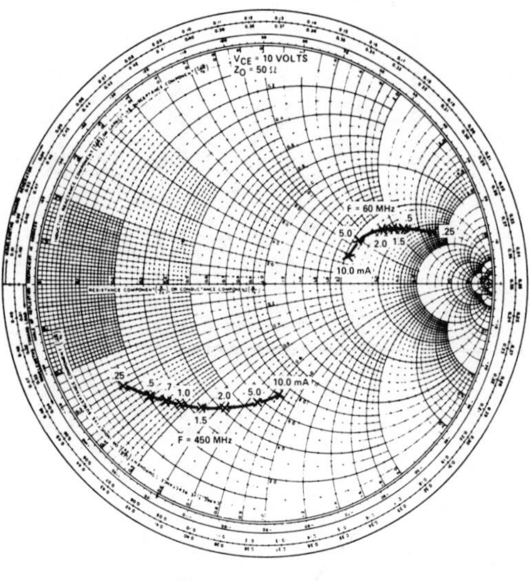

TYPICAL OPTIMUM NOISE SOURCE IMPEDANCE VS COLLECTOR CURRENT

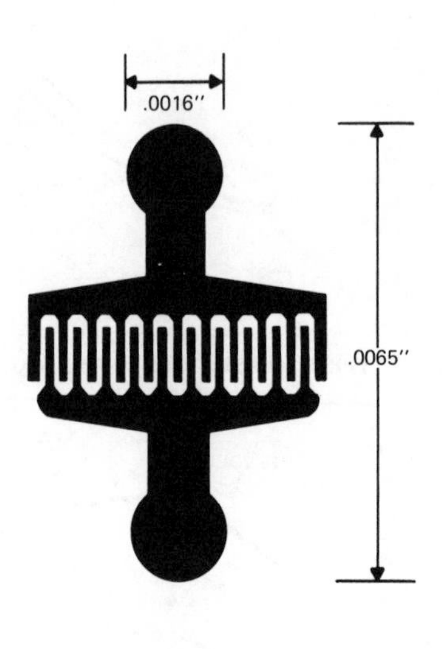

GEOMETRY 70

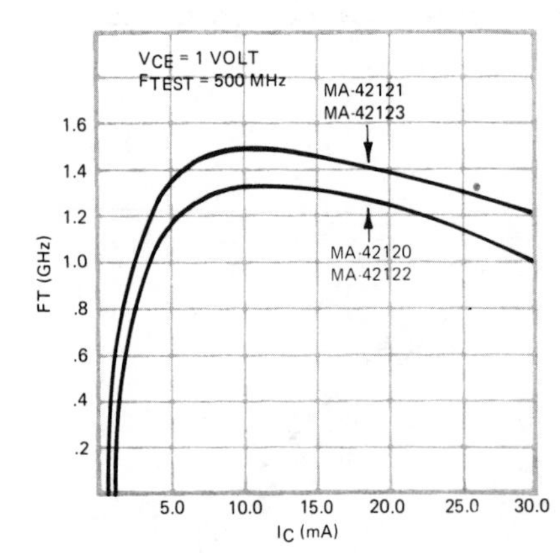

MA-42120 SERIES TYPICAL F_T VS COLLECTOR CURRENT

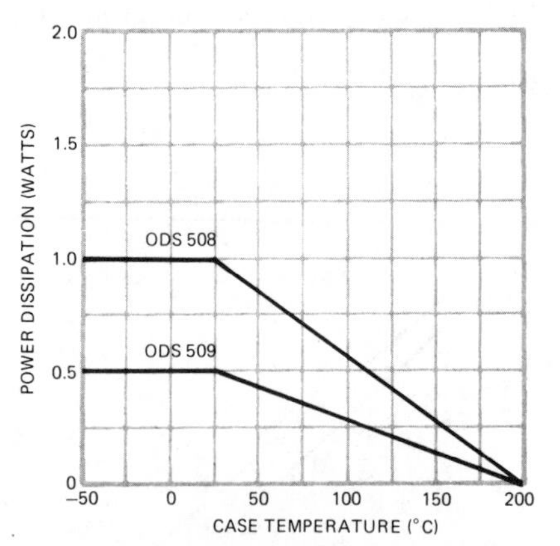

MA-42120 SERIES POWER DISSIPATION VS. CASE TEMPERATURE

Cont. on next page

Fig. 6-20.—Cont. Data sheet for Microwave Associates' MA-42120 series of transistors. (*Courtesy Microwave Associates*)

MA-42120 SERIES

specification sheet

ELECTRICAL CHARACTERISTICS (25°C Ambient Temperature Unless Otherwise Noted)

Symbol	Characteristics	MA42120 MA42122 Min	Typ	Max	MA42121 MA42123 Min	Typ	Max	Units	Test Conditions
h_{FE}	DC Current Gain	20	45	110	40	75	185		I_C = 5.0 mA, V_{CE} = 5.0
V_{CE} (sat)	Pulsed Collector Saturation Voltage[1]		0.30	0.38		0.25	0.35	V	I_C = 80 mA, I_B = 8.0 mA
V_{BE} (sat)	Pulsed Base Saturation Voltage[1]		0.95	0.98		0.93	0.96	V	I_C = 40 mA, I_B = 20 mA
BV_{CBO}	Collector to Base Breakdown	30	35		30	35		V	I_C = 10 μA, I_E = 0
V_{CEO} (sus)	Collector to Emitter Sustaining Voltage	14	16.5		14	16.5		V	I_C = 1.0 mA, I_B = 0
I_{EBO}	Emitter Cutoff Current		20	100		20	100	μA	I_C = 0, V_{EB} = 4.0V
I_{CBO}	Collector Cutoff Current		0.01	50		0.01	50	nA	V_{CB} = 10V, I_E = 0
I_{CBO}	Collector Cutoff Current		0.3	1.0		0.3	1.0	μA	V_{CB} = 10V, I_E = 0, T_A = 125°C
C_{cb}	Collector to Base Capacitance (MA42120, MA42121)		1.0	1.4		1.0	1.4	pF	V_{CB} = 10V, I_E = 0
C_{eb}	Collector to Base Capacitance (MA42122, MA42123)		0.85	1.0		0.85	1.0	pF	V_{CB} = 10V, I_E = 0
C_{eb}	Emitter to Base Capacitance		1.5	3.0		1.5	3.0	pF	V_{EB} = 0.5V, I_C = 0
$\|h_{fe}\|$	Magnitude of High Frequency Current Gain	2.0	2.6		2.6	3.0			V_{CE} = 10V, I_C = 20 mA, f = 500 MHz

NOTE:
1. Pulse Conditions: Length = 300 μs; duty cycle = 1%

MAXIMUM RATING
(Case Temperature 25°C unless otherwise noted)

	Total Power Dissipation	508 case – 1.0 W 509 case – .5 W
V_{CBO}	Collector to Base Voltage	30 V
V_{EBO}	Emitter to Base Voltage	4.0 V
V_{CES}	Collector to Emitter Voltage	30 V
I_C	Collector Current	80 mA
	Storage Temperature	–65° to +200°C
	Operating Junction Temperature	+200°C
	Lead Temperature (Soldering – 10 seconds each lead)	+250°C
	Hermeticity	5 x (10)$^{-8}$ cc/sec of He

ENVIRONMENTAL RATINGS PER MIL-STD-750

	Method	Level
Storage Temperature	1031	–65 to +200°C
Temperature Cycle	1051	10 cycles –65 to +200°C
Shock	2016	500 g's
Vibration	2056	15 g's
Constant Acceleration	2006	20,000 g's
Humidity	1021	10 days

Cont. on next page

Fig. 6-20.–Cont. Data sheet for Microwave Associates' MA-42120 series of transistors. (*Courtesy Microwave Associates*)

MA-42120 SERIES

specification sheet

MA-42120-508

TYPICAL COMMON-EMITTER S-PARAMETERS AT 25°C LEAD TEMPERATURE

V_{CE} = 10 VOLTS, Z_G = Z_L = 50 Ω

FREQUENCY (MHz)	COLLECTOR CURRENT (mA)	INPUT REFLECTION COEFFICIENT		FORWARD TRANSMISSION COEFFICIENT		REVERSE TRANSMISSION COEFFICIENT		OUTPUT REFLECTION COEFFICIENT	
		$\lvert S_{11E}\rvert$	ϕS_{11E}	$\lvert S_{21E}\rvert$	ϕS_{21E}	$\lvert S_{12E}\rvert$	ϕS_{12E}	$\lvert S_{22E}\rvert$	ϕS_{22E}
100	1.5	.89	−44°	3.8	146°	.05	65°	.94	−13°
	5.0	.69	−79°	8.6	126°	.03	55°	.81	−18°
	20.0	.50	−127°	12.0	107°	.02	52°	.67	−16°
200	1.5	.80	−72°	3.3	128°	.06	51°	.86	−20°
	5.0	.60	−113°	6.1	107°	.04	49°	.71	−21°
	20.0	.50	−150°	7.4	92°	.02	62°	.61	−16°
300	1.5	.72	−95°	2.7	112°	.07	43°	.82	−25°
	5.0	.55	−134°	4.3	95°	.04	50°	.68	−23°
	20.0	.51	−161°	5.0	83°	.03	68°	.61	−18°
400	1.5	.66	−115°	2.2	101°	.08	38°	.79	−29°
	5.0	.53	−148°	3.4	86°	.05	54°	.66	−25°
	20.0	.52	−169°	3.8	75°	.04	72°	.62	−20°
450	1.5	.64	−123°	2.1	95°	.07	38°	.77	−30°
	5.0	.53	−154°	3.1	81°	.05	58°	.66	−27°
	20.0	.53	−172°	3.4	71°	.04	77°	.63	−21°
500	1.5	.63	−130°	1.9	90°	.07	39°	.76	−32°
	5.0	.53	−159°	2.7	77°	.05	62°	.66	−28°
	20.0	.54	−175°	3.0	67°	.05	79°	.63	−23°
600	1.5	.61	−143°	1.6	83°	.07	40°	.73	−35°
	5.0	.53	−167°	2.3	72°	.06	66°	.65	−31°
	20.0	.55	178°	2.5	63°	.06	82°	.63	−26°
700	1.5	.61	−153°	1.5	76°	.07	46°	.70	−39°
	5.0	.55	−174°	2.0	65°	.06	74°	.63	−34°
	20.0	.57	174°	2.2	56°	.06	87°	.62	−29°
800	1.5	.61	−162°	1.3	70°	.06	55°	.68	−44°
	5.0	.56	179°	1.8	61°	.07	79°	.62	−38°
	20.0	.59	170°	1.8	52°	.08	91°	.62	−34°
900	1.5	.61	−169°	1.2	66°	.07	64°	.68	−50°
	5.0	.56	174°	1.6	57°	.08	84°	.62	−44°
	20.0	.60	166°	1.6	49°	.09	94°	.62	−41°
1000	1.5	.61	−176°	1.1	60°	.07	77°	.68	−55°
	5.0	.56	169°	1.4	52°	.09	91°	.63	−50°
	20.0	.61	160°	1.5	44°	10	99°	.63	−47°

MA 42121-508

TYPICAL COMMON-EMITTER S-PARAMETERS AT 25°C LEAD TEMPERATURE

V_{CE} = 10 VOLTS, Z_G = Z_L = 50

FREQUENCY (MHz)	COLLECTOR CURRENT (mA)	INPUT REFLECTION COEFFICIENT		FORWARD TRANSMISSION COEFFICIENT		REVERSE TRANSMISSION COEFFICIENT		OUTPUT REFLECTION COEFFICIENT	
		$\lvert S_{11E}\rvert$	ϕS_{11E}	$\lvert S_{21E}\rvert$	ϕS_{21E}	$\lvert S_{12E}\rvert$	ϕS_{12E}	$\lvert S_{22E}\rvert$	ϕS_{22E}
100	1.5	.92	−36°	3.8	150°	.04	67°	.95	−13°
	5.0	.74	−67°	9.1	131°	.03	59°	.82	−19°
	20.0	.51	−118°	13	109°	.02	55°	.66	−18°
200	1.5	.83	−59°	3.4	133°	.07	55°	.86	−21°
	5.0	.61	−98°	6.7	111°	.04	48°	.69	−24°
	20.0	.49	−142°	8.0	93°	.03	59°	.59	−18°
300	1.5	.74	−81°	2.9	117°	.08	47°	.81	−27°
	5.0	.54	−120°	4.9	99°	.05	50°	.65	−26°
	20.0	.49	−156°	5.5	84°	.03	65°	.58	−19°
400	1.5	.67	−99°	2.5	106°	.09	41°	.77	−31°
	5.0	.51	−136°	3.9	89°	.06	51°	.62	−28°
	20.0	.50	−164°	4.2	77°	.04	67°	.59	−22°
450	1.5	.65	−106°	2.3	100°	.09	40°	.75	−33°
	5.0	.50	−142°	3.5	85°	.06	53°	.62	−29°
	20.0	.51	−167°	3.7	73°	.05	71°	.59	−23°
500	1.5	.62	−114°	2.1	95°	.09	40°	.74	−35°
	5.0	.50	−147°	3.1	81°	.06	55°	.62	−30°
	20.0	.51	−170°	3.3	69°	.05	74°	.60	−24°
600	1.5	.59	−127°	1.8	88°	.09	38°	.71	−37°
	5.0	.49	−157°	2.6	75°	.06	58°	.61	−32°
	20.0	.53	−176°	2.7	64°	.06	76°	.60	−26°
700	1.5	.58	−138°	1.6	80°	.09	39°	.68	−41°
	5.0	.50	−164°	2.3	68°	.07	63°	.59	−35°
	20.0	.55	179°	2.4	58°	.06	81°	.59	−30°
800	1.5	.58	−147°	1.5	74°	.08	43°	.65	−45°
	5.0	.51	−170°	2.0	63°	.08	68°	.57	−39°
	20.0	.57	175°	2.0	54°	.07	84°	.59	−35°
900	1.5	.58	−154°	1.3	69°	.08	47°	.65	−51°
	5.0	.52	−174°	1.8	60°	.08	71°	.57	−44°
	20.0	.58	171°	1.8	51°	.08	88°	.59	−41°
1000	1.5	.57	−161°	1.2	64°	.08	56°	.65	−56°
	5.0	.52	−179°	1.6	55°	.09	78°	.59	−49°
	20.0	.59	166°	1.6	46°	.09	93°	.61	−46°

Cont. on next page

Fig. 6-20.—Cont. Data sheet for Microwave Associates' MA-42120 series of transistors. (*Courtesy Microwave Associates*)

MA-42120 SERIES

specification sheet

MA 42122-509
TYPICAL COMMON-EMITTER S-PARAMETERS AT 25°C LEAD TEMPERATURE
V_{CE} = 10 VOLTS, $Z_G = Z_L = 50\ \Omega$

FREQUENCY (MHz)	COLLECTOR CURRENT (mA)	INPUT REFLECTION COEFFICIENT		FORWARD TRANSMISSION COEFFICIENT		REVERSE TRANSMISSION COEFFICIENT		OUTPUT REFLECTION COEFFICIENT	
		$\lvert S_{11E}\rvert$	ϕS_{11E}	$\lvert S_{21E}\rvert$	ϕS_{21E}	$\lvert S_{12E}\rvert$	ϕS_{12E}	$\lvert S_{22E}\rvert$	ϕS_{22E}
100	1.5	.87	−43°	3.7	145°	.03	63°	.96	−10°
	5.0	.66	−70°	8.1	127°	.02	53°	.87	−12°
	20.0	.43	−103°	11.2	109°	.02	55°	.78	−12°
200	1.5	.76	−69°	3.2	127°	.04	54°	.89	−15°
	5.0	.52	−102°	5.7	107°	.03	56°	.79	−16°
	20.0	.37	−132°	7.0	93°	.02	66°	.73	−13°
300	1.5	.66	−94°	2.6	110°	.05	47°	.86	−19°
	5.0	.45	−126°	4.1	94°	.03	58°	.77	−18°
	20.0	.36	−149°	4.8	83°	.03	70°	.73	−15°
400	1.5	.69	−115°	2.2	99°	.05	45°	.84	−23°
	5.0	.41	−144°	3.3	84°	.04	62°	.76	−21°
	20.0	.37	−162°	3.7	75°	.04	73°	.73	−17°
450	1.5	.56	−124°	2.0	93°	.05	47°	.83	−25°
	5.0	.40	−151°	2.9	79°	.04	66°	.76	−22°
	20.0	.37	−167°	3.3	70°	.04	76°	.73	−19°
500	1.5	.54	−132°	1.8	88°	.05	49°	.82	−26°
	5.0	.40	−157°	2.6	75°	.05	69°	.75	−23°
	20.0	.37	−172°	2.9	66°	.05	79°	.73	−21°
600	1.5	.61	−148°	1.6	80°	.05	53°	.80	−29°
	5.0	.40	−169°	2.2	69°	.05	74°	.74	−26°
	20.0	.39	179°	2.4	60°	.06	81°	.73	−23°
700	1.5	.50	−160°	1.4	72°	.05	63°	.77	−33°
	5.0	.40	−179°	2.0	62°	.06	79°	.72	−29°
	20.0	.41	172°	2.1	53°	.06	84°	.72	−27°
800	1.5	.50	−171°	1.3	66°	.08	72°	.74	−38°
	5.0	.41	173°	1.7	57°	.07	83°	.71	−34°
	20.0	.43	166°	1.8	49°	.08	87°	.70	−32°
900	1.5	.49	178°	1.2	61°	.07	80°	.74	−43°
	5.0	.41	166°	1.6	52°	.08	85°	.70	−39°
	20.0	.43	159°	1.6	45°	.09	88°	.70	−37°
1000	1.5	.48	168°	1.1	55°	.08	89°	.74	−48°
	5.0	.41	167°	1.4	47°	.10	90°	.71	−45°
	20.0	.44	151°	1.5	39°	.11	93°	.70	−42°

MA 42123-509
TYPICAL COMMON-EMITTER S-PARAMETERS AT 25°C LEAD TEMPERATURE
V_{CE} = 10 VOLTS, $Z_G = Z_L = 50\ \Omega$

FREQUENCY (MHz)	COLLECTOR CURRENT (mA)	INPUT REFLECTION COEFFICIENT		FORWARD TRANSMISSION COEFFICIENT		REVERSE TRANSMISSION COEFFICIENT		OUTPUT REFLECTION COEFFICIENT	
		$\lvert S_{11E}\rvert$	ϕS_{11E}	$\lvert S_{21E}\rvert$	ϕS_{21E}	$\lvert S_{12E}\rvert$	ϕS_{12E}	$\lvert S_{22E}\rvert$	ϕS_{22E}
100	1.5	.89	−39°	3.8	147°	.04	54°	.95	−11°
	5.0	.68	−85°	8.6	128°	.03	56°	.85	−15°
	20.0	.43	−121°	10.9	104°	.02	57°	.67	−16°
200	1.5	.79	−63°	3.3	129°	.05	56°	.87	−17°
	5.0	.53	−95°	6.1	108°	.04	55°	.75	−17°
	20.0	.40	−144°	6.6	89°	.03	66°	.61	−16°
300	1.5	.69	−86°	2.7	113°	.06	48°	.84	−21°
	5.0	.45	−118°	4.4	95°	.05	57°	.72	−20°
	20.0	.41	−158°	4.5	80°	.04	70°	.61	−18°
400	1.5	.61	−105°	2.3	101°	.07	44°	.81	−25°
	5.0	.41	−135°	3.5	86°	.05	58°	.71	−22°
	20.0	.41	−168°	3.5	72°	.06	72°	.61	−21°
450	1.5	.58	−114°	2.2	96°	.07	45°	.79	−27°
	5.0	.40	−141°	3.2	81°	.05	62°	.70	−23°
	20.0	.42	−172°	3.1	67°	.06	73°	.61	−22°
500	1.5	.55	−121°	2.0	91°	.07	45°	.79	−28°
	5.0	.39	−148°	2.8	77°	.06	65°	.70	−24°
	20.0	.43	−176°	2.7	64°	.07	76°	.62	−24°
600	1.5	.52	−136°	1.7	83°	.07	46°	.76	−31°
	5.0	.39	−159°	2.4	71°	.06	66°	.69	−27°
	20.0	.45	177°	2.3	58°	.08	77°	.62	−27°
700	1.5	.50	−148°	1.6	74°	.07	51°	.73	−34°
	5.0	.39	−168°	2.1	64°	.07	70°	.67	−29°
	20.0	.47	171°	2.0	51°	.09	80°	.61	−31°
800	1.5	.50	−159°	1.4	68°	.07	57°	.71	−39°
	5.0	.39	−176°	1.9	59°	.08	74°	.66	−33°
	20.0	.48	166°	1.7	47°	.11	83°	.60	−36°
900	1.5	.49	−167°	1.3	63°	.07	64°	.70	−43°
	5.0	.39	178°	1.7	55°	.09	76°	.65	−38°
	20.0	.49	160°	1.5	43°	.12	84°	.59	−43°
1000	1.5	.47	−176°	1.2	57°	.08	74°	.70	−49°
	5.0	.39	171°	1.5	49°	.10	81°	.66	−43°
	20.0	.50	154°	1.4	37°	.14	87°	.60	−49°

Fig. 6-20.—Cont. Data sheet for Microwave Associates' MA-42120 series of transistors. (*Courtesy Microwave Associates*)

resistance on the data sheet, such as in the case of the 2N5179 transistor given in Chapter 5. Others will specify an optimum source-reflection coefficient. Such is the case for the Microwave Associates' MA-42120-Series transistor data sheet that is shown in Fig. 6-20. Note the Smith Chart, on page 3 of the data sheet, labeled "Typical Optimum Noise Source Impedance vs. Collector Current." Obviously, as shown on the chart, if you were planning to use the transistor at some frequency other than 60 MHz or 450 MHz, you would be out of luck as far as optimum noise-figure design is concerned. Typically, most data sheets are incomplete like this. There is just not enough space in a typical data book to provide the user with all of the information that he needs in order to design amplifiers at every possible frequency and bias point. The data sheet is meant only as a starting point in any design. Chances are you will end up making many of your own measurements on a device before it becomes a part of the design.

On page 2 of the data sheet, you will find a set of curves labeled "Typical Optimum N.F. vs. Collector Current." Note that for this particular device, at 450 MHz, the optimum collector current for minimum noise figure is approximately 1.5 mA. This value of collector current should result in a noise figure of just above 2 dB. Again, the data is presented for only 60 MHz and 450 MHz.

Designing amplifiers for a minimum noise figure is simply a matter of determining, either experimentally or from the data sheet, the source resistance and the bias point that produce the minimum noise figure for the device (Example 6-8). Once determined, the actual source impedance is simply forced to "look like" the optimum value. Of course, all stability considerations still apply. If the Rollett stability factor (K) calculates to be less than 1, then you must be careful in your choice of source- and load-reflection coefficients. It is best, in this case, to draw the stability circles for an accurate graphical indication of where the unstable regions lie.

After providing the transistor with its optimum source impedance, the next step is to determine the optimum load-reflection coefficient needed to properly terminate the transistor's output. This is given by:

$$\Gamma_L = \left[S_{22} + \frac{S_{12}S_{21}\Gamma_S}{1 - S_{11}\Gamma_S} \right]^* \quad \text{(Eq. 6-33)}$$

where,

Γ_S is the source-reflection coefficient for minimum noise figure.

EXAMPLE 6-8

It has been determined that the optimum bias point for minimum noise figure for a transistor is $V_{CE} = 10$ V and $I_C = 5$ mA. Its optimum source-reflection coefficient, as given on the data sheet, is:

$$\Gamma_S = 0.7 \angle 140^\circ$$

The S parameters for the transistor, under the given bias conditions at 200 MHz, are:

$$S_{11} = 0.4 \angle 162^\circ$$
$$S_{22} = 0.35 \angle -39^\circ$$
$$S_{12} = 0.04 \angle 60^\circ$$
$$S_{21} = 5.2 \angle 63^\circ$$

Design a low-noise amplifier to operate between a 75-ohm source and a 100-ohm load at 200 MHz. What gain can you expect from the amplifier when it is built?

Solution

The Rollett stability factor (K) calculates to be 1.74 which indicates unconditional stability (Equation 6-15). Therefore, we may proceed with the design. The design values of the input-matching network are shown in Fig. 6-21. Here the normalized 75-ohm source resistance is transformed to Γ_S using two components.

Arc AB = Shunt C = j1.7 mhos
Arc BC = Series L = j0.86 ohm

Using Equations 4-11 through 4-14, the component values are calculated to be:

$$C_1 = \frac{1.7}{(50)(2\pi)(200 \times 10^6)} = 27 \text{ pF}$$

$$L_1 = \frac{(0.86)(50)}{2\pi(200 \times 10^6)} = 34 \text{ nH}$$

The load-reflection coefficient needed to properly terminate the transistor is then found using Equation 6-33.

Continued on next page

EXAMPLE 6-8—Cont.

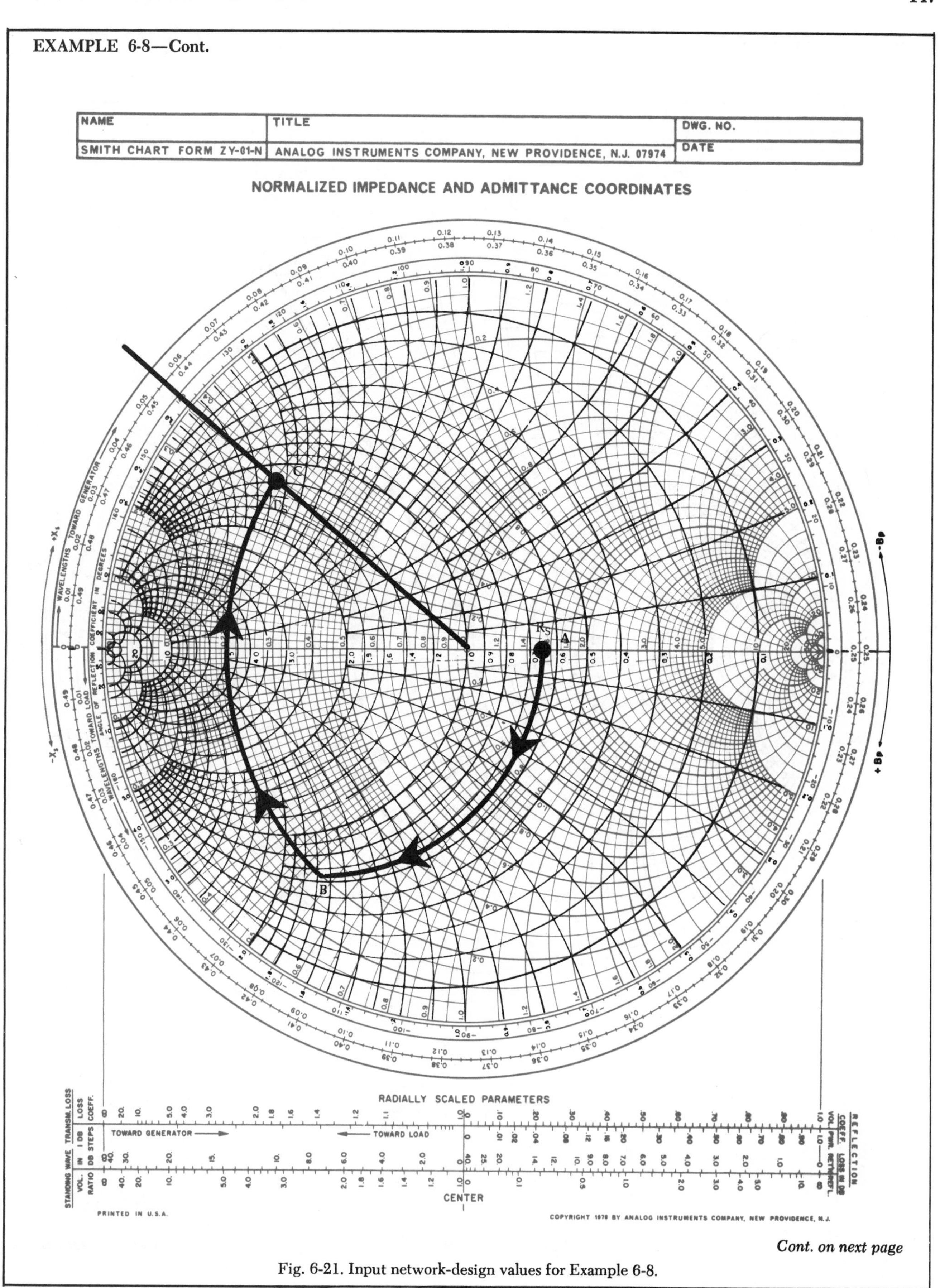

Cont. on next page

Fig. 6-21. Input network-design values for Example 6-8.

EXAMPLE 6-8—Cont.

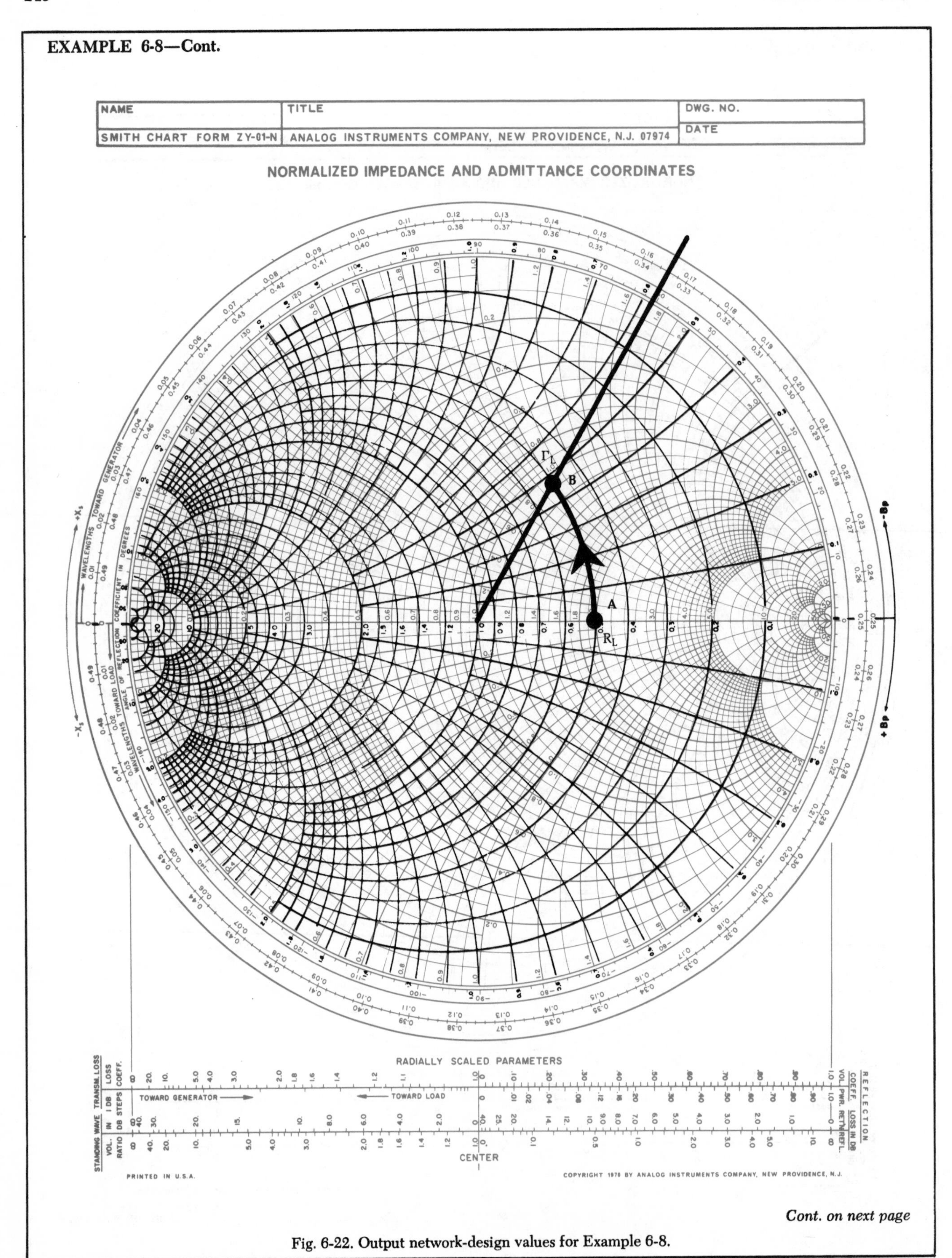

Cont. on next page

Fig. 6-22. Output network-design values for Example 6-8.

EXAMPLE 6-8—Cont.

$$\Gamma_L = \left[0.35 \angle -39^\circ + \frac{(0.04 \angle 60^\circ)(5.2 \angle 63^\circ)(0.7 \angle 140^\circ)}{1 - (0.4 \angle 162^\circ)(0.7 \angle 140^\circ)}\right]^*$$

$$= 0.427 \angle 60.7^\circ$$

This value, along with the normalized load-resistance value, is plotted in Fig. 6-22. The 100-ohm load must be transformed into Γ_L. One possible method is shown in Fig. 6-22. Note that a single shunt inductor provides the necessary impedance transformation:

$$\text{Arc AB} = \text{Shunt L} = -j0.48 \text{ mho}$$

Again using Equations 4-11 through 4-14, the inductor's value is found to be:

$$L_2 = \frac{50}{2\pi(200 \times 10^6)(0.48)}$$

$$= 83 \text{ nH}$$

The final design, including a typical bias network, is shown in Fig. 6-23. The 0.1-μF capacitors are used only as bypass and coupling elements. The gain of the amplifier, as calculated with Equation 6-22, is 13.3 dB.

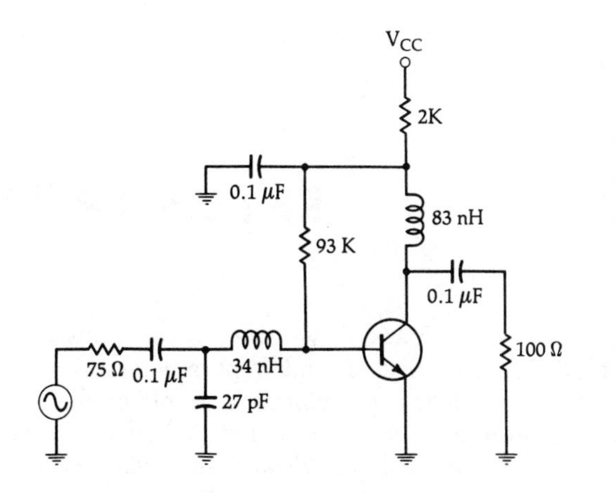

Fig. 6-23. Final circuit for Example 6-8.

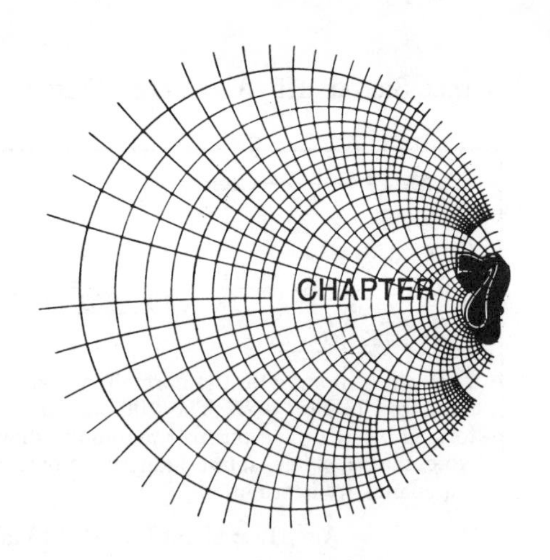

RF POWER AMPLIFIERS

In Chapters 5 and 6, we studied the transistor as a small-signal device. Y and S parameters were introduced as a means of facilitating amplifier design, and design equations were provided. When the transistor is used as a *large-signal device,* however, these equations are no longer valid. In fact, both Y and S parameters are called *small-signal* parameters, and should not be considered in the design of rf power amplifiers.

RF POWER TRANSISTOR CHARACTERISTICS

Instead of specifying the Y and S parameters for a power transistor, manufacturers will typically specify the *large-signal input impedance* and the *large-signal output impedance* for the device. These parameters are typically measured on the device when it is operating as a matched amplifier at the desired dc supply voltage and rf power output level. A matched amplifier, in this case, refers to a condition in which the input and output impedances are conjugately matched to the source and load, respectively.

The Rf Power Transistor Data Sheet

Pertinent design information for rf power transistors is usually presented in the form of large-signal input and output impedances, as shown in Fig. 7-1. Fig. 7-1 is a data sheet for the Motorola MRF233 rf power transistor. This particular data sheet was chosen for instructional purposes because it includes both series- and shunt-impedance information. This gives the circuit designer the opportunity of using an impedance format with which he is accustomed—without the need of converting from one format to the other.

Figure 5, on page 3 of the data sheet, is a Smith Chart representation of the *series* input and output impedance of the transistor (between 40 and 100 MHz). The information is also tabulated on the right side of the chart for your convenience. Note that the impedance is presented in the form $Z = R \pm jX$. Thus, at 100 MHz, the input impedance of the transistor is found to be $Z_{in} = 1.7 - j2.7$ ohms, while the output impedance is $Z_{out} = 5 - j5.6$ ohms. This equivalent-series representation for the transistor is shown in Fig. 7-2.

Figures 6, 7, 8, and 9, of the data sheet, present the same impedance information in parallel form. The input and output impedance of the transistor are presented as a shunt resistance in parallel with a capacitor. Thus, referring to Figures 6 and 7 of the data sheet, the input impedance of the transistor is represented by a 6-ohm shunt resistor in parallel with a 422-pF capacitor. The curves of Figures 8 and 9 indicate an equivalent parallel output impedance for the transistor, which includes an 11.3-ohm resistor in parallel with a 158-pF capacitor, at 100 MHz. These shunt combinations are shown in the equivalent circuit of Fig. 7-3.

Note that you can perform your own transformation from series to shunt, and back again, by using Equations 2-6 and 2-7 and, then, following the procedure of Example 2-2.

Figure 2, on page 3 of the data sheet, is useful in helping you determine how much input signal power you will need to produce a given output power. Note that as the frequency of operation increases, the required input drive level increases. An input power to the transistor of 1 watt will produce a 20-watt output signal at 50 MHz (13-dB Gain), while, at 90 MHz, that same input level will produce only 14 watts out (11.5-dB Gain).

Figure 3 presents the same basic information as Figure 2, but in a different format. Note that the output power decreases as the frequency of operation increases when given a constant input power level.

The remainder of the data sheet is straightforward and resembles that of any typical small-signal transistor.

TRANSISTOR BIASING

The type of bias applied to an rf power transistor is determined by the "class" of amplification that the designer wishes. There are many different classes of amplification available for the designer to choose from. The particular class chosen for a design will depend upon the application at hand.

The primary emphasis of this chapter will be on class-C amplifiers. However, class-A and class-B amplifier bias arrangements will also be covered.

Class-A Amplifiers and Linearity

A class-A amplifier is defined as an amplifier that is biased so that the output current flows at all times.

MRF233

The RF Line

15 W – 90 MHz
RF POWER
TRANSISTOR
NPN SILICON

NPN SILICON RF POWER TRANSISTORS

. . . designed for 12.5 Volt, mid-band large-signal amplifier applications in industrial and commercial FM equipment operating in the 40 to 100 MHz range.

- Specified 12.5 Volt, 90 MHz Characteristics –
  Output Power = 15 Watts
  Minimum Gain = 10 dB
  Efficiency = 55%
- 100% Tested for Load Mismatch at all Phase Angles with 30:1 VSWR
- Characterized with Series Equivalent Large-Signal Impedance Parameters
- Characterized with Parallel Equivalent Large-Signal Impedance Parameters

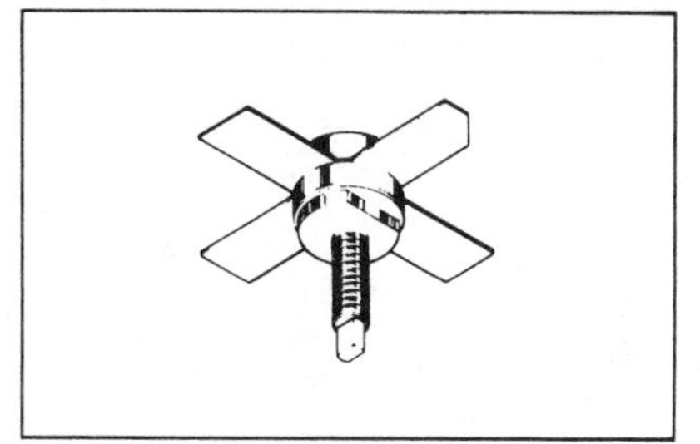

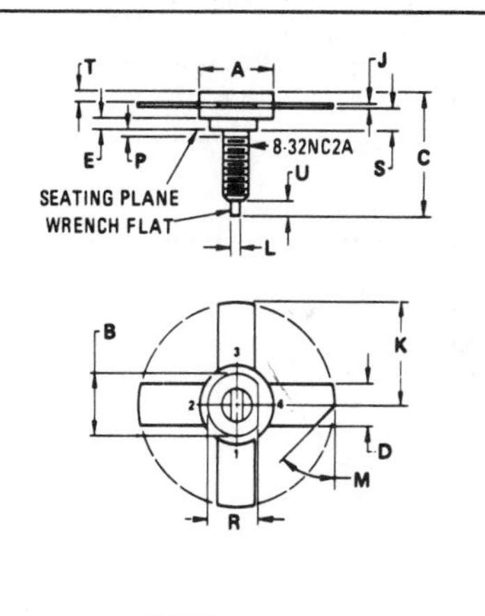

STYLE 1:
PIN 1. EMITTER
2. BASE
3. EMITTER
4. COLLECTOR

DIM	MILLIMETERS MIN	MILLIMETERS MAX	INCHES MIN	INCHES MAX
A	9.40	9.78	0.370	0.385
B	8.13	8.38	0.320	0.330
C	17.02	20.07	0.670	0.790
D	5.46	5.97	0.215	0.235
E	1.78	–	0.070	–
J	0.08	0.18	0.003	0.007
K	12.45	–	0.490	–
L	1.40	1.78	0.055	0.070
M	45° NOM		45° NOM	
P	–	1.27	–	0.050
R	7.59	7.80	0.299	0.307
S	4.01	4.52	0.158	0.178
T	2.11	2.54	0.083	0.100
U	2.49	3.35	0.098	0.132

145A-09

MAXIMUM RATINGS

Rating	Symbol	Value	Unit
Collector-Emitter Voltage	V_{CEO}	18	Vdc
Collector-Base Voltage	V_{CBO}	36	Vdc
Emitter-Base Voltage	V_{EBO}	4.0	Vdc
Collector Current – Continuous	I_C	3.5	Adc
Total Device Dissipation @ $T_C = 25°C$ (1) Derate Above 25°C	P_D –	50 285	Watts mW/°C
Storage Temperature Range	T_{stg}	-65 to +200	°C
Stud Torque (2)	–	6.5	In-lb

THERMAL CHARACTERISTICS

Characteristic	Symbol	Max	Unit
Thermal Resistance, Junction to Case	$R_{\theta JC}$	3.5	°C/W

(1) These devices are designed for RF operation. The total device dissipation rating applies only when the devices are operated as Class C RF amplifiers.
(2) For Repeated Assembly use 5 In. Lb.

Cont. on next page

Fig. 7-1. Data sheet. (*Courtesy Motorola Semiconductor Products Inc.*)

MRF233

ELECTRICAL CHARACTERISTICS (T_C = 25°C unless otherwise noted).

Characteristic	Symbol	Min	Typ	Max	Unit
OFF CHARACTERISTICS					
Collector-Emitter Breakdown Voltage (I_C = 100 mAdc, I_B = 0)	BV_{CEO}	18	–	–	Vdc
Collector-Emitter Breakdown Voltage (I_C = 50 mAdc, V_{BE} = 0)	BV_{CES}	36	–	–	Vdc
Emitter-Base Breakdown Voltage (I_E = 5.0 mAdc, I_C = 0)	BV_{EBO}	4.0	–	–	Vdc
Collector Cutoff Current (V_{CB} = 15 Vdc, I_E = 0)	I_{CBO}	–	–	1.0	mAdc
ON CHARACTERISTICS					
DC Current Gain (I_C = 1.0 Adc, V_{CE} = 5.0 Vdc)	h_{FE}	5.0	–	–	–
DYNAMIC CHARACTERISTICS					
Output Capacitance (V_{CB} = 12.5 Vdc, I_E = 0, f = 1.0 MHz)	C_{ob}	–	100	120	pF
FUNCTIONAL TESTS (Figure 1)					
Common-Emitter Amplifier Power Gain (V_{CC} = 12.5 Vdc, P_{out} = 15 W, f = 90 MHz)	G_{PE}	10	–	–	dB
Collector Efficiency (V_{CC} = 12.5 Vdc, P_{out} = 15 W, f = 90 MHz)	η	55	–	–	%
Load Mismatch (V_{CC} = 12.5 Vdc, P_{out} = 15 W, f = 90 MHz, $T_C \leqslant$ 25°C)	–	VSWR > 30:1 Through All Phase Angles in a 3 Second Interval After Which Devices Will Meet G_{PE} Test Limits			

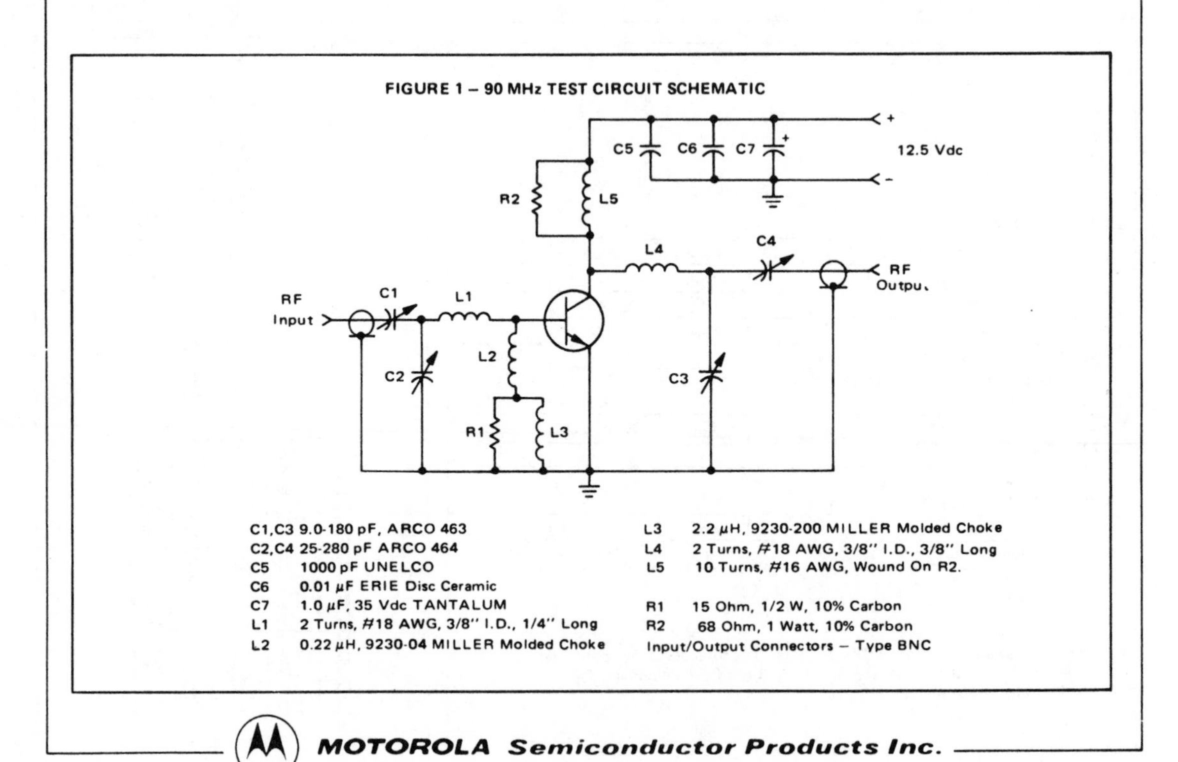

MOTOROLA Semiconductor Products Inc.

Cont. on next page

Fig. 7-1.—Cont. Data sheet. (*Courtesy Motorola Semiconductor Products Inc.*)

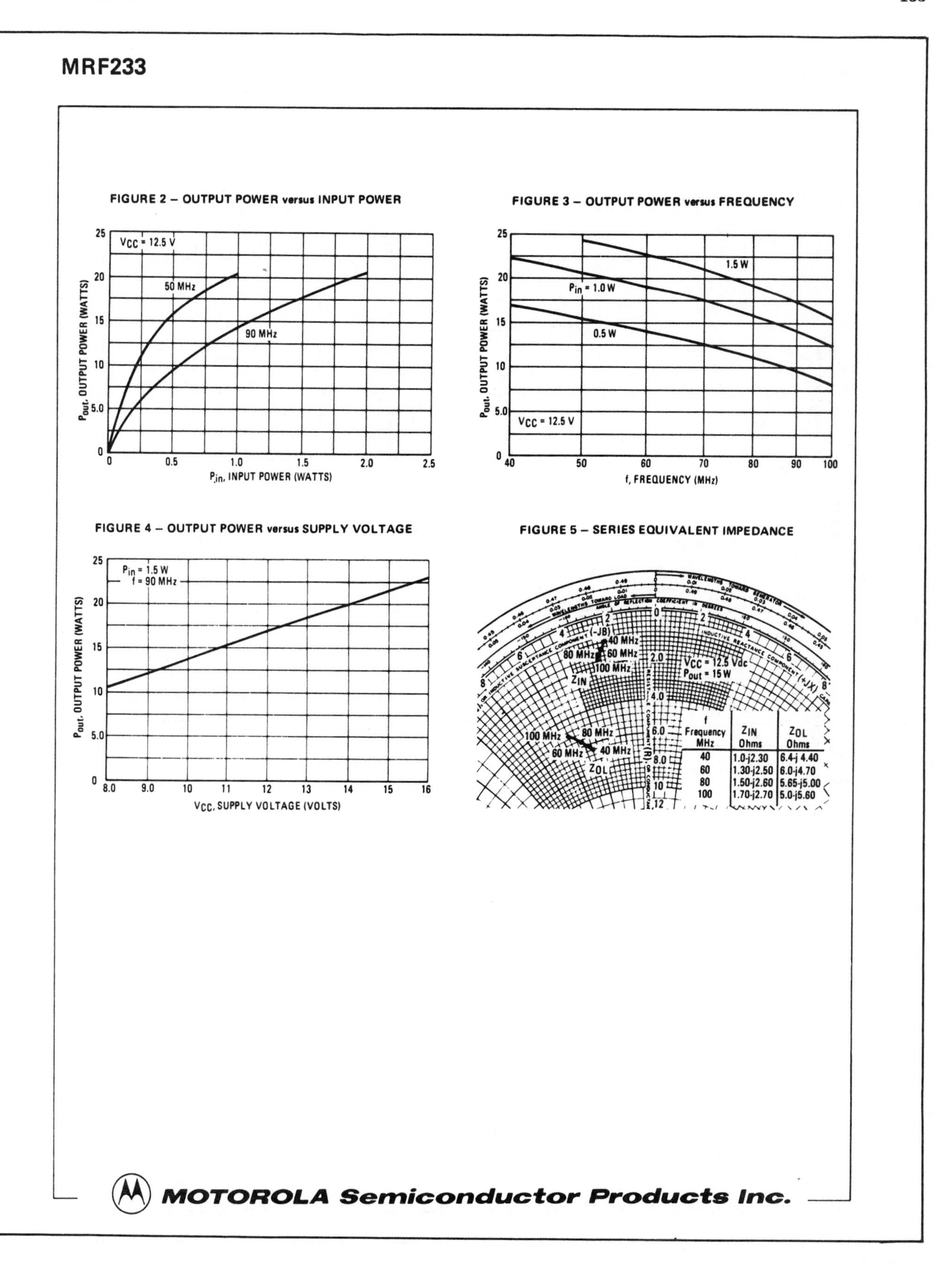

f Frequency MHz	Z_{IN} Ohms	Z_{OL} Ohms
40	1.0-j2.30	6.4-j 4.40
60	1.30-j2.50	6.0-j4.70
80	1.50-j2.60	5.65-j5.00
100	1.70-j2.70	5.0-j5.60

Cont. on next page

Fig. 7-1.—Cont. Data sheet. (*Courtesy Motorola Semiconductor Products Inc.*)

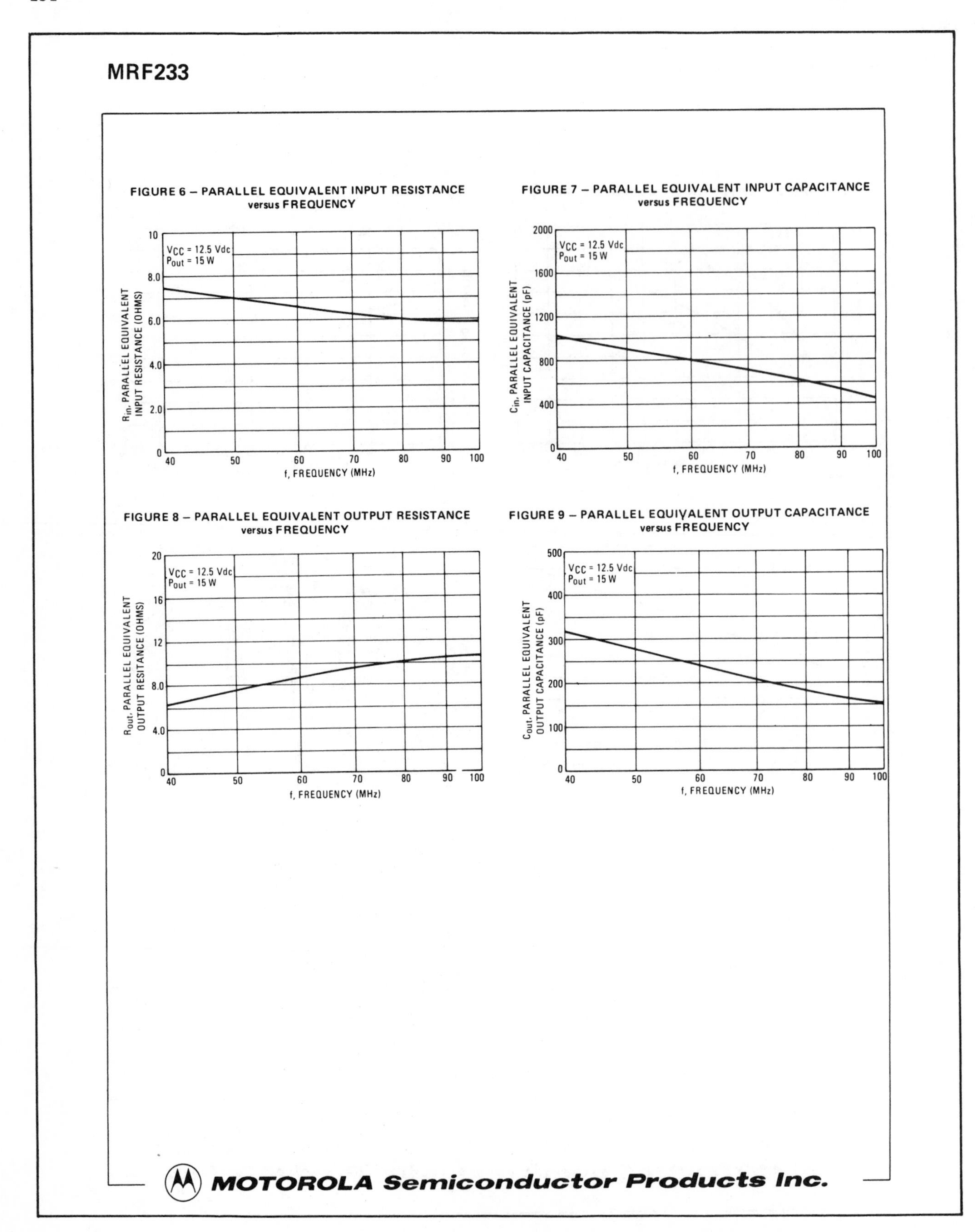

Fig. 7-1.—Cont. Data sheet. (*Courtesy Motorola Semiconductor Products Inc.*)

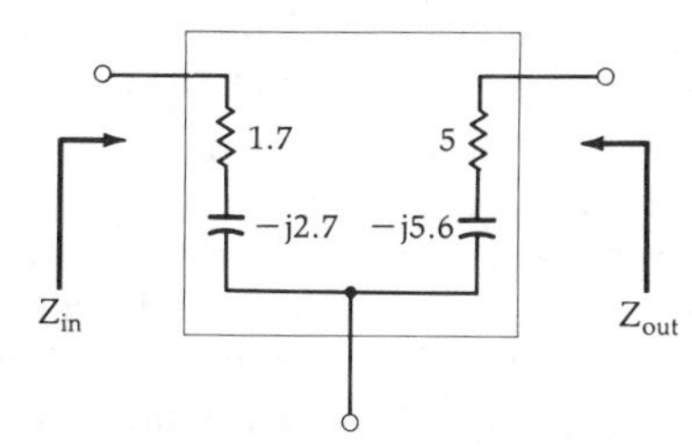

Fig. 7-2. Equivalent circuit for series input and output impedance at 100 MHz.

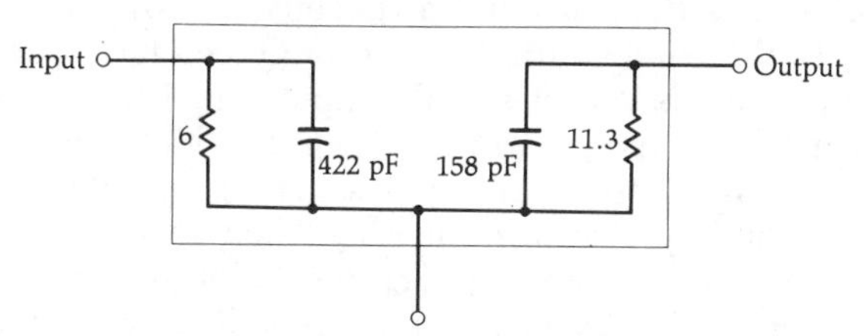

Fig. 7-3. Equivalent circuit for parallel input and output impedance at 100 MHz.

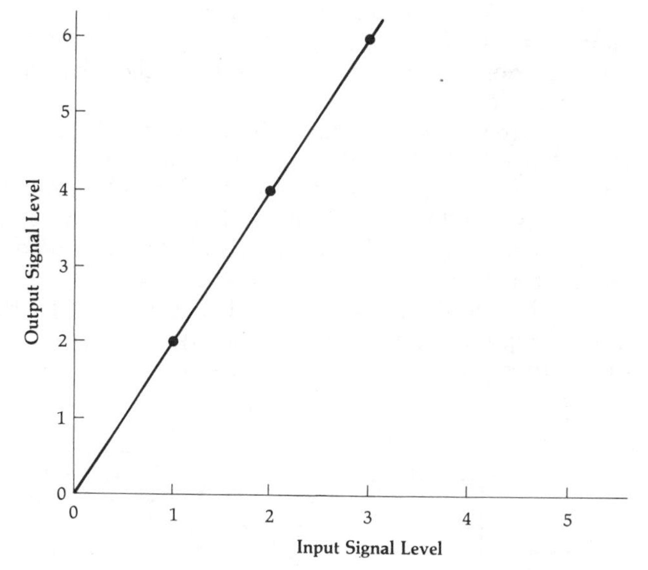

Fig. 7-4. Transfer characteristic for a linear amplifier.

Thus, the input signal-drive level to the amplifier is kept small enough to avoid driving the transistor into cutoff. Another way of stating this is to say that the *conduction angle* of the transistor is 360°—meaning that the transistor conducts for the full cycle of the input signal.

The class-A amplifier is the most *linear* of all amplifier types. Linearity is simply a measure of how closely the output signal of the amplifier resembles the input signal. A linear amplifier is one in which the output signal is proportional to the input signal, as shown in Fig. 7-4. Notice that in this case, the output signal level is equal to twice the input signal level, and the transfer function from input to output is a straight line.

No transistor is perfectly linear, however, and, therefore, the output signal of an amplifier is never an exact replica of the input signal. There are always spurious components added to a signal in the form of *harmonic* generation or *intermodulation distortion* (IMD). These types of nonlinearities in transistors produce amplifier transfer functions which no longer resemble straight lines. Instead, a curved characteristic appears, as shown in Fig. 7-5A. The distortion caused to an input signal of such an amplifier is shown at Fig. 7-5B. Notice the *flat topping* of the output signal that occurs due to the second-harmonic content generated by the amplifier. This type of distortion is called harmonic distortion and is expressed by the equation:

$$V_{out} = AV_{in} + BV_{in}^2 + CV_{in}^3 + \ldots\ldots \quad \text{(Eq. 7-1)}$$

The second term of Equation 7-1 is known as the *second harmonic* or *second-order distortion.* The third term is called the *third harmonic* or *third-order distortion.* Of course, a perfectly linear amplifier will produce no second, third, or higher order products to distort the signal.

Notice in Fig. 7-5, where the amplifier's transfer function is given as $V_{out} = 5V_{in} + 2V_{in}^2$, that the second-order distortion component increases as the square of the input signal. Thus, with increasing input-signal levels, the second-order component will increase much faster than the fundamental component in the output signal. Eventually, the second-order content in the output signal will equal the amplitude of the fundamental. This effect is shown graphically in Fig. 7-6. The point at which the second-order and first-order content of the output signal are equal is called the *second-*

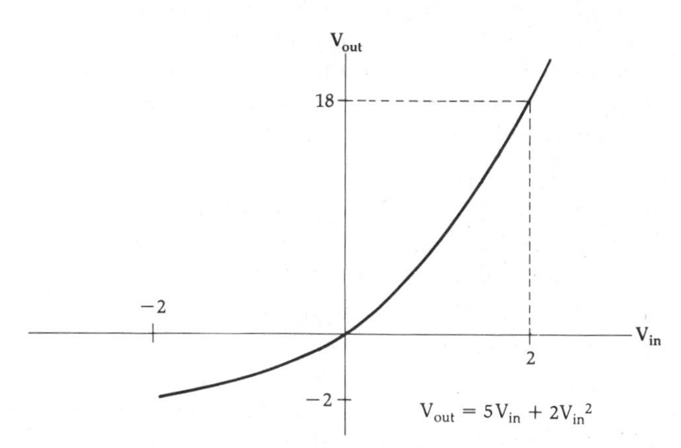

(*A*) *Transfer characteristic.*

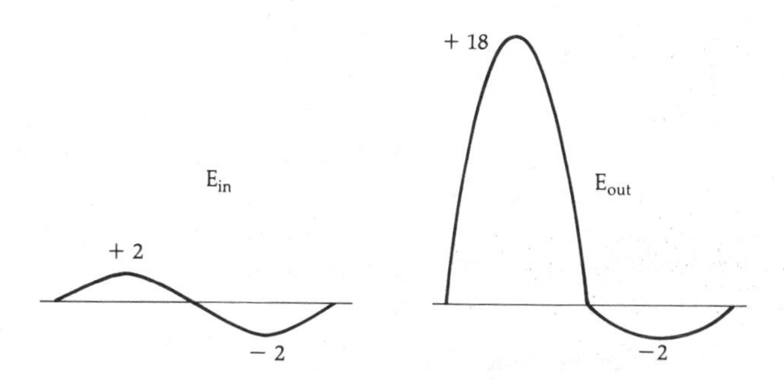

(*B*) *Resulting waveforms.*

Fig. 7-5. Nonlinear amplifier characteristics.

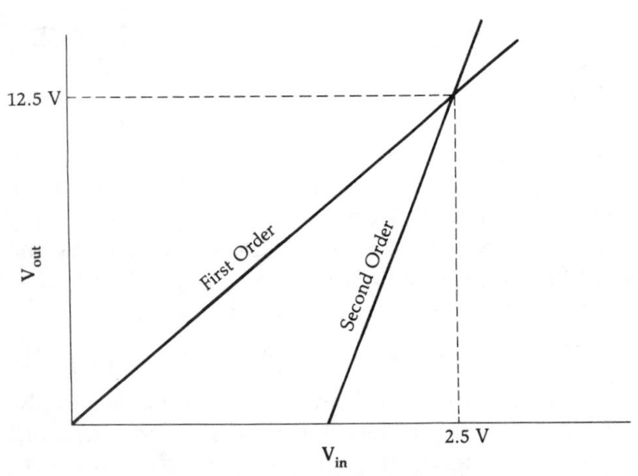

Fig. 7-6. Second-order intercept point.

order intercept point. A similar graph may be drawn for an amplifier which exhibits a third-order distortion characteristic. In this case, the third-order term is plotted along with the fundamental gain term of the amplifier. In this manner, the third-order intercept may be determined. The second- and third-order intercept of an amplifier are often used as figures of merit. The higher the intercept point, the better the amplifier is at amplifying large signals.

When two or more signals are input to an amplifier simultaneously, the second-, third-, and higher-order *intermodulation* components are caused by the sum and difference products of each of the fundamental input signals and their associated harmonics. For example, when two perfect sinusoidal signals, at frequencies f_1 and f_2, are input to any nonlinear amplifier, the following output components will result:

fundamental: f_1, f_2
second order: $2f_1, 2f_2, f_1 + f_2, f_1 - f_2$
third order: $3f_1, 3f_2, 2f_1 \pm f_2, 2f_2 \pm f_1$
+ higher order terms

Under normal circuit operation, the second-, third-, and higher-order terms are usually at a much smaller signal level than the fundamental component and, in the time domain, this is seen as distortion. Note that, if f_1 and f_2 are very close in frequency, the $2f_1 - f_2$ and $2f_2 - f_1$ terms fall very close to the two fundamental terms. Third-order distortion products are, therefore, much more difficult to eliminate through filtering once they are generated within an amplifier.

The bias requirements for a class-A power amplifier are the same as those for the small-signal amplifiers presented in Chapter 6. In fact, the distinction between a class-A power amplifier and its small-signal counterpart is a hazy one at best. For all practical purposes, they are equivalent except for input and output signal levels.

Class-B Power Amplifiers

A class-B amplifier is one in which the conduction angle for the transistor is approximately 180°. Thus, the transistor conducts only half the time—either on the positive or negative half cycle of the input signal. Again, it is the dc bias applied to the transistor that determines the class-B operation.

Class-B amplifiers are more efficient than class-A amplifiers (70% vs. less than 50%). However, they are much less linear. Therefore, a typical class-B amplifier will produce quite a bit of harmonic distortion that must be filtered from the amplified signal.

Probably the most common configuration of a class-B amplifier is the push-pull arrangement shown in Fig. 7-7. In this configuration, transistor Q_1 conducts during the positive half cycles of the input signal while transistor Q_2 conducts during the negative half cycles. In this manner, the entire input signal is reproduced at the secondary of transformer T_2. Thus, neither device by itself produces an amplified replica of the input signal. Instead, the signal is actually split in half. Each half is then amplified and reassembled at the output.

Of course, a single transistor may be used in a class-B configuration. The only requirement is that a resonant circuit must be placed in the output network of the transistor in order to reproduce the "other" half of the input signal.

There are several methods of biasing a transistor for class-B operation. One of the most widely used methods is shown in Fig. 7-8. This method simply establishes a base voltage of 0.7 volt on the transistor, using an external silicon diode. Often, this diode is mounted on the transistor itself to help prevent *thermal runaway*, which is often a problem with incorrectly biased power amplifiers. Diode CR1 is usually of the heavy-duty variety because the value of resistor R is usually

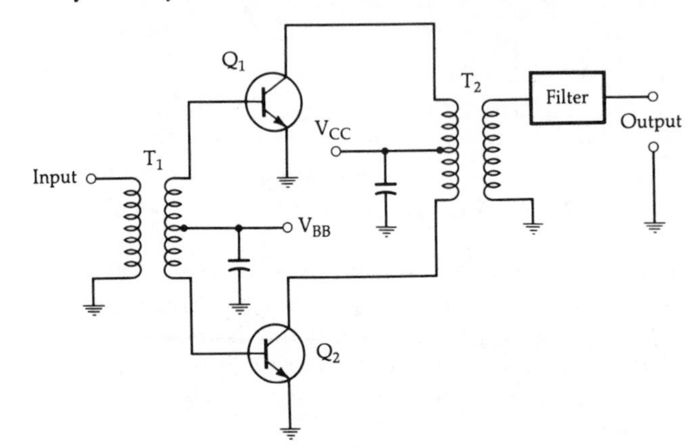

Fig. 7-7. Push-pull class-B amplifier.

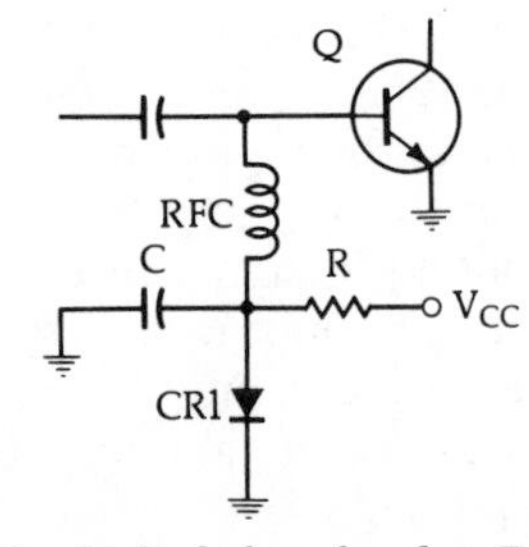

Fig. 7-8. Simple diode bias for class-B operation.

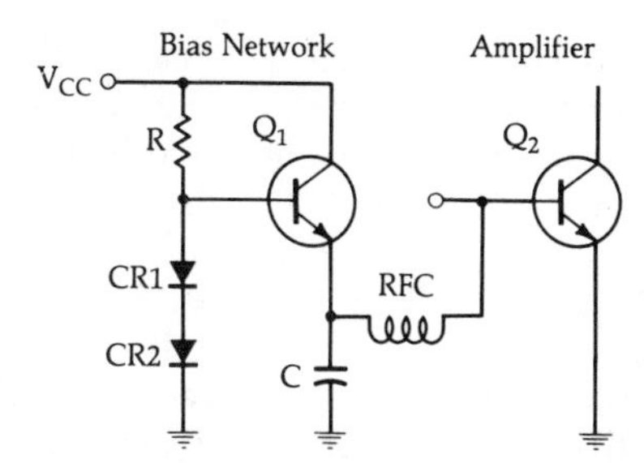

Fig. 7-9. Emitter-follower bias for class-B operation.

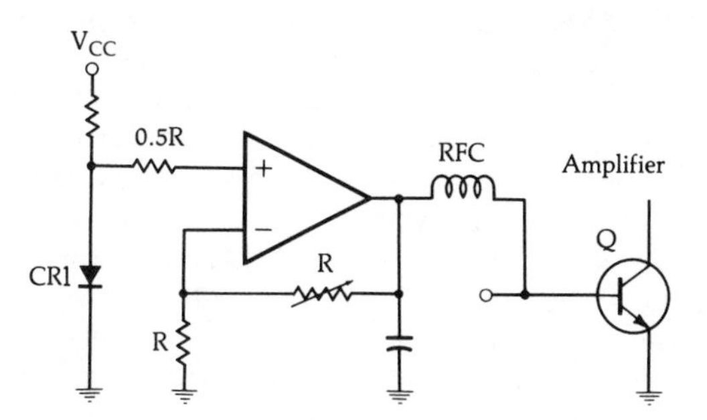

Fig. 7-10. Operational amplifier bias for class-B operation.

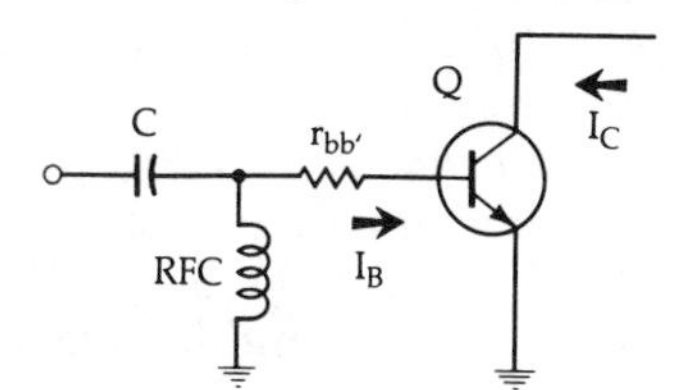

Fig. 7-11. Class-C self bias.

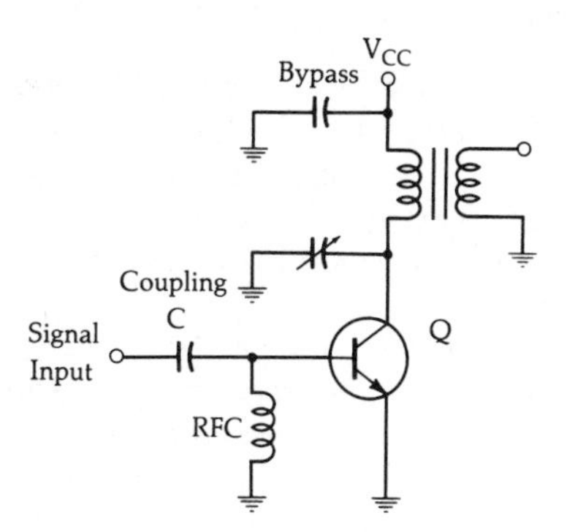

Fig. 7-12. Circuit for class-C self bias.

chosen so that the current through CR1 is rather high. This ensures that the bias to the transistor is stable. An alternative bias network is shown in Fig. 7-9. Here, two silicon diodes are used to forward bias an emitter-follower, which is used as a current amplifier. The voltage at the emitter of Q_1 and, hence, at the base of Q_2, is still 0.7 volt due to the V_{BE} drop across transistor Q_1. The rf choke and capacitor shown in both Figs. 7-8 and 7-9 are there only to prevent the flow of rf into the bias network.

Still another bias arrangement for class-B operation is shown in Fig. 7-10. Here the bias voltage is made variable so that an optimum solution may be found for best IMD performance. Care must be taken in all three bias arrangements to ensure that the RFC is a low-Q choke for optimum operation.

Class-C Power Amplifiers

A class-C amplifier is one in which the conduction angle for the transistor is significantly less than 180°. The transistor is biased such that under steady-state conditions no collector current flows. The transistor idles at cutoff. Linearity of the class-C amplifier is the poorest of the classes of amplifiers. Its efficiency can approach 85%, however, which is much better than either the class-B or the class-A amplifier.

In order to bias a transistor for class-C operation, it is necessary to *reverse bias* the base-emitter junction. External biasing is usually not needed, however, because it is possible to force the transistor to provide its own bias. This is shown in Fig. 7-11. If the base of the transistor is returned to ground through an rf choke (RFC), the base current flowing through the internal base-spreading resistance ($r_{bb'}$) tends to reverse bias the base-emitter junction. This is exactly the effect you would like to achieve. Of course, it is possible to provide an external dc voltage to reverse bias the junction, but why bother with the extra time and expense if the transistor will provide everything you need. Fig. 7-12 shows a typical class-C amplifier bias arrangement.

POWER AMPLIFIER DESIGN

At the beginning of this chapter, you learned that the important design information for rf power transistors is presented in the form of large-signal impedance parameters. The formulas presented in Chapter 6 for small-signal transistor design, using Y and S parameters, are no longer valid. Instead, the designer must model with the help of the data sheet, deciding what the input and output impedance of the transistor looks like at the frequency of interest. With this information in hand, the designer needs only to match the input and output impedance of the device to the source and load, respectively. These two steps require only that the designer read the input and output impedances off of the data sheet, and then apply the principles of Chapter 4 to complete the matching network. Care must be taken to ensure that the information extracted from the data sheet is of the proper format—series or shunt information.

Often, instead of supplying complete output information for a transistor in the form of a series or shunt resistance and capacitance output, manufacturers will supply output capacitance information only. This is because the optimum load resistance for the transistor is very easily calculated using a very simple formula, as we shall soon see.

Optimum Collector Load Resistance

In the absence of collector output *resistance* information on the data sheet, it becomes necessary for the designer to make a very simple calculation to determine the optimum load resistance for the transistor (Example 7-1). This value of load resistance is dependent upon the output power level required and is given by:

$$R_L = \frac{(V_{CC} - V_{sat})^2}{2P} \qquad \text{(Eq. 7-2)}$$

where,
V_{CC} = the supply voltage,
V_{sat} = the saturation voltage of the transistor,
P = the output power level required.

EXAMPLE 7-1

What value of load resistance is required to obtain 2.0 watts of rf output from a transistor if the supply voltage is 12 volts and the saturation voltage of the transistor is 2 volts?

Solution

Using Equation 7-2, we have

$$R_L = \frac{(12 - 2)^2}{2(2)}$$
$$= 25 \text{ ohms}$$

Note that Equation 7-2 provides you with a value of load *resistance* only. It does not indicate anything about the reactive portion of the load. On the data sheet, however, the manufacturer typically provides values of shunt output capacitance versus frequency for the transistor. The designer's job is to provide a load for the transistor which absorbs this stray or parasitic capacitance so that the transistor may be matched to its load (see Chapter 4). Example 7-2 may illustrate this point.

Keep in mind that if the output *resistance* information had not been provided in the data sheet, it would have been a simple matter to calculate the required R_L using Equation 7-2. Once this calculation is made, the output matching network is designed in the same manner as was done in Example 7-2. The 50-ohm load is simply transformed into the impedances that the transistor would like to see for the specified power output.

Driver Amplifiers and Interstage Matching

Often it is required that power gain be distributed throughout several amplifier stages in order to produce a specified output power into a load. This is especially true in transmitter applications that require a substantial amount of power into an antenna.

The typical procedure for such a design involves finding, first, an output transistor that will handle the required *output* power and, then, designing driver amplifiers that will provide the necessary *drive* power to the final transistor. This type of gain distribution is shown in Fig. 7-17. Note that the required output-power level from the final amplifier is 15 watts. A final transistor was chosen which will handle the required output power and which will provide a gain of 10 dB. The required drive level to the stage is, therefore, 1.5 watts, and is supplied by a transistor with a gain of 15 dB. The signal source must, therefore, supply the driver with a signal level of 47 milliwatts, which is within the capabilities of most oscillators.

Let's examine the interstage match between Stage A and Stage B in a little more detail. Often, in dealing with power amplifiers, it is unclear whether or not a true *impedance match* occurs between the power amplifier and its load. A true impedance match for an amplifier would involve providing a load for the transistor that is the complex conjugate of its output impedance. In designing power amplifiers, however, we speak of providing a load resistance (Equation 7-2) for the transistor in order to extract a specified power gain from the stage. This is simply a matter of semantics and, from a circuit-design viewpoint, it doesn't really matter how you look at it. Fig. 7-18 illustrates this point. Suppose the transistor of Stage B has an input impedance as shown ($Z_{in} = 1.7 - j2.7$ ohms). Also, suppose that Stage A, in order to supply the required 1.5 watts of rf drive, requires a load resistance of 25 ohms. The role of the *impedance-matching* network, then, is to *transform* the low-input impedance of Stage B up to the 25-ohm level required by Stage A. In addition, the matching network must absorb or resonate out the 15-pF output capacitance of Stage A.

MATCHING TO COAXIAL FEEDLINES

The T and Pi networks studied in Chapter 4 are excellent candidates for use in matching coaxial feedlines to power amplifiers. Often such a network will serve a dual purpose, especially when configured as a low-pass filter, in providing harmonic suppression for a transmitter.

Fig. 7-19 is a diagram of a coaxial feed to an antenna at the antenna's resonant frequency. Resistance R_a is the antenna's radiation resistance. A quarter-wavelength vertical antenna operating against a very good ground plane has a *radiation resistance* of about 35 ohms while a half-wave center-fed dipole has a radiation resistance of about 70 ohms—at its resonant frequency. This is simply the resistance that the coaxial cable sees at the antenna terminals. Above and below the resonant frequency of the antenna, its radiation resistance begins to show a reactive component. This is illustrated in Fig. 7-20. Above resonance (Fig. 7-20A), the antenna looks inductive, and below resonance (Fig. 7-20B), the antenna looks capacitive.

At the transmitter end of the coaxial feedline, the impedance that the output transistor actually sees is not only a function of the antenna's radiation resistance, but also a function of the length of the coaxial feedline. The impedance along the line varies sinusoidally as you move away from the antenna. Thus, at a distance of one half-wavelength back from the an-

EXAMPLE 7-2

Using the data sheet of Fig. 7-1, design a class-C power amplifier that will deliver 15 watts between a 50-ohm source and load at 100 MHz.

Solution

The data sheet for the MRF233 transistor provides input and output impedance information for the transistor in both series and parallel form. The designer is, therefore, left with a choice as to which he prefers. The input-matching design may proceed as in the following method.

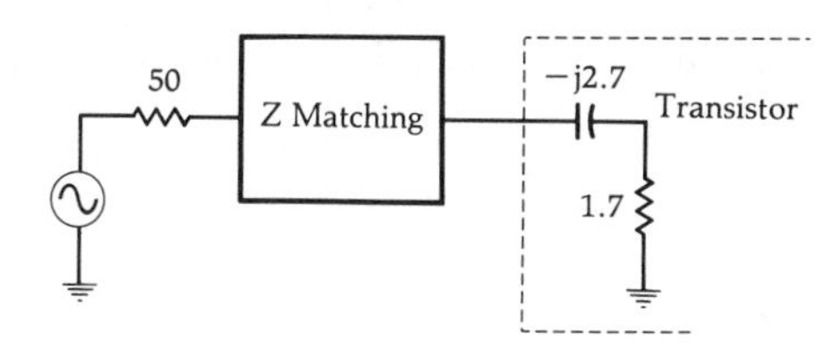

Fig. 7-13. Transistor input impedance for Example 7-2.

The input impedance of the transistor appears as shown in the diagram of Fig. 7-13. This information was taken from Figure 5 of the data sheet. Note that the object of the input-matching network is to transform the input impedance of the transistor up to 50 ohms to provide an optimum load for the source.

Using the techniques of Chapter 4, first resonate the series capacitance with an equal and opposite series inductance of +j2.7 ohms. Then, match the remaining 1.7-ohm resistive load to the source as follows. Using Equation 4-1 through 4-3, for the L-network:

$$Q_S = Q_P = \sqrt{\frac{R_P}{R_S} - 1} \quad \text{(Eq. 4-1)}$$

$$= \sqrt{\frac{50}{1.7} - 1}$$

$$= 5.33$$

$$X_S = R_S Q_S \quad \text{(Eq. 4-2)}$$

$$= (1.7)(5.33)$$

$$= 9.06 \text{ ohms}$$

$$X_P = \frac{R_P}{Q_P} \quad \text{(Eq. 4-3)}$$

$$= \frac{50}{5.33}$$

$$= 9.38 \text{ ohms}$$

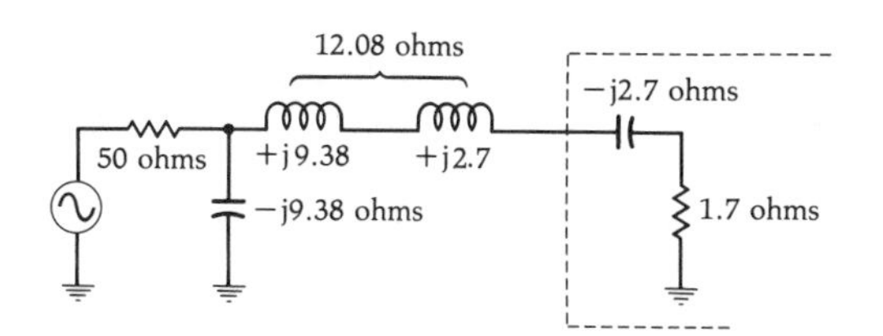

Fig. 7-14. Input matching network for Example 7-2.

This matching network is shown in Fig. 7-14. Note that it is convenient to use a shunt-C series-L matching network in order to easily absorb the +j2.7-ohm inductor that was needed earlier. Thus, the two inductors may be combined into a single component in the actual design.

Since complete output-impedance information is provided for the transistor (Figure 5 of the data sheet), it is only necessary to match this output impedance to the 50-ohm load. Proceeding as before, first resonate the −j5.6-ohm series capacitance with an inductor of equal value. Then, match the remaining 5-ohm resistive portion of the transistor output to the load as follows:

$$Q_S = Q_P = \sqrt{\frac{R_P}{R_S} - 1}$$

$$= \sqrt{\frac{50}{5} - 1}$$

$$= 3$$

$$X_S = Q_S R_S$$

$$= (3)(5)$$

$$= 15 \text{ ohms}$$

$$X_P = \frac{R_P}{Q_P}$$

$$= \frac{50}{3}$$

$$= 16.7 \text{ ohms}$$

The output network is shown in Fig. 7-15. Note that it is again convenient to use the series-L shunt-C arrangement so that we may absorb the +j5.6-ohm inductor used previously.

A practical circuit for this design might appear as shown in Fig. 7-16.

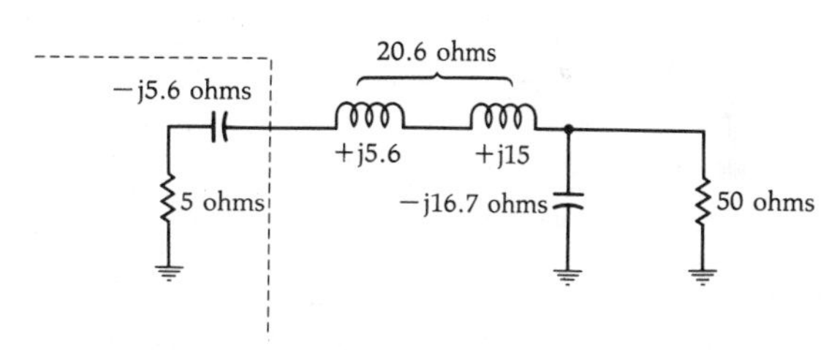

Fig. 7-15. Output matching network for Example 7-2.

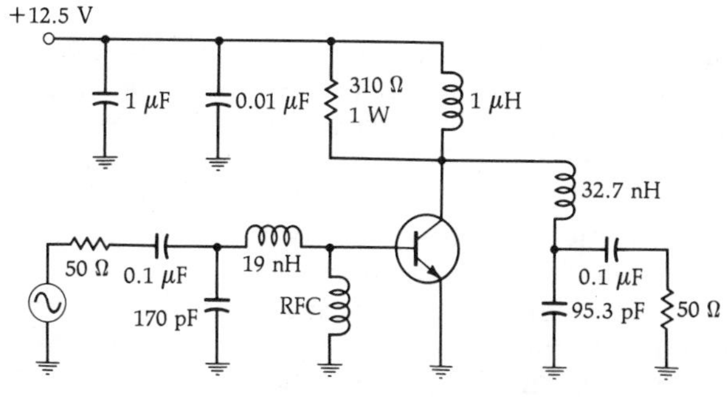

Fig. 7-16. Final circuit realization for Example 7-2.

tenna, the impedance looking into the coax would appear to be equal to the antenna's radiation resistance. At other distances removed from the antenna, however, the coax would appear to have a much different input impedance depending upon the degree of mismatch between the antenna and the feedline. Therefore, it is extremely difficult to estimate the actual input impedance of any transmission line *unless the line*

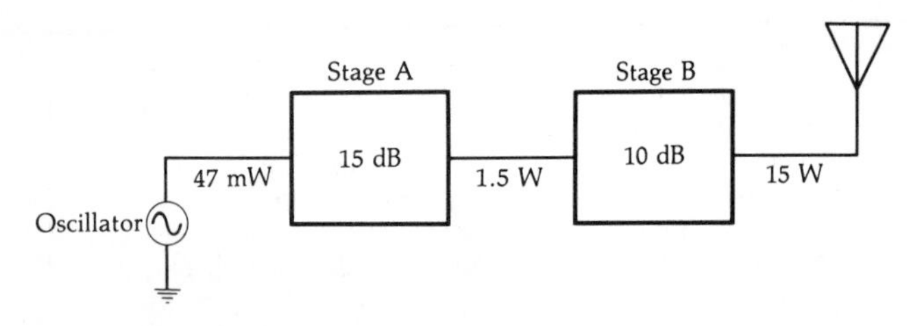

Fig. 7-17. System drive requirements for a 15-watt transmitter.

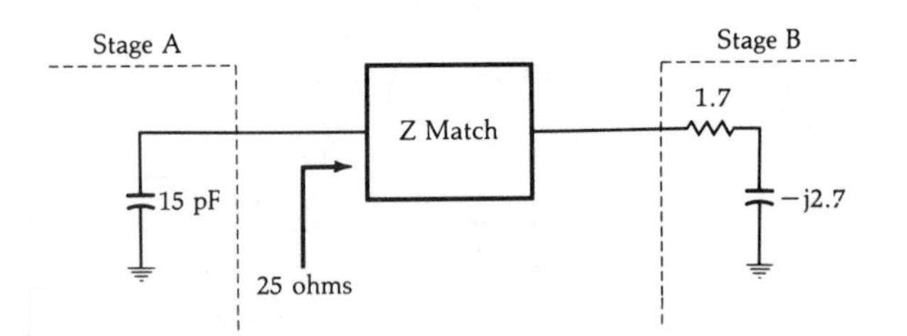

Fig. 7-18. Requirement for an interstage impedance-matching network.

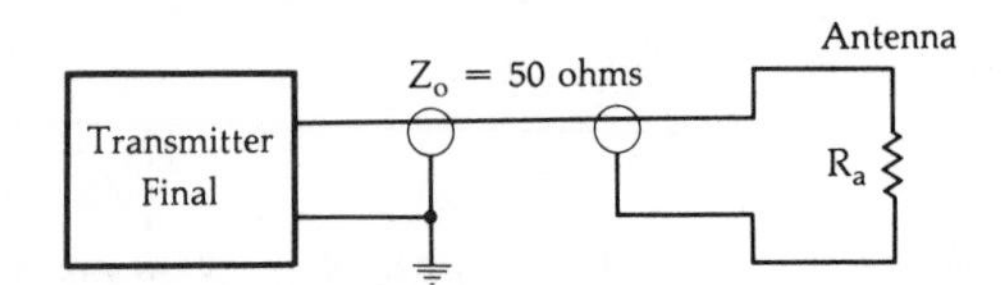

Fig. 7-19. Antenna radiation resistance at resonance.

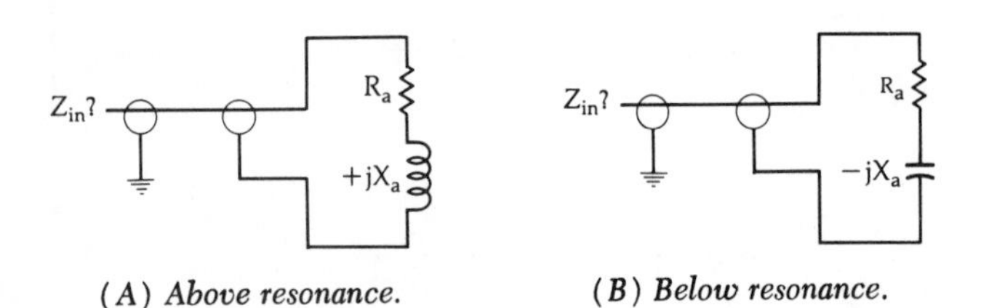

(*A*) *Above resonance.* (*B*) *Below resonance.*

Fig. 7-20. Radiation resistance of an antenna.

is terminated in its characteristic impedance. That is to say, a 50-ohm coaxial cable will not look like 50 ohms at its input unless there is a 50-ohm load at the other end of the cable. Since this is hardly ever the case when driving practical antenna systems, it is not very practical to design a matching network unless the network is tunable. In addition, many antenna installations operate over quite a range of frequencies. Since the radiation resistance of the antenna varies with frequency, the input impedance of the coaxial cable must also vary, and the matching network must be able to compensate for these variations.

Fig. 7-21 indicates two possible methods of providing a tunable impedance-matching network for a transmission line. The T network of Fig. 7-21A uses both tapped inductors and a tunable capacitor. The Pi network of Fig. 7-21B uses only tunable capacitors. Note that, in both cases, the low-pass configuration is used to aid in suppressing harmonics of the transmitted signal. The circuits of Fig. 7-21 are designed in the same manner as those shown in Chapter 4. Of course, if you had a requirement that the harmonics of the transmitted signal were required to be at a certain level below the fundamental, say 50 dB, then the filter-design approach used in Chapter 3 might be the best approach to take.

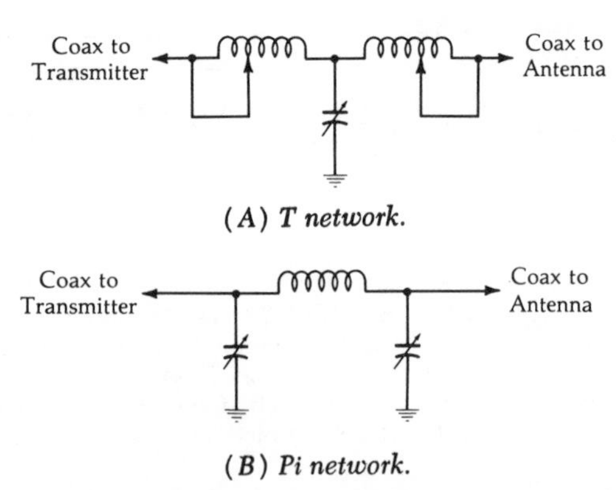

(*A*) *T network.*

(*B*) *Pi network.*

Fig. 7-21. Variable coaxial feedline matching networks.

AUTOMATIC SHUTDOWN CIRCUITRY

Since power amplifiers are designed to supply a considerable amount of power to an antenna system, an impedance mismatch presented to the amplifier could cause very severe problems. As we learned in Chapter 4, an impedance mismatch between a source and its load causes reflection of some of the signal incident upon that load. This reflected signal will eventually make its way back to the source and, in high-power transmitters, can cause serious side-effects, such as transistor damage in the form of secondary breakdown. For this reason, many manufacturers of power amplifiers include a VSWR monitoring circuit in the transmitter, which monitors the *standing wave ratio* of the output circuit and, in the event that the VSWR becomes excessive, indicating a severe mismatch, the circuit automatically decreases the rf drive to the final amplifier, thereby reducing the transmitter's output power. The reduction in output power subsequently reduces the reflected power from the load, thus, protecting the output transistor. A simplified diagram of such a system appears in Fig. 7-22.

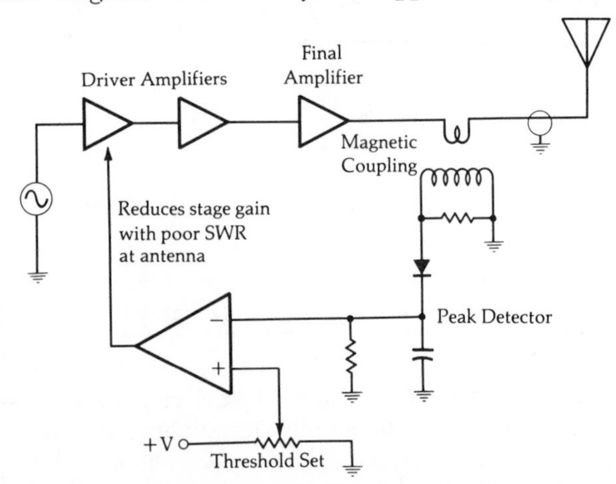

Fig. 7-22. Automatic shutdown circuitry.

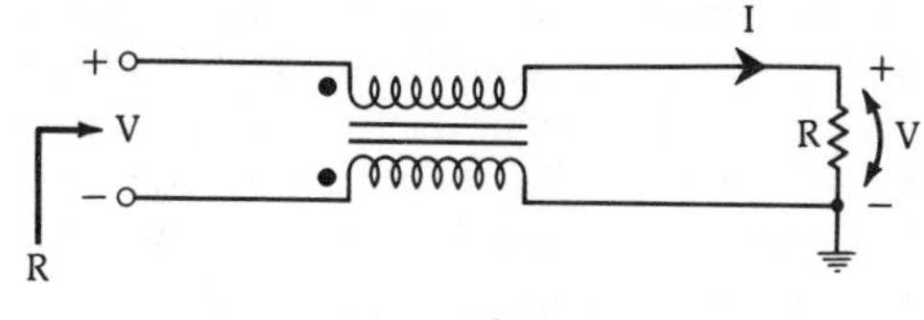

(A) 1:1 balun.

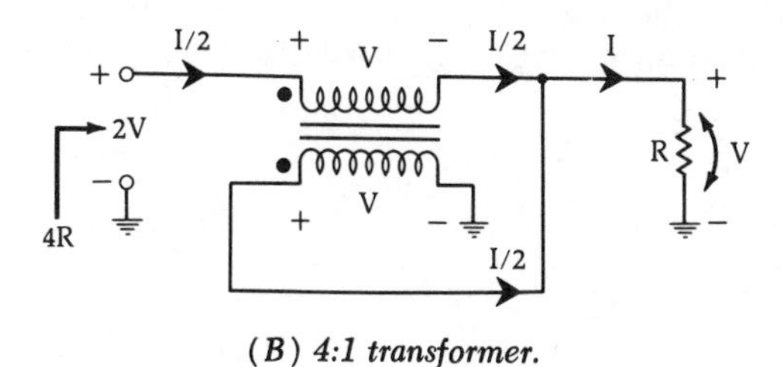

(B) 4:1 transformer.

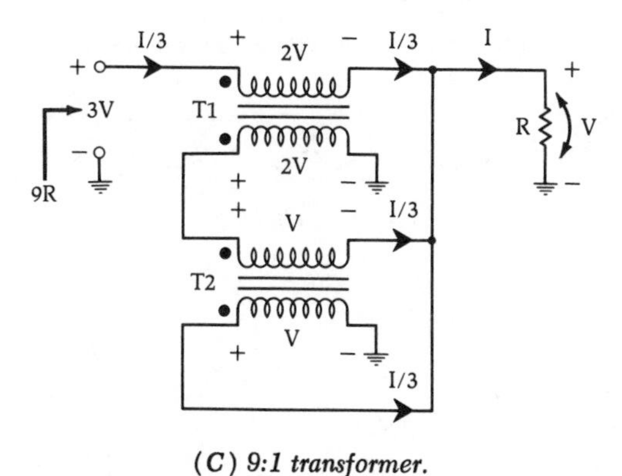

(C) 9:1 transformer.

Fig. 7-23. Types of broadband transformers.

BROADBAND TRANSFORMERS

Several types of broadband transformers, which are often used in power-amplifier design, are illustrated in Fig. 7-23. Fig. 7-23A is known as a 1:1 *balun*. It is used mainly to connect a *bal*anced source to an *un*balanced load, or vice-versa, and it provides no impedance-transforming function.

Fig. 7-23B is a 4:1 transformer. That is to say, it will transform an impedance of 4R down to an impedance of R, or vice-versa. The small dot located next to each winding indicates polarity. A detailed look at Fig. 7-23B will explain how the 4:1 transformation occurs. First, suppose that a voltage (V) has been impressed across the load resistor and the voltage across the same voltage must also be impressed across the lower winding of the transformer since the two are in parallel. The voltage on the lower winding impresses the same voltage (V) on the *upper* winding with the polarity indicated. This is true because each winding has the same number of turns. The voltage at the input terminals is, therefore, equal to the sum of the voltage across the load resistor and the voltage across the upper winding, or 2V. Suppose now that a current of I/2 is injected at the input terminals of the transformer. This current flowing in the top winding of the transformer induces a current of I/2 in the bottom winding in the direction shown. The current in the load resistor is, therefore, equal to I/2 + I/2 or I. Therefore, if the load resistor is equal to 1 ohm, the resistance seen looking into the input terminals of the transformer, then, must be:

$$\text{Resistance} = \frac{\text{Voltage}}{\text{Current}} = \frac{2V}{I/2} = 4R$$

A 9:1 transformer is illustrated in Fig. 7-23C. Note that this transformer is actually made up of two separate transformers, T1 and T2. A similar analysis may be made of this transformer to confirm the 9:1 transformation. Voltages and currents are included in the diagrams to aid in the analysis process. Obviously, several transformers may be included in such a configuration to produce other transformation ratios. Fig. 7-24, for example, includes three separate transformers that are used to produce a 16:1 impedance transformation.

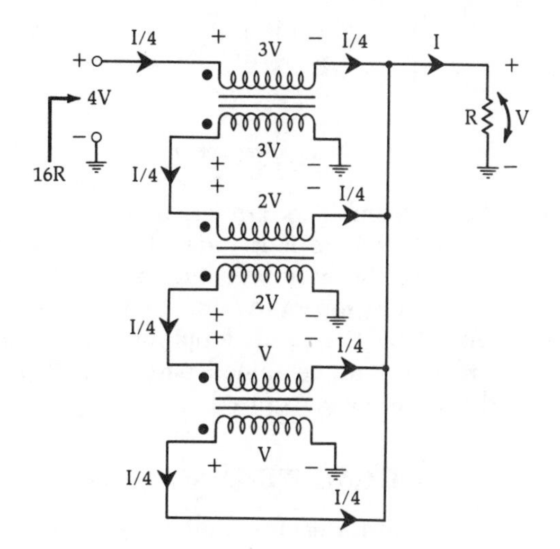

Fig. 7-24. A 16:1 broadband transformer.

Power Splitters

A basic power splitter is shown in Fig. 7-25. Ideally, the power into the primary of the transformer is split evenly between amplifier No. 1 and amplifier No. 2. However, due to input-impedance variations between the two amplifiers, this is rarely the case. Instead, one amplifier is usually provided with a bit more drive power than the other. To aid in equalizing the power split, resistor R at the center tap of the secondary is often left out of the circuit. (Once again, the small dots are used to indicate polarity.)

Power Combiners

A typical power combiner is shown in Fig. 7-26. Here, the power output of each amplifier is combined in transformer T1 to provide an output power of

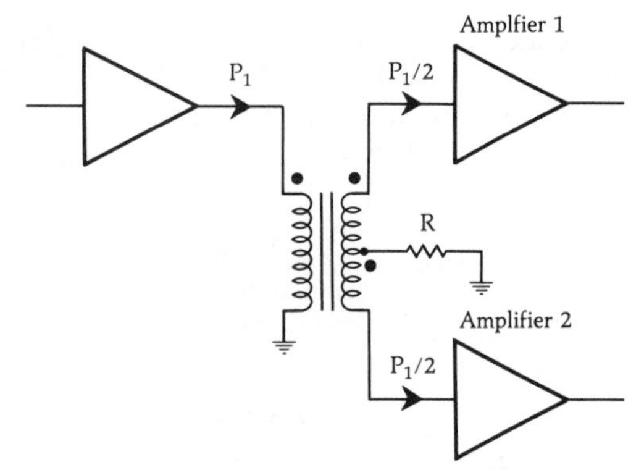

Fig. 7-25. A power splitter.

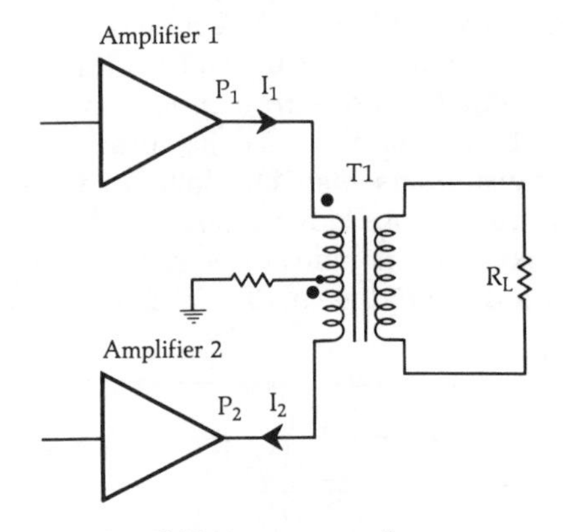

Fig. 7-26. A power combiner.

$P_1 + P_2$. Power combiners are often used in power-amplifier designs where it is impractical for a single stage to produce the necessary output power. In this case, two amplifiers, operating 180° out-of-phase, will each provide half the needed output power to the power combiner. The combined output is, therefore, equal to the power level required.

PRACTICAL WINDING HINTS

Broadband transformers are often called transmission-line transformers because they make use of the transmission-line properties of the windings. This is done by using *bifilar-* or *trifilar*-type windings rather than the conventional type of winding.

A conventional transformer usually has two entirely separate windings. That is, one of the windings is usually wound onto the core first and, then, the other winding is wound on top of the first winding. Typically, the larger winding is wound first for convenience. This winding technique is shown in the toroidal transformer diagram of Fig. 7-27. Note that an impedance transformation occurs between the primary and secondary of the transformer. The value of the transformation is dependent upon the turns ratio from the primary to the secondary. Transmission-line transformers use an entirely different technique for the windings, as shown in Fig. 7-28. First, the primary and secondary windings are made by twisting the wires together for a certain number of turns per inch (Fig. 7-28A). This produces a certain characteristic impedance for the resulting "transmission line" in much the same manner that a coaxial cable exhibits a certain characteristic impedance which is dependent upon the spacing of its center conductor to its outer conductor. The actual characteristic impedance of the twisted pair is dependent upon the number of turns per inch, the shape of the windings, and the size wire used. For low-impedance lines, tight twists (many turns per

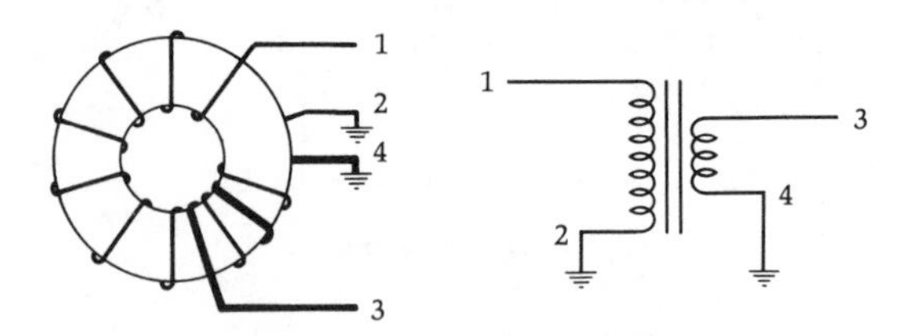

Fig. 7-27. A conventional transformer.

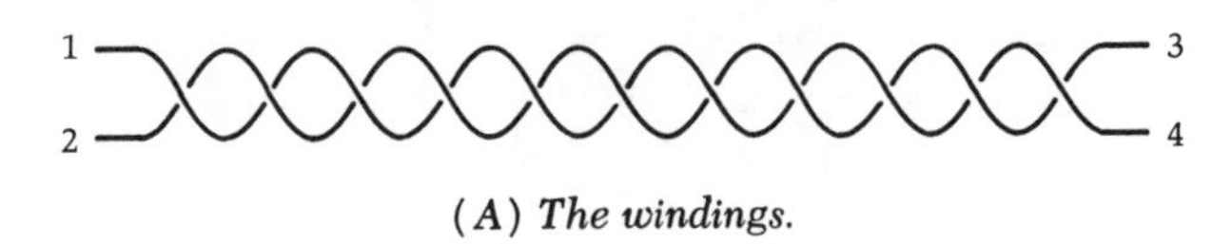

(*A*) *The windings.*

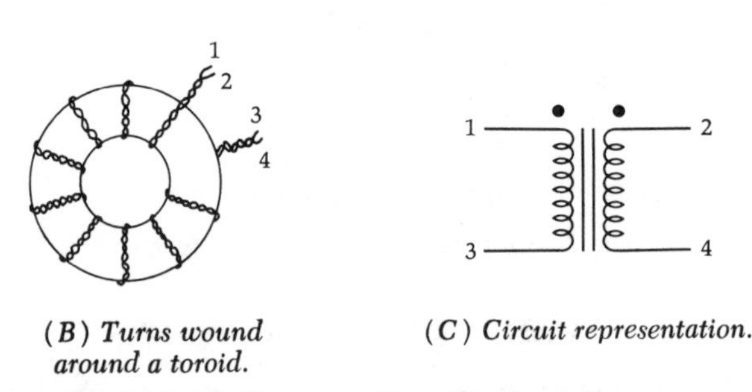

(*B*) *Turns wound around a toroid.* (*C*) *Circuit representation.*

Fig. 7-28. A bifilar-wound broadband transformer.

inch) are used while high-impedance lines may not be twisted at all. Instead, the windings will be placed side-by-side around the core. For optimum operation, the characteristic impedance of the winding should be equal to:

$$Z_O = \sqrt{R_S R_L} \qquad \text{(Eq. 7-3)}$$

where,

R_S = the primary impedance,
R_L = the secondary impedance.

Typically, Z_O must be found experimentally.

The transformer of Fig. 7-28 is called a *bifilar-wound* transformer because it uses two conductors in the twisted-winding arrangement. A *trifilar* transformer is one that uses three conductors, and so on.

As shown, Fig. 7-28 is simply a 1:1 broadband transformer. If connected as shown in Fig. 7-23B, however, this arrangement could be used to produce a 4:1 broadband transformer. This may be done by simply connecting lines 2 and 3 together and using that junction as your output port. Line 4 is connected to ground and line 1 is then the input port.

Often, instead of using a twisted pair, the designer may instead use coaxial cable for the windings. The center conductor and outer conductor of the coax, then, take the place of each conductor in the twisted pair. This is done primarily because each type of coax has a very well-defined and consistent characteristic impedance and this eliminates the experimentation involved in defining the characteristic impedance of the twisted pair. Of course, the use of standard coaxial cable does not allow for trifilar-wound transformers. Typically, broadband transformers are wound on low-Q, high-permeability, ferrite toroidal cores (see Chapter 1). The high permeability is needed at the low end of the frequency spectrum where, for a given inductance, fewer turns would be needed.

SUMMARY

The power-amplifier design process is not as well defined as that of the small-signal amplifier. Thus, considerable experimentation may be necessary in order to optimize a design. Standard Y and S parameters are not used in power-amplifier design. Instead, large-signal impedance parameters are typically provided by the transistor manufacturers to aid in the design process. Following the impedance-matching procedures outlined in Chapter 4, the designer must match the source to the transistor's input impedance, and transform the load impedance to a value that is dependent upon the required output-power level from the stage. The source must be capable of providing the required rf drive level, or the calculated rf power output from the stage will never be achieved.

VECTOR ALGEBRA

Many of the design equations contained in earlier chapters require that the user be familiar with vector algebra. It is the intent of this appendix to provide, for those who are unfamiliar with this subject, a working knowledge of vector addition, subtraction, multiplication, and division.

As illustrated in Fig. A-1, a vector may be expressed in either rectangular or polar form. In *rectangular* form, the vector quantity is expressed as a sum of its coordinate parts. Thus, the vector A shown in Fig. A-1 can be expressed as the sum of 5 units in the x direction and 5 units in the y direction, or $A = 5 + j5$. That same vector may be expressed in *polar* notation as a distance (R) from the point of origin at an angle (θ) from the x axis. If vector A were measured, its length would be found to be 7.07 units at an angle of 45° from the x axis. Thus,

$$A = 5 + j5 \quad \text{or} \quad A = 7.07 \angle 45°$$

Similarly, vector B can be expressed in rectangular form as $5 - j10$ or in polar form as $11.18 \angle -63.4°$. Note that negative angles are measured clockwise from the x axis while positive angles are measured counterclockwise.

Rectangular/Polar and Polar/Rectangular Conversion

Rather than plotting a vector to graphically determine its component parts, it is more convenient to perform a few simple mathematical calculations. Any vector expressed in rectangular form may be converted to polar form (Example A-1) using the following formulas:

$$R = \sqrt{x^2 + y^2} \quad \text{and} \quad \theta = \arctan \frac{y}{x}$$

The conversion from polar to rectangular notation (Example A-2) can be made by using the following formulas:

$$x = R \cos \theta \quad \text{and} \quad y = R \sin \theta$$

Vector Addition

Two vector quantities may be added by performing two separate additions—one for the respective x components and one for the respective y components (Ex-

EXAMPLE A-1

The input impedance of a transistor is found to be $Z = 25 - j10$ ohms. Express this impedance in polar notation.

Solution

The magnitude of the resulting vector (R) is found as:

$$R = \sqrt{x^2 + y^2} = \sqrt{625 + 100} = 26.9$$

The resulting angle from the x axis is found to be:

$$\theta = \arctan \frac{y}{x} = \arctan \frac{-10}{25} = -21.8°$$

Thus, $Z = 25 - j10$ ohms can also be expressed as $Z = 26.9 \angle -21.8°$ ohms.

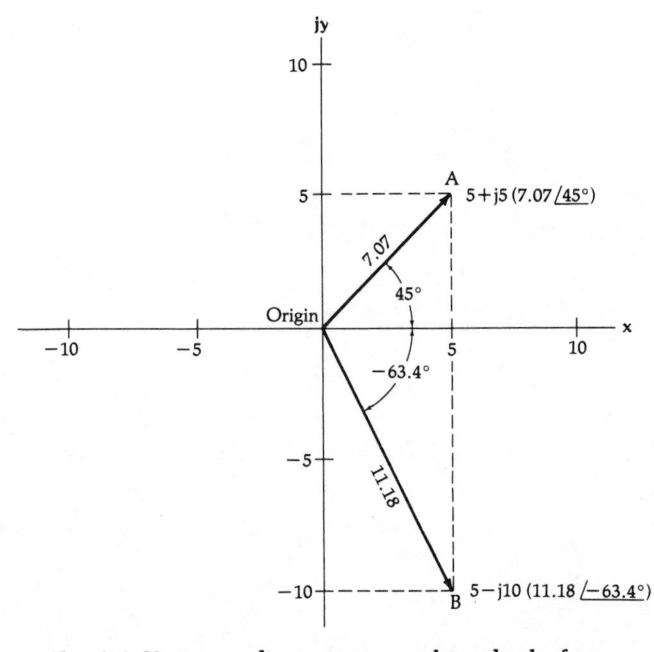

Fig. A-1. Vector coordinates in rectangular and polar form.

EXAMPLE A-2

The input impedance of a transistor is found to be $Z = 26.9 \angle -21.8°$. Express this impedance in rectangular form.

Solution

First:

$$\begin{aligned} x &= R \cos\theta \\ &= 26.9 \cos(-21.8°) \\ &= 26.9(0.9285) \\ &= 25 \end{aligned}$$

and, then,

$$\begin{aligned} y &= R \sin\theta \\ &= 26.9 \sin(-21.8°) \\ &= 26.9(-0.3714) \\ &= -10 \end{aligned}$$

Thus, $Z = 25 - j10$ ohms.

ample A-3). Of course, the resultant may be expressed in either rectangular or polar form.

Vector Subtraction

Vector subtraction is performed in a similar manner to that of addition (Example A-4). The two vector quantities must first be expressed in rectangular form, and their respective x and y components may then be subtracted.

EXAMPLE A-3

An impedance of $Z_1 = 11.18 \angle 63.40°$ ohms is added in series with an impedance of $Z_2 = 18.03 \angle -56.3°$ ohms. What is the resulting series impedance (Z_T) expressed in rectangular form?

Solution

Before the addition can be performed, the polar quantities of the problem must be transformed to rectangular notation. For Z_1:

$$\begin{aligned} x_1 &= R_1 \cos\theta_1 \\ &= 11.18 \cos(63.4°) \\ &= 5 \\ y_1 &= R_1 \sin\theta_1 \\ &= 11.18 \sin(63.4°) \\ &= 10 \end{aligned}$$

Thus,

$$Z_1 = 5 + j10 \text{ ohms}$$

For Z_2:

$$\begin{aligned} x_2 &= R_2 \cos\theta_2 \\ &= 18.03 \cos(-56.3°) \\ &= 10 \\ y_2 &= R_2 \sin\theta_2 \\ &= 18.03 \sin(-56.3°) \\ &= -15 \end{aligned}$$

Thus, $Z_2 = 10 - j15$ ohms.

To perform the addition, add the respective x components and the respective y components.

$$\begin{aligned} x_T &= x_1 + x_2 \\ &= 5 + 10 \\ &= 15 \\ y_T &= y_1 + y_2 \\ &= 10 - 15 \\ &= -5 \end{aligned}$$

Thus, $Z_T = 15 - j5$ ohms.

EXAMPLE A-4

Using the following values:

$$V_1 = 11.18 \angle 63.4°$$
$$V_2 = 18.03 \angle -56.3°$$

perform the calculation, $V_T = V_1 - V_2$.

Solution

Both quantities must first be expressed in rectangular form. For V_1:

$$\begin{aligned} x_1 &= R_1 \cos\theta_1 \\ &= 11.18 \cos(63.4°) \\ &= 5 \\ y_1 &= R_1 \sin\theta_1 \\ &= 11.18 \sin(63.4°) \\ &= 10 \end{aligned}$$

and, then, for V_2:

$$\begin{aligned} x_2 &= R_2 \cos\theta_2 \\ &= 18.03 \cos(-56.3°) \\ &= 10 \\ y_2 &= R_2 \sin\theta_2 \\ &= 18.03 \sin(-56.3°) \\ &= -15 \end{aligned}$$

Subtracting the x and y components, we get:

$$\begin{aligned} x_T &= x_1 - x_2 \\ &= 5 - 10 \\ &= -5 \\ y_T &= y_1 - y_2 \\ &= 10 - (-15) \\ &= 25 \end{aligned}$$

Therefore, $V_T = -5 + j25$.

Vector Multiplication

Multiplication of two vectors is accomplished by first converting both vectors to polar form. The magnitudes (R) of the vectors are then multiplied and their angles are added (Example A-5). Thus,

$$R_T = R_1R_2 \quad \text{and} \quad \theta_T = \theta_1 + \theta_2$$

Vector Division

Vector division is performed by first converting both vectors to polar form. The vector quotient is then found by *dividing* the magnitudes and subtracting the angles (Example A-6). Use the formulas:

$$R_T = \frac{R_1}{R_2} \quad \text{and} \quad \theta_T = \theta_1 - \theta_2$$

Real, Imaginary, and Magnitude Components

Several references are made throughout the text to "the real part of," "the imaginary part of," and "the

EXAMPLE A-5

For a transistor, $S_{21} = 5.6 \angle 60°$ and $S_{12} = 0.1 \angle 30°$. Find the product $S_{21}S_{12}$.

Solution

Both S parameters are already in polar form, therefore:

$$R_T = R_1R_2 = (5.6)(0.1) = 0.56$$

and,

$$\theta_T = \theta_1 + \theta_2 = 60° + 30° = 90°$$

Thus, the product $S_{21}S_{12}$ is equal to $0.56 \angle 90°$.

EXAMPLE A-6

Perform the following vector division:

$$V_T = \frac{V_1}{V_2}$$

where,

$V_1 = 40 \angle 60°$
$V_2 = 5 + j5$

Solution

V_1 is already in polar form. Convert V_2 to polar form.

$$V_2 = 7.071 \angle 45°$$

Divide the magnitudes.

$$R_T = \frac{R_1}{R_2} = \frac{40}{7.071} = 5.66$$

Subtract the angles.

$$\theta_T = \theta_1 - \theta_2 = 60° - 45° = 15°$$

Therefore, the quotient is equal to $5.66 \angle 15°$.

magnitude of" a complex vector (Example A-7). These components are described as follows:

When given the complex vector V, where

$$V = R \angle \theta = x + jy$$

The *real part* of the vector V is given as:

$$Re(V) = x$$

the imaginary part of the vector V is given as:

$$Im(V) = jy$$

and the magnitude of the vector V is, then, given as:

$$|V| = R$$

EXAMPLE A-7

Given the complex vector $V = 10 \angle 60°$, find Re(V), Im(V), and |V|.

Solution

First, express the vector in rectangular form.

$$x = R \cos \theta = 10 \cos(60°) = 5$$

$$y = R \sin \theta = 10 \sin(60°) = 8.66$$

Therefore, $V = 5 + j8.66$ and

$$Re(V) = 5$$
$$Im(V) = j8.66$$
$$|V| = 10$$

NOISE CALCULATIONS

Noise can be defined as any undesired disturbance, be it man-made or natural, in any dynamic electrical or electronic system. It may take the form of atmospheric noise caused by the immense energy of the sun or it may be something as trivial as the thermal noise associated with a carbon resistor. Whatever the source, noise is an obstacle that man has been trying to overcome since time began.

The purpose of this appendix is to simply pinpoint a few of the many sources of noise in the electronic systems of today. We will examine noise in its relationship with amplifier design and receiver systems design. The emphasis of this appendix will be placed on the practical aspects of noise rather than on its probabilistic nature.

TYPES OF NOISE

Basically, we are concerned with two types of noise —thermal noise and shot noise.

Thermal Noise

In any conducting medium whose temperature is above absolute zero (0 Kelvin), the *random* motion of charge carriers within the conductor produces random voltages and currents. These voltages and currents produce noise. As the temperature of the conductor increases, the random motion and the velocity of the charge carriers increase; hence, the noise voltage increases.

The open-circuit noise voltage across the terminals of any conductor is given by:

$$V = \sqrt{4kTRB} \qquad \text{(Eq. B-1)}$$

where,

V = the rms noise voltage in volts,
k = Boltzmann's constant (1.38×10^{-23} j/Kelvin),
T = the absolute temperature in Kelvin (°C + 273),
R = the resistance of the conductor in ohms,
B = the bandwidth in hertz.

Notice that the amount of noise voltage present in the conductor is dependent upon the system bandwidth. The narrower the system bandwidth, the less thermal noise that is introduced. Consequently, for optimum noise performance, the bandwidth of any circuit should never be wider than that required to transmit the desired signal.

EXAMPLE B-1

What noise voltage is produced in a 10K resistor at room temperature (293 Kelvin) over an effective bandwidth of 10 MHz?

Solution

Using Equation B-1, we find that:

$$V = \sqrt{4kTRB}$$
$$= \sqrt{4(1.38 \times 10^{-23})(293)(10{,}000)(10 \times 10^{6})}$$
$$= 40.22 \text{ microvolts}$$

Thermal noise is also known as *Johnson noise* and *white noise.*

Shot Noise

Shot noise is a type of noise that is common to the particle-like nature of the charge carriers. It is often thought that a dc current flow in any semiconductor material is constant at every instant. In fact, however, since the current flow is made up of individual electrons and holes, it is only the time-average flow of these charge carriers that is seen as a constant current. Any fluctuation in the number of charge carriers at any instant produces a random current change at that instant. This random current change is known as noise.

Shot noise is also often called *Schottky noise* and is found by the formula:

$$I_n^2 = 2qI_{dc}B \qquad \text{(Eq. B-2)}$$

where,

I_n^2 = the mean square noise current,
q = the electron charge (1.6×10^{-19} coulombs),
I_{dc} = the direct current in amperes,
B = the bandwidth in hertz.

NOISE FIGURE

The *noise figure,* or NF, of a network is a quantity used as a "figure-of-merit" to compare the noise in a network with the noise in an ideal or noiseless network. It is a measure of the degradation in signal-to-noise ratio (SNR) between the input and output ports of the network. *Noise factor* (F) is the numerical ratio of NF, where NF is expressed in dB. Thus,

$$NF = 10 \log_{10} F \qquad \text{(Eq. B-3)}$$

and,

$$F = \frac{\text{Input SNR}}{\text{Output SNR}} \qquad \text{(Eq. B-4)}$$

Cascaded Devices

Often, it is necessary to calculate the noise figure of a group of amplifiers that are connected in cascade (Example B-2). This is easily done if the noise figure of each individual amplifier in the cascade configuration is known.

$$F_{TOTAL} = F_1 + \frac{F_2 - 1}{G_1} + \frac{F_3 - 1}{G_1 G_2} + \frac{F_4 - 1}{G_1 G_2 G_3} \cdots \qquad \text{(Eq. B-5)}$$

where,

F_n = the noise factor of each stage,
G_n = the numerical gain of each stage (not in dB).

EXAMPLE B-2

What is the noise figure of the three cascade-connected amplifiers diagrammed in Fig. B-1?

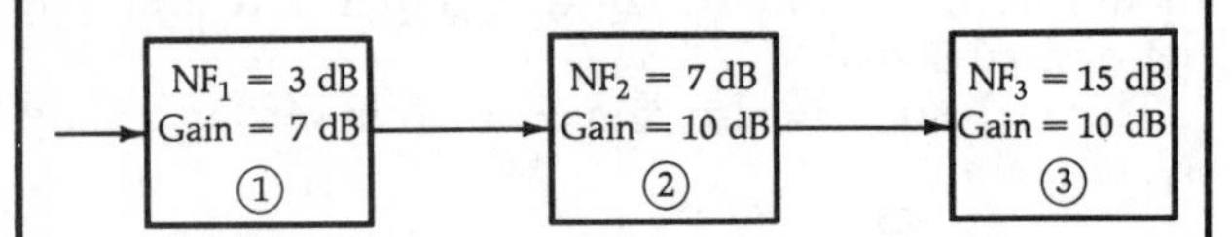

Fig. B-1. Block diagram for Example B-2.

Solution

Since the gains and noise figures of each stage are given in dB, they must first be changed to numerical ratios.

$$F_1 = 2, F_2 = 5, F_3 = 31.6$$
$$G_1 = 5, G_2 = 10, G_3 = 10$$

The overall noise factor is then given by:

$$F_{TOTAL} = 2 + \frac{5 - 1}{5} + \frac{31.6 - 1}{(5)(10)}$$
$$= 3.4$$
$$NF = 10 \log_{10} 3.4$$
$$= 5.3 \text{ dB}$$

Another look at Equation B-5 will reveal a very interesting point. If the gain of the first stage is sufficiently high, the denominators of the second and succeeding terms will force those terms to very small values leaving only F_1 in the equation. Hence, the NF of the first stage will typically determine the NF of the cascade configuration (Example B-3).

Lossy Networks

The NF of a lossy network is equal to the loss of the network in dB. For example, a mixer with a conversion

EXAMPLE B-3

If the gain of the first stage in Example B-2 were 25 dB, what would be the NF of the entire cascade?

Solution

We know that 25 dB is a numerical ratio of 316. Therefore, using Equation B-5:

$$F_{TOTAL} = 2 + \frac{5 - 1}{316} + \frac{31.6 - 1}{(316)(10)}$$
$$= 2.022$$
$$NF = 10 \log_{10} 2.022$$
$$= 3.06 \text{ dB}$$

Note that the NF of the entire cascade is approximately equal to the NF of the first stage.

EXAMPLE B-4

Find the NF of the receiver whose block diagram is shown in Fig. B-2.

Solution

For analysis purposes, the last three blocks of Fig. B-2 may be replaced with a single block having a noise figure of:

$$NF_c = 10 \text{ dB} + 7 \text{ dB} + 4 \text{ dB}$$
$$= 21 \text{ dB}$$

or,

$$F_c = 126$$

The noise factor looking into the preamplifier is, therefore, equal to:

$$F_{preamp} = F_p + \frac{F_c - 1}{G_{preamp}}$$
$$= 2 + \frac{126 - 1}{10}$$
$$= 2 + 12.5$$
$$= 14.5$$

Therefore, the NF at the preamplifier is equal to:

$$NF_{preamp} = 10 \log_{10} 14.5$$
$$= 11.6 \text{ dB}$$

The noise figure of the entire receiver is, thus, equal to the sum of 11.6 dB and the noise figure of the two-pole filter.

$$NF_{rcvr} = 11.6 \text{ dB} + 6 \text{ dB}$$
$$= 17.6 \text{ dB}$$

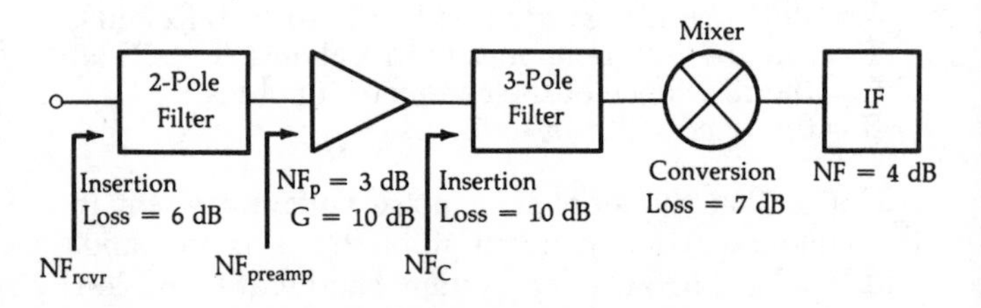

Fig. B-2. Receiver block diagram for Examples B-3 and B-4.

loss of 10 dB has an NF of 10 dB. Similarly, a filter with an insertion loss of 5 dB has an NF of 5 dB.

The NF of two or more cascaded lossy networks can be found simply by adding the losses (in dB) of each network. Thus, if a mixer with a conversion loss of 10 dB is followed by a filter with an insertion loss of 3.5 dB, the combined noise figure of the cascaded network is equal to 13.5 dB (Example B-4).

RECEIVER SYSTEMS CALCULATIONS

The thermal noise that is added to a signal while passing through a system can be calculated as:

$$n_o = kTB \qquad \text{(Eq. B-6)}$$

where,
n_o = the noise power in watts,
k = Boltzmann's constant,
T = the temperature in Kelvins,
B = the noise bandwidth of the system.

Expressed as noise power in dBm, we have:

$$n_o = 10 \log_{10} \frac{kTB}{1 \times 10^{-3}} \qquad \text{(Eq. B-7)}$$

If n_o and the NF of a receiver are known (or can be calculated), the required input signal level to the receiver for a given output signal-to-noise ratio can then be calculated (Example B-5).

$$S_i = NF + n_o + S/N \qquad \text{(Eq. B-8)}$$

where,
S_i = the required input signal level in dBm,
NF = the noise figure of the receiver,
n_o = the thermal noise power of the receiver in dBm,
S/N = the required output signal-to-noise ratio in dB.

EXAMPLE B-5

Using the block diagram shown in Fig. B-2, calculate the required input signal level for a 10-dB signal-to-noise ratio at the output of the if stage. The noise bandwidth (B) is 1.25 MHz.

Solution

The NF of the receiver was determined in Example B-4 to be 17.6 dB.

Calculate n_o using Equation B-7. (Assume a room temperature of 293 K.)

$$n_o = 10 \log_{10} \frac{(1.38 \times 10^{-23})(293)(1.25 \times 10^{6})}{1 \times 10^{-3}}$$
$$= -113 \text{ dBm}$$

Therefore, the required input signal level is calculated as:

$$S_i = NF + n_o + S/N$$
$$= 17.6 - 113 + 10$$
$$= -85.4 \text{ dBm}$$

BIBLIOGRAPHY

TECHNICAL PAPERS

Anderson, L. H., "How to Design Matching Networks," *Ham Radio,* April 1978.

Besser, L., "Take the Hassle Out of FET Amp Design," *MSN,* September 1977.

Besser, L., "Update Amplifier Design With Network Synthesis," *Microwaves,* October 1977.

Brown, W. L., "Don't Guess at Bias Circuit Design," *Electronic Design,* May 9, 1968.

Burwasser, A., "Wideband Monofilar Autotransformers," *RF Design,* January/February 1981.

Burwasser, A., "How to Design Broadband JFET Amplifiers to Provide Top Performance From VLF to Over 100 MHz," *Ham Radio,* November 1979.

Cohn, S. B., "Direct-Coupled-Resonator Filters," *Proceedings of the IRE,* February 1957.

Cohn, S. B., "Dissipation Loss in Multiple-Coupled Resonator Filters," *Proceedings of the IRE,* August 1959.

DeMaw, M. F., "The Practical Side of Toroids," *QST Magazine,* June 1979.

DeMaw, M. F., "Magnetic Cores in RF Circuits," *RF Design,* April 1980.

Dishal, M., "Exact Design and Analysis of Double and Triple Tuned Band-Pass Amplifiers," *Proceedings of the IRE,* June 1947.

Fisk, J., "How to Use the Smith Chart," *Ham Radio,* March 1978.

Grammer, G., "Simplified Design of Impedance Matching Networks," *QST Magazine,* March 1957.

Grossman, M., "Focus on Ferrite Materials: They Star as HF Magnetic Cores," *Electronic Design,* April 30, 1981.

Hall, C., "RF Chokes, Their Performance Above and Below Resonance," *Ham Radio,* June 1978.

Hejhall, Roy, "Systemizing RF Power Amplifier Design," *Motorola AN-282A,* 1975.

Hejhall, Roy, "RF Small Signal Design Using Two-Port Parameters," *Motorola AN-215A,* February 1977.

Hejhall, Roy, "Understanding Transistor Response Parameters," *Motorola AN-139A,* 1974.

Hewlett-Packard Components, "Transistor Parameter Measurements," *Hewlett-Packard AN-77-1,* February 1967.

Hewlett-Packard Components, "A Low Noise 4 GHz Transistor Amplifier Using the HXTR-6101 Silicon Bipolar Transistor," *Hewlett-Packard AN-967,* May 1975.

Hewlett-Packard Components, "A 6 GHz Amplifier Using the HFET-1101 GaAs FET," *Hewlett-Packard AN-970,* February 1978.

Hewlett-Packard Components, "S-Parameter Techniques for Faster, More Accurate Network Design," *Hewlett-Packard AN-95-1.*

Hewlett-Packard Components, "S-Parameter Design," *Hewlett-Packard AN-154,* April 1972.

Hewlett-Packard Components, "Microwave Transistor Bias Considerations," *Hewlett-Packard AN-944-1,* October 1973.

Hoff, I. M., "Pi-Network Design for High-Frequency Power Amplifiers," *Ham Radio,* June 1978.

Howe, H. Jr., "Basic Parameters Hold the Key to Microstrip-Amplifier Design," *EDN,* October 5, 1980.

Kraus, H. and Allen, C., "Designing Toroidal Transformers to Optimize Wideband Performance," *Electronics,* August 16, 1973.

Litty, T. P., "Microstrip RF Amplifier Design," *RF Design,* January/February 1979.

Motorola Semiconductor Products Inc., "Semiconductor Noise Figure Considerations," *Motorola AN-421,* August 1976.

Ruthroff, C., "Some Broadband Transformers," *Proceedings of the IRE,* August 1959.

Sando, S., "Very Low-Noise GaAs FET Preamp for 432 MHz," *Ham Radio,* April 1978.

Scherer, D., "Today's Lesson—Learn About Low-Noise Design," *Microwaves,* April 1979.

Sevick, J., "Broadband Matching Transformers Can Handle Many Kilowatts," *Electronics,* November 25, 1976.

Shuch, Paul, "Solid State Microwave Amplifier Design," *Ham Radio,* October 1976.

Shulamn, J. M., "T-Network Impedance Matching to Coaxial Feedlines," *Ham Radio,* September 1978.

Trout, B. "Small-Signal RF Design With Dual-Gate MOSFETs," *Motorola AN-478A,* 1971.

Wetherhold, E. E., "Design 7-Element Low-Pass Filters Using Standard Value Capacitors," *EDN,* January 7, 1981.

Zverev, A., "Introduction to Filters," *Electro-Technology,* June 1964.

BOOKS

Carlson, A., *Communications Systems,* McGraw-Hill, New York, NY, 1975.

Carson, R., *High Frequency Amplifiers,* John Wiley & Sons, New York, NY, 1975.

DeMaw, M. F., *Practical RF Communications Data for Engineers and Technicians,* Howard W. Sams & Co., Inc., Indianapolis, IN, 1978.

Fink, D. G., *Electronics Engineers' Handbook,* McGraw-Hill, New York, NY, 1975.

Geffe, P., *Simplified Modern Filter Design,* John F. Rider, New York, NY, 1963.

Gupta, S., Bayless, J., and Peikari, B., *Circuit Analysis,* Intext Educational Pub., Scranton, PA, 1972.

Hardy, J., *High Frequency Circuit Design,* Reston Publishing Co., Inc., Reston, VA, 1979.

Harper, C. A., *Handbook of Components for Electronics,* McGraw-Hill, New York, NY.

Kaufman, M. and Seidman, A., *Handbook of Electronics Calculations for Engineers and Technicians,* McGraw-Hill, New York, NY, 1979.

Krauss, H., Bostian, C., and Raak, F., *Solid State Radio Engineering,* John Wiley & Sons, New York, NY, 1980.

Lenk, J., *Handbook of Simplified Solid State Circuit Design,* Prentice-Hall, Englewood Cliffs, NJ, 1971.

Miller, J., *Solid State Communications: Design of Communications Equipment Using Semiconductors,* McGraw-Hill, New York, NY, 1966.

Millman, J. and Halkias, C., *Integrated Electronics: Analog Digital Circuits & Systems,* McGraw-Hill, New York, NY, 1972.

Reference Data for Radio Engineers, Howard W. Sams & Co., Inc., Indianapolis, IN, 1975.

Sentz, R., *Voltage and Power Amplifiers,* Holt, Rinehart & Winston, Inc., New York, NY, 1968.

Smith, P. H., *Electronic Applications of the Smith Chart: In Waveguide, Circuit, & Component Analysis,* McGraw-Hill, New York, NY, 1969.

Williams, A. B., *Electronic Filter Design Handbook,* McGraw-Hill, New York, NY, 1981.

Ziemer, R. E. and Tranter, W., *Principles of Communications: Systems, Modulation & Noise,* Houghton Mifflin Co., Boston, MA, 1976.

Zverev, A., *Handbook of Filter Synthesis,* John Wiley & Sons, New York, NY, 1967.

INDEX